Mathematical Thought
from Ancient to Modern Times

Mathematical Thought
from Ancient to Modern Times

Volume 3

MORRIS KLINE

New York   Oxford   OXFORD UNIVERSITY PRESS

Oxford University Press

Oxford New York Toronto
Delhi Bombay Calcutta Madras Karachi
Petaling Jaya Singapore Hong Kong Tokyo
Nairobi Dar es Salaam Cape Town
Melbourne Auckland

and associated companies in
Berlin Ibadan

First published in 1972, in one volume, by Oxford University Press, Inc.,
200 Madison Avenue, New York, New York 10016

First issued as an Oxford University Press paperback, 1990

Oxford is a registered trademark of Oxford University Press

Library of Congress Cataloging-in-Publication Data
Kline, Morris, 1908–
Mathematical thought from ancient to modern times / Morris Kline.
p. cm. Includes bibliographical references.
ISBN 0-19-506137-3 (PBK) (v. 3)
1. Mathematics—History. I. Title.
QA21.K516 1990 510'.9—dc20 89-25520

Printed in the United States of America

Preface to the Three-Volume Paperback Edition of Mathematical Thought

The reception accorded the original edition of this book is most gratifying. I am flattered, if not a penny richer for it, by a pirated Chinese translation. Even more satisfying is a forthcoming authorized Spanish translation.

This work is part of my long-time efforts to humanize the subject of mathematics. At the very beginning of my career I banded with a few colleagues to produce a freshman text that departed from the traditional dry-as-dust mathematics textbook. Later, I wrote a calculus text with the same end in view. While I was directing a research group in electromagnetic theory and doing research myself, I still made time to write *Mathematics In Western Culture*, which is partly history and partly an exploration of the influence of mathematics upon philosophy, religion, literature, art, music, economic theory, and political thought. More recently I have written with the general reader in mind a book on the philosophical foundations of mathematics and a book on the underlying mathematical structure of a good deal of science, most especially cosmogony and physics.

I hope that students, teachers, as well as the general reader will profit from this more affordable and accessible three-volume paperback edition of *Mathematical Thought.* I wish to acknowledge the helpful suggestions made by Harold Edwards, Donald Gillis, and Robert Schlapp among others. My very special thanks go to Fred Pohle for his time, interest, and generosity. Having over the years taught a course based on this book, he saw a need for a multi-volume paperback version and provided the impetus for this edition. Beyond this he gave unstintingly of his time and knowledge in helping me correct errors. I am truly in his debt, as I am to my wife Helen, who undertook much of the work involved in preparing this edition.

Preface

If we wish to foresee the future of mathematics our proper
course is to study the history and present condition of the
science.   HENRI POINCARÉ

This book treats the major mathematical creations and developments
from ancient times through the first few decades of the twentieth century.
It aims to present the central ideas, with particular emphasis on those
currents of activity that have loomed largest in the main periods of the life
of mathematics and have been influential in promoting and shaping subsequent mathematical activity. The very concept of mathematics, the
changes in that concept in different periods, and the mathematicians' own
understanding of what they were achieving have also been vital concerns.

This work must be regarded as a survey of the history. When one
considers that Euler's works fill some seventy volumes, Cauchy's twenty-six
volumes, and Gauss's twelve volumes, one can readily appreciate that a
one-volume work cannot present a full account. Some chapters of this work
present only samples of what has been created in the areas involved, though
I trust that these samples are the most representative ones. Moreover, in
citing theorems or results, I have often omitted minor conditions required for
strict correctness in order to keep the main ideas in focus. Restricted as this
work may be, I believe that some perspective on the entire history has been
presented.

The book's organization emphasizes the leading mathematical themes
rather than the men. Every branch of mathematics bears the stamp of its
founders, and great men have played decisive roles in determining the course
of mathematics. But it is their ideas that have been featured; biography is
entirely subordinate. In this respect, I have followed the advice of Pascal:
"When we cite authors we cite their demonstrations, not their names."

To achieve coherence, particularly in the period after 1700, I have
treated each development at that stage where it became mature, prominent,
and influential in the mathematical realm. Thus non-Euclidean geometry is
presented in the nineteenth century even though the history of the efforts to

replace or prove the Euclidean parallel axiom date from Euclid's time onward. Of course, many topics recur at various periods.

To keep the material within bounds I have ignored several civilizations such as the Chinese,[1] Japanese, and Mayan because their work had no material impact on the main line of mathematical thought. Also some developments in mathematics, such as the theory of probability and the calculus of finite differences, which are important today, did not play major roles during the period covered and have accordingly received very little attention. The vast expansion of the last few decades has obliged me to include only those creations of the twentieth century that became significant in that period. To continue into the twentieth century the extensions of such subjects as ordinary differential equations or the calculus of variations would call for highly specialized material of interest only to research men in those fields and would have added inordinately to the size of the work. Beyond these considerations, the importance of many of the more recent developments cannot be evaluated objectively at this time. The history of mathematics teaches us that many subjects which aroused tremendous enthusiasm and engaged the attention of the best mathematicians ultimately faded into oblivion. One has but to recall Cayley's dictum that projective geometry is all geometry, and Sylvester's assertion that the theory of algebraic invariants summed up all that is valuable in mathematics. Indeed one of the interesting questions that the history answers is what survives in mathematics. History makes its own and sounder evaluations.

Readers of even a basic account of the dozens of major developments cannot be expected to know the substance of all these developments. Hence except for some very elementary areas the contents of the subjects whose history is being treated are also described, thus fusing exposition with history. These explanations of the various creations may not clarify them completely but should give some idea of their nature. Consequently this book may serve to some extent as a historical introduction to mathematics. This approach is certainly one of the best ways to acquire understanding and appreciation.

I hope that this work will be helpful to professional and prospective mathematicians. The professional man is obliged today to devote so much of his time and energy to his specialty that he has little opportunity to familiarize himself with the history of his subject. Yet this background is important. The roots of the present lie deep in the past and almost nothing in that past is irrelevant to the man who seeks to understand how the present came to be what it is. Moreover, mathematics, despite the proliferation into hundreds of branches, is a unity and has its major problems and goals. Unless the various specialties contribute to the heart of mathematics they are likely to be

1. A fine account of the history of Chinese mathematics is available in Joseph Needham's *Science and Civilization in China*, Cambridge University Press, 1959, Vol. 3, pp. 1–168.

sterile. Perhaps the surest way to combat the dangers which beset our fragmented subject is to acquire some knowledge of the past achievements, traditions, and objectives of mathematics so that one can direct his research into fruitful channels. As Hilbert put it, "Mathematics is an organism for whose vital strength the indissoluble union of the parts is a necessary condition."

For students of mathematics this work may have other values. The usual courses present segments of mathematics that seem to have little relationship to each other. The history may give perspective on the entire subject and relate the subject matter of the courses not only to each other but also to the main body of mathematical thought.

The usual courses in mathematics are also deceptive in a basic respect. They give an organized logical presentation which leaves the impression that mathematicians go from theorem to theorem almost naturally, that mathematicians can master any difficulty, and that the subjects are completely thrashed out and settled. The succession of theorems overwhelms the student, especially if he is just learning the subject.

The history, by contrast, teaches us that the development of a subject is made bit by bit with results coming from various directions. We learn, too, that often decades and even hundreds of years of effort were required before significant steps could be made. In place of the impression that the subjects are completely thrashed out one finds that what is attained is often but a start, that many gaps have to be filled, or that the really important extensions remain to be created.

The polished presentations in the courses fail to show the struggles of the creative process, the frustrations, and the long arduous road mathematicians must travel to attain a sizable structure. Once aware of this, the student will not only gain insight but derive courage to pursue tenaciously his own problems and not be dismayed by the incompleteness or deficiencies in his own work. Indeed the account of how mathematicians stumbled, groped their way through obscurities, and arrived piecemeal at their results should give heart to any tyro in research.

To cover the large area which this work comprises I have tried to select the most reliable sources. In the pre-calculus period these sources, such as T. L. Heath's *A History of Greek Mathematics*, are admittedly secondary, though I have not relied on just one such source. For the subsequent development it has usually been possible to go directly to the original papers, which fortunately can be found in the journals or in the collected works of the prominent mathematicians. I have also been aided by numerous accounts and surveys of research, some in fact to be found in the collected works. I have tried to give references for all of the major results; but to do so for all assertions would have meant a mass of references and the consumption of space that is better devoted to the account itself.

The sources have been indicated in the bibliographies of the various chapters. The interested reader can obtain much more information from these sources than I have extracted. These bibliographies also contain many references which should not and did not serve as sources. However, they have been included either because they offer additional information, because the level of presentation may be helpful to some readers, or because they may be more accessible than the original sources.

I wish to express thanks to my colleagues Martin Burrow, Bruce Chandler, Martin Davis, Donald Ludwig, Wilhelm Magnus, Carlos Moreno, Harold N. Shapiro, and Marvin Tretkoff, who answered numerous questions, read many chapters, and gave valuable criticisms. I am especially indebted to my wife Helen for her critical editing of the manuscript, extensive checking of names, dates, and sources, and most careful reading of the galleys and page proofs. Mrs. Eleanore M. Gross, who did the bulk of the typing, was enormously helpful. To the staff of Oxford University Press, I wish to express my gratitude for their scrupulous production of this work.

New York   M. K.
May 1972

Contents

49. The Emergence of Abstract Algebra, 1136

50. The Beginnings of Topology, 1158

51. The Foundations of Mathematics, 1182

Publisher's Note

to this Three-Volume Paperback Edition

Mathematical Thought from Ancient to Modern Times was first published by Oxford University Press as a one-volume cloth edition. In publishing this three-volume paperback edition we have retained the same pagination as the cloth in order to maintain consistency within the Index, Subject Index, and Notes. These volumes are paginated consecutively and, for the reader's convenience, both Indexes appear at the end of each volume.

34
The Theory of Numbers in the Nineteenth Century

> It is true that Fourier had the opinion that the principal object of mathematics was public use and the explanation of natural phenomena; but a philosopher like him ought to know that the sole object of the science is the honor of the human spirit and that under this view a problem of [the theory of] numbers is worth as much as a problem on the system of the world.   C. G. J. JACOBI

1. *Introduction*

Up to the nineteenth century the theory of numbers was a series of isolated though often brilliant results. A new era began with Gauss's *Disquisitiones Arithmeticae*[1] which he composed at the age of twenty. This great work had been sent to the French Academy in 1800 and was rejected but Gauss published it on his own. In this book he standardized the notation; he systematized the existing theory and extended it; and he classified the problems to be studied and the known methods of attack and introduced new methods. In Gauss's work on the theory of numbers there are three main ideas: the theory of congruences, the introduction of algebraic numbers, and the theory of forms as the leading idea in Diophantine analysis. This work not only began the modern theory of numbers but determined the directions of work in the subject up to the present time. The *Disquisitiones* is difficult to read but Dirichlet expounded it.

Another major nineteenth-century development is analytic number theory, which uses analysis in addition to algebra to treat problems involving the integers. The leaders in this innovation were Dirichlet and Riemann.

2. *The Theory of Congruences*

Though the notion of congruence did not originate with Gauss—it appears in the work of Euler, Lagrange, and Legendre—Gauss introduced the notation

1. Published 1801 = *Werke*, 1.

in the first section of *Disquisitiones* and used it systematically thereafter. The basic idea is simple. The number 27 is congruent to 3 modulo 4,

$$27 \equiv 3 \text{ modulo } 4,$$

because $27 - 3$ is exactly divisible by 4. (The word modulo is often abbreviated to mod.) In general, if a, b, and m are integers

$$a \equiv b \text{ modulo } m$$

if $a - b$ is (exactly) divisible by m or if a and b have the same remainders on division by m. Then b is said to be a residue of a modulo m and a is a residue of b modulo m. As Gauss shows, all the residues of a modulo m, for fixed a and m, are given by $a + km$ where $k = 0, \pm 1, \pm 2, \ldots$.

Congruences with respect to the same modulus can be treated to some extent like equations. Such congruences can be added, subtracted, and multiplied. One can also ask for the solution of congruences involving unknowns. Thus, what values of x satisfy

$$2x \equiv 25 \text{ modulo } 12?$$

This equation has no solutions because $2x$ is even and $2x - 25$ is odd. Hence $2x - 25$ cannot be a multiple of 12. The basic theorem on polynomial congruences, which Gauss re-proves in the second section, had already been established by Lagrange.[2] A congruence of the nth degree

$$Ax^n + Bx^{n-1} + \cdots + Mx + N \equiv 0 \text{ modulo } p$$

whose modulus is a prime number p which does not divide A cannot have more than n noncongruent roots.

In the third section Gauss takes up residues of powers. Here he gives a proof in terms of congruences of Fermat's minor theorem, which, stated in terms of congruences, reads: If p is a prime and a is not a multiple of p then

$$a^{p-1} \equiv 1 \text{ modulo } p.$$

The theorem follows from his study of congruences of higher degree, namely,

$$x^n \equiv a \text{ modulo } m$$

where a and m are relatively prime. This subject was continued by many men after Gauss.

The fourth section of *Disquisitiones* treats quadratic residues. If p is a prime and a is not a multiple of p and if there exists an x such that $x^2 \equiv a$ mod p, then a is a quadratic residue of p; otherwise a is a quadratic non-residue of p. After proving some subordinate theorems on quadratic residues Gauss gave the first rigorous proof of the law of quadratic reciprocity (Chap.

2. *Hist. de l'Acad. de Berlin*, 24, 1768, 192 ff., pub. 1770 = *Œuvres*, 2, 655–726.

25, sec. 4). Euler had given a complete statement much like Gauss's in one paper of his *Opuscula Analytica* of 1783 (Chap. 25, sec. 4). Nevertheless in article 151 of his *Disquisitiones* Gauss says that no one had presented the theorem in as simple a form as he had. He refers to other work of Euler including another paper in the *Opuscula* and to Legendre's work of 1785. Of these papers Gauss says rightly that the proofs were incomplete.

Gauss is supposed to have discovered a proof of the law in 1796 when he was nineteen. He gave another proof in the *Disquisitiones* and later published four others. Among his unpublished papers two more were found. Gauss says that he sought many proofs because he wished to find one that could be used to establish the biquadratic reciprocity theorem (see below). The law of quadratic reciprocity, which Gauss called the gem of arithmetic, is a basic result on congruences. After Gauss gave his proofs, more than fifty others were given by later mathematicians.

Gauss also treated congruences of polynomials. If A and B are two polynomials in x with, say, real coefficients then one knows that one can find unique polynomials Q and R such that

$$A = B \cdot Q + R,$$

where the degree of R is less than the degree of B. One can then say that two polynomials A_1 and A_2 are congruent modulo a third polynomial P if they have the same remainder R on division by P.

Cauchy used this idea[3] to define complex numbers by polynomial congruences. Thus if $f(x)$ is a polynomial with real coefficients then under division by $x^2 + 1$

$$f(x) \equiv a + bx \bmod x^2 + 1$$

because the remainder is of lower degree than the divisor. Here a and b are necessarily real by virtue of the division process. If $g(x)$ is another such polynomial then

$$g(x) \equiv c + dx \bmod x^2 + 1.$$

Cauchy now points out that if A_1, A_2, and B are any polynomials and if

$$A_1 = BQ_1 + R_1 \quad \text{and} \quad A_2 = BQ_2 + R_2,$$

then

$$A_1 + A_2 \equiv R_1 + R_2 \bmod B, \quad \text{and} \quad A_1 A_2 \equiv R_1 R_2 \bmod B.$$

We can now see readily that

$$f(x) + g(x) \equiv (a + c) + (b + d)x \bmod x^2 + 1$$

3. *Exercices d'analyse et de physique mathématique*, 4, 1847, 84 ff. = *Œuvres*, (1), 10, 312–23 and (2), 14, 93–120.

and since $x^2 \equiv -1 \bmod x^2 + 1$ that

$$f(x)g(x) \equiv (ac - bd) + (ad + bc)x \bmod x^2 + 1.$$

Thus the numbers $a + bx$ and $c + dx$ combine like complex numbers; that is, they have the formal properties of complex numbers, x taking the place of i. Cauchy also proved that every polynomial $g(x)$ not congruent to 0 modulo $x^2 + 1$ has an inverse, that is, a polynomial $h(x)$ such that $h(x)g(x)$ is congruent to 1 modulo $x^2 + 1$.

Cauchy did introduce i for x, i being for him a real indeterminate quantity. He then showed that for any

$$f(i) = a_0 + a_1 i + a_2 i^2 + \cdots$$

that

$$f(i) \equiv a_0 - a_2 + a_4 - \cdots + (a_1 - a_3 + a_5 - \cdots)i \text{ modulo } i^2 + 1.$$

Hence any expression involving complex numbers behaves as one of the form $c + di$ and one has all the apparatus needed to work with complex expressions. For Cauchy, then, the polynomials in i, with his understanding about i, take the place of complex numbers and one can put into one class all those polynomials having the same residue modulo $i^2 + 1$. These classes are the complex numbers.

It is interesting that in 1847 Cauchy still had misgivings about $\sqrt{-1}$. He says, "In the theory of algebraic equivalences substituted for the theory of imaginary numbers the letter i ceases to represent the symbolic sign $\sqrt{-1}$, which we repudiate completely and which we can abandon without regret since one does not know what this supposed sign signifies nor what sense to attribute to it. On the contrary we represent by the letter i a real quantity but indeterminate and in substituting the sign $\equiv$ for $=$ we transform what has been called an imaginary equation into an algebraic equivalence relative to the variable i and to the divisor $i^2 + 1$. Since this divisor remains the same in all the formulas one can dispense with writing it."

In the second decade of the century Gauss proceeded to search for reciprocity laws applicable to congruences of higher degree. These laws again involve residues of congruences. Thus for the congruence

$$x^4 \equiv q \bmod p$$

one can define q as a biquadratic residue of p if there is an integral value of x satisfying the equation. He did arrive at a law of biquadratic reciprocity (see below) and a law of cubic reciprocity. Much of this work appeared in papers from 1808 to 1817 and the theorem proper on biquadratic residues was given in papers of 1828 and 1832.[4]

4. *Comm. Soc. Gott.*, 6, 1828, and 7, 1832 = *Werke*, 2, 65–92 and 93–148; also pp. 165–78.

To attain elegance and simplicity in his theory of cubic and biquadratic residues Gauss made use of complex integers, that is, numbers of the form $a + bi$ with a and b integral or 0. In Gauss's work on biquadratic residues it was necessary to consider the case where the modulus p is a prime of the form $4n + 1$ and Gauss needed the complex factors into which prime numbers of the form $4n + 1$ can be decomposed. To obtain these Gauss realized that one must go beyond the domain of the ordinary integers to introduce the complex integers. Though Euler and Lagrange had introduced such integers into the theory of numbers it was Gauss who established their importance.

Whereas in the ordinary theory of integers the units are $+1$ and -1 in Gauss's theory of complex integers the units are ± 1 and $\pm i$. A complex integer is called composite if it is the product of two such integers neither of which is a unit. If such a decomposition is not possible the integer is called a prime. Thus $5 = (1 + 2i)(1 - 2i)$ and so is composite, whereas 3 is a complex prime.

Gauss showed that complex integers have essentially the same properties as ordinary integers. Euclid had proved (Chap. 4, sec. 7) that every integer is uniquely decomposable into a product of primes. Gauss proved that this unique decomposition, which is often referred to as the fundamental theorem of arithmetic, holds also for complex integers provided we do not regard the four unit numbers as different factors. That is, if $a = bc = (ib)(-ic)$, the two decompositions are the same. Gauss also showed that Euclid's process for finding the greatest common divisor of two integers is applicable to the complex integers.

Many theorems for ordinary primes carry over to the complex primes. Thus Fermat's theorem carries over in the form: If p be a complex prime $a + bi$ and k any complex integer not divisible by p then

$$k^{Np-1} \equiv 1 \text{ modulo } p$$

where Np is the norm $(a^2 + b^2)$ of p. There is also a law of quadratic reciprocity for complex integers, which was stated by Gauss in his 1828 paper.

In terms of complex integers Gauss was able to state the law of biquadratic reciprocity rather simply. One defines an uneven integer as one not divisible by $1 + i$. A primary uneven integer is an uneven integer $a + bi$ such that b is even and $a + b - 1$ is even. Thus -7 and $-5 + 2i$ are primary uneven numbers. The law of reciprocity for biquadratic residues states that if α and β are two primary uneven primes and A and B are their norms, then

$$\left(\frac{\alpha}{\beta}\right)_4 = (-1)^{(1/4)(A-1)(1/4)(B-1)}\left(\frac{\beta}{\alpha}\right)_4 .$$

The symbol $(\alpha/\beta)_4$ has the following meaning: If p is any complex prime and k is any biquadratic residue not divisible by p, then $(k/p)_4$ is the power i^e of i which satisfies the congruence

$$k^{(Np-1)/4} \equiv 1 \text{ modulo } p$$

wherein Np stands for the norm of p. This law is equivalent to the statement: The biquadratic characters of two primary uneven prime numbers with respect to one another are identical, that is, $(\alpha/\beta)_4 = (\beta/\alpha)_4$, if either of the primes is congruent to 1 modulo 4; but if neither of the primes satisfies the congruence, then the two biquadratic characters are opposite, that is, $(\alpha/\beta)_4 = -(\beta/\alpha)_4$.

Gauss stated this reciprocity theorem but did not publish his proof. This was given by Jacobi in his lectures at Königsberg in 1836-37. Ferdinand Gotthold Eisenstein (1823-52), a pupil of Gauss, published five proofs of the law, of which the first two appeared in 1844.[5]

For cubic reciprocity Gauss found that he could obtain a law by using the "integers" $a + b\rho$ where ρ is a root of $x^2 + x + 1 = 0$ and a and b are ordinary (rational) integers but Gauss did not publish this result. It was found in his papers after his death. The law of cubic reciprocity was first stated by Jacobi[6] and proved by him in his lectures at Königsberg. The first published proof is due to Eisenstein.[7] Upon noting this proof Jacobi claimed[8] that it was precisely the one given in his lectures but Eisenstein indignantly denied any plagiarism.[9] There are also reciprocity laws for congruences of degree greater than four.

3. Algebraic Numbers

The theory of complex integers is a step in the direction of a vast subject, the theory of algebraic numbers. Neither Euler nor Lagrange envisioned the rich possibilities which their work on complex integers opened up. Neither did Gauss.

The theory grew out of the attempts to prove Fermat's assertion about $x^n + y^n = z^n$. The cases $n = 3$, 4, and 5 have already been discussed (Chap. 25, sec. 4). Gauss tried to prove the assertion for $n = 7$ but failed. Perhaps because he was disgusted with his failure, he said in a letter of 1816 to Heinrich W. M. Olbers (1758-1840), "I confess indeed that Fermat's theorem as an isolated proposition has little interest for me, since a multitude of such propositions, which one can neither prove nor refute, can easily be

5. *Jour. für Math.*, 28, 1844, 53-67 and 223-45.
6. *Jour. für Math.*, 2, 1827, 66-69 = *Werke*, 6, 233-37.
7. *Jour. für Math.*, 27, 1844, 289-310.
8. *Jour. für Math.*, 30, 1846, 166-82, p. 172 = *Werke*, 6, 254-74.
9. *Jour. für Math.*, 35, 1847, 135-274 (p. 273).

formulated." This particular case of $n = 7$ was disposed of by Lamé in 1839,[10] and Dirichlet established the assertion for $n = 14$.[11] However, the general proposition was unproven.

It was taken up by Ernst Eduard Kummer (1810–93), who turned from theology to mathematics, became a pupil of Gauss and Dirichlet, and later served as a professor at Breslau and Berlin. Though Kummer's major work was in the theory of numbers, he made beautiful discoveries in geometry which had their origin in optical problems; he also made important contributions to the study of refraction of light by the atmosphere.

Kummer took $x^p + y^p$ where p is prime, and factored it into

$$(x + y)(x + \alpha y) \cdots (x + \alpha^{p-1}y),$$

where α is an imaginary pth root of unity. That is, α is a root of

(1) $$\alpha^{p-1} + \alpha^{p-2} + \cdots + \alpha + 1 = 0.$$

This led him to extend Gauss's theory of complex integers to algebraic numbers insofar as they are introduced by equations such as (1), that is, numbers of the form

$$f(\alpha) = a_0 + a_1\alpha + \cdots + a_{p-2}\alpha^{p-2},$$

where each a_i is an ordinary (rational) integer. (Since α satisfies (1), terms in α^{p-1} can be replaced by terms of lower power.) Kummer called the numbers $f(\alpha)$ complex integers.

By 1843 Kummer made appropriate definitions of integer, prime integer, divisibility, and the like (we shall give the standard definitions in a moment) and then made the mistake of assuming that unique factorization holds in the class of algebraic numbers that he had introduced. He pointed out while transmitting his manuscript to Dirichlet in 1843 that this assumption was necessary to prove Fermat's theorem. Dirichlet informed him that unique factorization holds only for certain primes p. Incidentally, Cauchy and Lamé made the same mistake of assuming unique factorization for algebraic numbers. In 1844 Kummer[12] recognized the correctness of Dirichlet's criticism.

To restore unique factorization Kummer created a theory of ideal numbers in a series of papers starting in 1844.[13] To understand his idea let us consider the domain of $a + b\sqrt{-5}$, where a and b are integers. In this domain

$$6 = 2 \cdot 3 = (1 + \sqrt{-5})(1 - \sqrt{-5})$$

10. *Jour. de Math.*, 5, 1840, 195–211.
11. *Jour. für Math.*, 9, 1832, 390–93 = *Werke*, 1, 189–94.
12. *Jour. de Math.*, 12, 1847, 185–212.
13. *Jour. für Math.*, 35, 1847, 319–26, 327–67.

and all four factors can readily be shown to be prime integers. Then unique decomposition does not hold. Let us introduce, for this domain, the ideal numbers $\alpha = \sqrt{2}$, $\beta_1 = (1 + \sqrt{-5})/\sqrt{2}$, $\beta_2 = (1 - \sqrt{-5})/\sqrt{2}$. We see that $6 = \alpha^2 \beta_1 \beta_2$. Thus 6 is now uniquely expressed as the product of four factors, all ideal numbers as far as the domain $a + b\sqrt{-5}$ is concerned.[14] In terms of these ideals and other primes factorization in the domain is unique (apart from factors consisting of units). With ideal numbers one can prove some of the results of ordinary number theory in all domains that previously lacked unique factorization.

Kummer's ideal numbers, though ordinary numbers, do not belong to the class of algebraic numbers he had introduced. Moreover, the ideal numbers were not defined in any general way. As far as Fermat's theorem is concerned, with his ideal numbers Kummer did succeed in showing that it was correct for a number of prime numbers. In the first hundred integers only 37, 59, and 67 were not covered by Kummer's demonstration. Then Kummer in a paper of 1857[15] extended his results to these exceptional primes. These results were further extended by Dimitry Mirimanoff (1861– 1945), a professor at the University of Geneva, by perfecting Kummer's method.[16] Mirimanoff proved that Fermat's theorem is correct for each n up to 256 if x, y, and z are prime to that exponent n.

Whereas Kummer worked with algebraic numbers formed out of the roots of unity, Richard Dedekind (1831–1916), a pupil of Gauss, who spent fifty years of his life as a teacher at a technical high school in Germany, approached the problem of unique factorization in an entirely new and fresh manner. Dedekind published his results in supplement 10 to the second edition of Dirichlet's *Zahlentheorie* (1871) which Dedekind edited. He extended these results in the supplements to the third and fourth editions of the same book.[17] Therein he created the modern theory of algebraic numbers.

Dedekind's theory of algebraic numbers is a generalization of Gauss's complex integers and Kummer's algebraic numbers but the generalization is somewhat at variance with Gauss's complex integers. A number r that is a root of

$$(2) \qquad a_0 x^n + a_1 x^{n-1} + \cdots + a_{n-1} x + a_n = 0,$$

where the a_i's are ordinary integers (positive or negative), and that is not a root of such an equation of degree less than n is called an algebraic number of degree n. If the coefficient of the highest power of x in (2) is 1, the solutions are called algebraic integers of degree n. The sum, difference, and product of

14. With the introduction of the ideal numbers, 2 and 3 are no longer indecomposable, for 2 = α^2 and 3 = $\beta_1 \beta_2$.

15. *Abh. König. Akad. der Wiss. Berlin*, 1858, 41–74.

16. *Jour. für Math.*, 128, 1905, 45–68.

17. 4th ed., 1894 = *Werke*, 3, 2–222.

algebraic integers are algebraic integers, and if an algebraic integer is a rational number it is an ordinary integer.

We should note that under the new definitions an algebraic integer can contain ordinary fractions. Thus $(-13 + \sqrt{-115})/2$ is an algebraic integer of the second degree because it is a root of $x^2 + 13x + 71 = 0$. On the other hand $(1 - \sqrt{-5})/2$ is an algebraic number of degree 2 but not an algebraic integer because it is a root of $2x^2 - 2x + 3 = 0$.

Dedekind introduced next the concept of a number field. This is a collection F of real or complex numbers such that if α and β belong to F then so do $\alpha + \beta$, $\alpha - \beta$, $\alpha\beta$ and, if $\beta \neq 0$, α/β. Every number field contains the rational numbers because if α belongs then so does α/α or 1 and consequently $1 + 1$, $1 + 2$, and so forth. It is not difficult to show that the set of all algebraic numbers forms a field.

If one starts with the rational number field and θ is an algebraic number of degree n then the set formed by combining θ with itself and the rational numbers under the four operations is also a field of degree n. This field may be described alternatively as the smallest field containing the rational numbers and θ. It is also called an extension field of the rational numbers. Such a field does not consist of all algebraic numbers and is a specific algebraic number field. The notation $R(\theta)$ is now common. Though one might expect that the members of $R(\theta)$ are the quotients $f(\theta)/g(\theta)$ where $f(x)$ and $g(x)$ are any polynomials with rational coefficients, one can prove that if θ is of degree n, then any member α of $R(\theta)$ can be expressed in the form

$$\alpha = a_0\theta^{n-1} + a_1\theta^{n-2} + \cdots + a_{n-1},$$

where the a_i are ordinary rational numbers. Moreover, there exist algebraic integers $\theta_1, \theta_2, \ldots, \theta_n$ of this field such that all the algebraic integers of the field are of the form

$$A_1\theta_1 + A_2\theta_2 + \cdots + A_n\theta_n,$$

where the A_i are ordinary positive and negative integers.

A ring, a concept introduced by Dedekind, is essentially any collection of numbers such that if α and β belong, so do $\alpha + \beta$, $\alpha - \beta$, and $\alpha\beta$. The set of all algebraic integers forms a ring as does the set of all algebraic integers of any specific algebraic number field.

The algebraic integer α is said to be divisible by the algebraic integer β if there is an algebraic integer γ such that $\alpha = \beta\gamma$. If j is an algebraic integer which divides every other integer of a field of algebraic numbers then j is called a unit in that field. These units, which include $+1$ and -1, are a generalization of the units $+1$ and -1 of ordinary number theory. The algebraic integer α is a prime if it is not zero or a unit and if any factorization of α into $\beta\gamma$, where β and γ belong to the same algebraic number field, implies that β or γ is a unit in that field.

Now let us see to what extent the fundamental theorem of arithmetic holds. In the ring of *all* algebraic integers there are no primes. Let us consider the ring of integers in a specific algebraic number field $R(\theta)$, say the field $a + b\sqrt{-5}$, where a and b are ordinary rational numbers. In this field unique factorization does not hold. For example,

$$21 = 3 \cdot 7 = (4 + \sqrt{-5})(4 - \sqrt{-5}) = (1 + 2\sqrt{-5})(1 - 2\sqrt{-5}).$$

Each of these last four factors is prime in the sense that it cannot be expressed as a product of the form $(c + d\sqrt{-5})(e + f\sqrt{-5})$ with $c, d, e,$ and f integral.

On the other hand let us consider the field $a + b\sqrt{6}$ where a and b are ordinary rational numbers. If one applies the four algebraic operations to these numbers one gets such numbers. If a and b are restricted to integers one gets the algebraic integers (of degree 2) of this domain. In this domain we can take as an equivalent definition of unit that the algebraic integer M is a unit if $1/M$ is also an algebraic integer. Thus 1, -1, $5 - 2\sqrt{6}$, and $5 + 2\sqrt{6}$ are units. Every integer is divisible by any one of the units. Further, an algebraic integer of the domain is prime if it is divisible only by itself and the units. Now

$$6 = 2 \cdot 3 = \sqrt{6} \cdot \sqrt{6}.$$

It would seem as though there is no unique decomposition into primes. But the factors shown are not primes. In fact

$$6 = 2 \cdot 3 = \sqrt{6} \cdot \sqrt{6} = (2 + \sqrt{6})(-2 + \sqrt{6})(3 + \sqrt{6})(3 - \sqrt{6}).$$

Each of the last four factors is a prime in the domain and unique decomposition does hold in this domain.

In the ring of integers of a specific algebraic number field factorization of the algebraic integers into primes is always possible but *unique* factorization does not generally hold. In fact for domains of the form $a + b\sqrt{-D}$, where D may have any positive integral value not divisible by a square, the unique factorization theorem is valid only when $D = 1, 2, 3, 7, 11, 19, 43, 67,$ and 163, at least for D's up to 10^9.[18] Thus the algebraic numbers themselves do not possess the property of unique factorization.

4. The Ideals of Dedekind

Having generalized the notion of algebraic number, Dedekind now undertook to restore unique factorization in algebraic number fields by a scheme quite different from Kummer's. In place of ideal numbers he introduced

18. H. M. Stark has shown that the above values of D are the only ones possible. See his "On the Problem of Unique Factorization in Complex Quadratic Fields," *Proceedings of Symposia in Pure Mathematics*, XII, 41–56, Amer. Math. Soc., 1969.

classes of algebraic numbers which he called ideals in honor of Kummer's ideal numbers.

Before defining Dedekind's ideals let us note the underlying thought. Consider the ordinary integers. In place of the integer 2, Dedekind considers the class of integers $2m$, where m is any integer. This class consists of all integers divisible by 2. Likewise 3 is replaced by the class $3n$ of all integers divisible by 3. The product 6 becomes the collection of all numbers $6p$, where p is any integer. Then the product $2 \cdot 3 = 6$ is replaced by the statement that the class $2m$ "times" the class $3n$ equals the class $6p$. Moreover, the class $2m$ is a factor of the class $6p$, despite the fact that the former class contains the latter. These classes are examples in the ring of ordinary integers of what Dedekind called ideals. To follow Dedekind's work one must accustom oneself to thinking in terms of classes of numbers.

More generally, Dedekind defined his ideals as follows: Let K be a specific algebraic number field. A set of integers A of K is said to form an ideal if when α and β are any two integers in the set, the integers $\mu\alpha + \nu\beta$, where μ and ν are any other algebraic integers in K, also belong to the set. Alternatively an ideal A is said to be generated by the algebraic integers $\alpha_1, \alpha_2, \ldots, \alpha_n$ of K if A consists of all sums

$$\lambda_1\alpha_1 + \lambda_2\alpha_2 + \cdots + \lambda_n\alpha_n,$$

where the λ_i are any integers of the field K. This ideal is denoted by $(\alpha_1, \alpha_2, \ldots, \alpha_n)$. The zero ideal consists of the number 0 alone and accordingly is denoted by (0). The unit ideal is that generated by the number 1, that is, (1). An ideal A is called principal if it is generated by the single integer α, so that (α) consists of all the algebraic integers divisible by α. In the ring of the ordinary integers every ideal is a principal ideal.

An example of an ideal in the algebraic number field $a + b\sqrt{-5}$, where a and b are ordinary rational numbers, is the ideal generated by the integers 2 and $1 + \sqrt{-5}$. This ideal consists of all integers of the form $2\mu + (1 + \sqrt{-5})\nu$, where μ and ν are arbitrary integers of the field. The ideal also happens to be a principal ideal because it is generated by the number 2 alone in view of the fact that $(1 + \sqrt{-5})2$ must also belong to the ideal generated by 2.

Two ideals $(\alpha_1, \alpha_2, \ldots, \alpha_p)$ and $(\beta_1, \beta_2, \ldots, \beta_q)$ are equal if every member of the former ideal is a member of the latter and conversely. To tackle the problem of factorization we must first consider the product of two ideals. The product of the ideal $A = (\alpha_1, \ldots, \alpha_s)$ and the ideal $B = (\beta_1, \ldots, \beta_t)$ of K is defined to be the ideal

$$AB = (\alpha_1\beta_1, \alpha_1\beta_2, \alpha_2\beta_1, \ldots, \alpha_i\beta_j, \ldots, \alpha_s\beta_t).$$

It is almost evident that this product is commutative and associative. With this definition we may say that A divides B if there exists an ideal C such that

$B = AC$. One writes $A|B$ and A is called a factor of B. As already suggested above by our example of the ordinary integers, the elements of B are *included* in the elements of A and ordinary divisibility is replaced by class inclusion.

The ideals that are the analogues of the ordinary prime numbers are called prime ideals. Such an ideal P is defined to be one which has no factors other than itself and the ideal (1), so that P is not contained in any other ideal of K. For this reason a prime ideal is also called maximal. All of these definitions and theorems lead to the basic theorems for ideals of an algebraic number field K. Any ideal is divisible by only a finite number of ideals and if a prime ideal divides the product AB of two ideals (of the same number class) it divides A or B. Finally the fundamental theorem in the theory of ideals is that every ideal can be factored uniquely into prime ideals.

In our earlier examples of algebraic number fields of the form $a + b\sqrt{D}$, D integral, we found that some permitted unique factorization of the algebraic integers of those fields and others did not. The answer to the question of which do or do not is given by the theorem that the factorization of the integers of an algebraic number field K into primes is unique if and only if all the ideals in K are principal.

From these examples of Dedekind's work we can see that his theory of ideals is indeed a generalization of the ordinary integers. In particular, it furnishes the concepts and properties in the domain of algebraic numbers which enable one to establish unique factorization.

Leopold Kronecker (1823–91), who was Kummer's favorite pupil and who succeeded Kummer as professor at the University of Berlin, also took up the subject of algebraic numbers and developed it along lines similar to Dedekind's. Kronecker's doctoral thesis "On Complex Units," written in 1845 though not published until much later,[19] was his first work in the subject. The thesis deals with the units that can exist in the algebraic number fields created by Gauss.

Kronecker created another theory of fields (domains of rationality).[20] His field concept is more general than Dedekind's because he considered fields of rational functions in any number of variables (indeterminates). Specifically Kronecker introduced (1881) the notion of an indeterminate adjoined to a field, the indeterminate being just a new abstract quantity. This idea of extending a field by adding an indeterminate he made the cornerstone of his theory of algebraic numbers. Here he used the knowledge that had been built up by Liouville, Cantor, and others on the distinction between algebraic and transcendental numbers. In particular he observed that if x is transcendental over a field K (x is an indeterminate) then the field

19. *Jour. für Math.*, 93, 1882, 1–52 = *Werke*, 1, 5–71.
20. "Grundzüge einer arithmetischen Theorie der algebraischen Grössen," *Jour. für Math.*, 92, 1882, 1–122 = *Werke*, 2, 237–387; also published separately by G. Reimer, 1882.

$K(x)$ obtained by adjoining the indeterminate x to K, that is, the smallest field containing K and x, is isomorphic to the field $K[x]$ of rational functions in one variable with coefficients in K.[21] He did stress that the indeterminate was merely an element of an algebra and not a variable in the sense of analysis.[22] He then showed in 1887[23] that to each ordinary prime number p there corresponds within the ring $Q(x)$ of polynomials with rational coefficients a prime polynomial $p(x)$ which is irreducible in the rational field Q. By considering two polynomials to be equal if they are congruent modulo a given prime polynomial $p(x)$, the ring of all polynomials in $Q(x)$ becomes a field of residue classes possessing the same algebraic properties as the algebraic number field $K(\delta)$ arising from the field K by adjoining a root δ of $p(x) = 0$. Here he used the idea Cauchy had already employed to introduce imaginary numbers by using polynomials congruent modulo $x^2 + 1$. In this same work he showed that the theory of algebraic numbers is independent of the fundamental theorem of algebra and of the theory of the complete real number system.

In his theory of fields (in the "Grundzüge") whose elements are formed by starting with a field K and then adjoining indeterminates $x_1, x_2, \ldots, x_n$, Kronecker introduced the notion of a modular system that played the role of ideals in Dedekind's theory. For Kronecker a modular system is a set M of those polynomials in n variables $x_1, x_2, \ldots, x_n$ such that if P_1 and P_2 belong to the set so does $P_1 + P_2$ and if P belongs so does QP where Q is any polynomial in $x_1, x_2, \ldots, x_n$.

A basis of a modular system M is any set of polynomials $B_1, B_2 \ldots$ of M such that every polynomial of M is expressible in the form

$$R_1 B_1 + R_2 B_2 + \cdots,$$

where $R_1, R_2, \ldots$ are constants or polynomials (not necessarily belonging to M). The theory of divisibility in Kronecker's general fields was defined in terms of modular systems much as Dedekind had done with ideals.

The work on algebraic number theory was climaxed in the nineteenth century by Hilbert's famous report on algebraic numbers.[24] This report is primarily an account of what had been done during the century. However, Hilbert reworked all of this earlier theory and gave new, elegant, and powerful methods of securing these results. He had begun to create new ideas in algebraic number theory from about 1892 on and one of the new creations on Galoisian number fields was also incorporated in the report. Subsequently Hilbert and many other men extended algebraic number theory vastly.

21. *Werke*, 2, 253.
22. *Werke*, 2, 339.
23. *Jour. für Math.*, 100, 1887, 490–510 = *Werke*, 3, 211–40.
24. "Die Theorie der algebraischen Zahlkörper" (The Theory of Algebraic Number Fields), *Jahres. der Deut. Math.-Ver.*, 4, 1897, 175–546 = *Ges. Abh.*, 1, 63–363.

However, these later developments, relative Galoisian fields, relative Abelian number fields and class fields, all of which stimulated an immense amount of work in the twentieth century, are of concern primarily to specialists.

Algebraic number theory, originally a scheme for investigating the solutions of problems in the older theory of numbers, has become an end in itself. It has come to occupy a position in between the theory of numbers and abstract algebra, and now number theory and modern higher algebra merge in algebraic number theory. Of course algebraic number theory has also produced new theorems in the ordinary theory of numbers.

5. *The Theory of Forms*

Another class of problems in the theory of numbers is the representation of integers by forms. The expression

$$(3) \qquad\qquad ax^2 + 2bxy + cy^2,$$

wherein a, b, and c are integral, is a binary form because two variables are involved, and it is a quadratic form because it is of the second degree. A number M is said to be represented by the form if for specific integral values of a, b, c, x, and y the above expression equals M. One problem is to find the set of numbers M that are representable by a given form or class of forms. The converse problem, given M and given a, b, and c or some class of a, b, and c, to find the values of x and y that represent M, is equally important. The latter problem belongs to Diophantine analysis and the former may equally well be considered part of the same subject.

Euler had obtained some special results on these problems. However, Lagrange made the key discovery that if a number is representable by one form it is also representable by many other forms, which he called equivalent. The latter could be obtained from the original form by a change of variables

$$(4) \qquad\qquad x = \alpha x' + \beta y', \qquad y = \gamma x' + \delta y'$$

wherein α, β, γ, and δ are integral and $\alpha\delta - \beta\gamma = 1$.[25] In particular, Lagrange showed that for a given discriminant (Gauss used the word determinant) $b^2 - ac$ there is a finite number of forms such that each form with that discriminant is equivalent to one of this finite number. Thus all forms with a given discriminant can be segregated into classes, each class consisting of forms equivalent to one member of that class. This result and some inductively established results by Legendre attracted Gauss's attention. In a bold step Gauss extracted from Lagrange's work the notion of equivalence of forms and concentrated on that. The fifth section of his *Disquisitiones*, by far the largest section, is devoted to this subject.

25. *Nouv. Mém. de l'Acad. de Berlin*, 1773, 263–312; and 1775, 323 ff. = *Œuvres*, 3, 693–795 .

Gauss systematized and extended the theory of forms. He first defined equivalence of forms. Let

$$F = ax^2 + 2bxy + cy^2$$

be transformed by means of (4) into the form

$$F' = a'x'^2 + 2b'x'y' + c'y'^2.$$

Then

$$b'^2 - a'c' = (b^2 - ac)(\alpha\delta - \beta\gamma)^2.$$

If now $(\alpha\delta - \beta\gamma)^2 = 1$, the discriminants of the two forms are equal. Then the inverse of the transformation (4) will also contain integral coefficients (by Cramer's rule) and will transform F' into F. F and F' are said to be equivalent. If $\alpha\delta - \beta\gamma = 1$, F and F' are said to be properly equivalent, and if $\alpha\delta - \beta\gamma = -1$, then F and F' are said to be improperly equivalent.

Gauss proved a number of theorems on the equivalence of forms. Thus if F is equivalent to F' and F' to F'' then F is equivalent to F''. If F is equivalent to F', any number M representable by F is representable by F' and in as many ways by one as by the other. He then shows, if F and F' are equivalent, how to find all the transformations from F into F'. He also finds all the representations of a given number M by a form F, provided the values of x and y are relatively prime.

By definition two equivalent forms have the same value for their discriminant $D = b^2 - ac$. However, two forms with the same discriminant are not necessarily equivalent. Gauss shows that all forms with a given D can be segregated into classes; the members of any one class are properly equivalent to each other. Though the number of forms with a given D is infinite, the number of classes for a given D is finite. In each class one form can be taken as representative and Gauss gives criteria for the choice of a simplest representative. The simplest form of all those with determinant D has $a = 1, b = 0, c = -D$. This he calls the principal form and the class to which it belongs is called the principal class.

Gauss then takes up the composition (product) of forms. If the form

$$F = AX^2 + 2BXY + CY^2$$

is transformed into the product of two forms

$$f = ax^2 + 2bxy + cy^2 \quad \text{and} \quad f' = a'x'^2 + 2b'x'y' + c'y'^2$$

by the substitution

$$X = p_1xx' + p_2xy' + p_3x'y + p_4yy'$$
$$Y = q_1xx' + q_2xy' + q_3x'y + q_4yy'$$

828 THE THEORY OF NUMBERS 1800–1900

then F is said to be transformable into ff'. If further the six numbers

$$p_1q_2 - q_1p_2, p_1q_3 - q_1p_3, p_1q_4 - q_1p_4, p_2q_3 - q_2p_3, p_2q_4 - q_2p_4, p_3q_4 - q_3p_4$$

do not have a common divisor, then F is called a composite of the forms f and f'.

Gauss was then able to prove an essential theorem: If f and g belong to the same class and f' and g' belong to the same class, then the form composed of f and f' will belong to the same class as the form composed of g and g'. Thus one can speak of a class of forms composed of two (or more) given classes. In this composition of classes, the principal class acts as a unit class; that is, if the class K is composed with a principal class, the resulting class will be K.

Gauss now turns to a treatment of ternary quadratic forms

$$Ax^2 + 2Bxy + Cy^2 + 2Dxz + 2Eyz + Fz^2,$$

where the coefficients are integers, and undertakes a study very much like what he has just done for binary forms. The goal as in the case of binary forms is the representation of integers. The theory of ternary forms was not carried far by Gauss.

The objective of the entire work on the theory of forms was, as already noted, to produce theorems in the theory of numbers. In the course of his treatment of forms Gauss shows how the theory can be used to prove any number of theorems about the integers including many that were previously proved by such men as Euler and Lagrange. Thus Gauss proves that any prime number of the form $4n + 1$ can be represented as a sum of squares in one and only one way. Any prime number of the form $8n + 1$ or $8n + 3$ can be represented in the form $x^2 + 2y^2$ (for positive integral x and y) in one and only one way. He shows how to find all the representations of a given number M by the given form $ax^2 + 2bxy + cy^2$ provided the discriminant D is a positive non-square number. Further if F is a primitive form (the values of a, b, and c are relatively prime) with discriminant D and if p is a prime number dividing D, then the numbers not divisible by p which can be represented by F agree in that they are all either quadratic residues of p or non-residues of p.

Among the results Gauss drew from his work on ternary quadratic forms is the first proof of the theorem that every number can be represented as the sum of three triangular numbers. These, we recall, are the numbers

$$1, 3, 6, 10, 15, \cdots, \frac{n^2 + n}{2}, \cdots.$$

He also re-proved the theorem already proved by Lagrange that any positive integer can be expressed as the sum of four squares. Apropos of the former result, it is worth noting that Cauchy read a paper to the Paris Academy in

1815 which established the general result first asserted by Fermat that every integer is the sum of k or fewer k-gonal numbers.[26] (The general k-gonal number is $n + (n^2 - n)(k - 2)/2$.)

The algebraic theory of binary and ternary quadratic forms as presented by Gauss has an interesting geometrical analogue which Gauss himself initiated. In a review which appeared in the *Göttingische Gelehrte Anzeigen* of 1830[27] of a book on ternary quadratic forms by Ludwig August Seeber (1793–1855) Gauss sketched the geometrical representation of his forms and classes of forms.[28] This work is the beginning of a development called the geometrical theory of numbers which first gained prominence when Hermann Minkowski (1864–1909), who served as professor of mathematics at several universities, published his *Geometrie der Zahlen* (1896).

The subject of forms became a major one in the theory of numbers of the nineteenth century. Further work was done by a host of men on binary and ternary quadratic forms and on forms with more variables and of higher degree.[29]

6. Analytic Number Theory

One of the major developments in number theory is the introduction of analytic methods and of analytic results to express and prove facts about integers. Actually, Euler had used analysis in number theory (see below) and Jacobi used elliptic functions to obtain results in the theory of congruences and the theory of forms.[30] However, Euler's uses of analysis in number theory were minor and Jacobi's number-theoretic results were almost accidental by-products of his analytic work.

The first deep and deliberate use of analysis to tackle what seemed to be a clear problem of algebra was made by Peter Gustav Lejeune-Dirichlet (1805–59), a student of Gauss and Jacobi, professor at Breslau and Berlin, and then successor to Gauss at Göttingen. Dirichlet's great work, the *Vorlesungen über Zahlentheorie*,[31] expounded Gauss's *Disquisitiones* and gave his own contributions.

The problem that caused Dirichlet to employ analysis was to show that every arithmetic sequence

$$a, a + b, a + 2b, a + 3b, \cdots, a + nb, \cdots,$$

26. *Mém. de l'Acad. des Sci.*, Paris, (1), 14, 1813–15, 177–220 = *Œuvres*, (2), 6, 320–53.
27. *Werke*, 2, 188–196.
28. Felix Klein in his *Entwicklung* (see the bibliography at the end of this chapter), pp. 35–39, expands on Gauss's sketch.
29. See works of Smith and Dickson, listed in the bibliography, for further details.
30. *Jour. für Math.*, 37, 1848, 61–94 and 221–54 = *Werke*, 2, 219–88.
31. Published 1863; the second, third, and fourth editions of 1871, 1879, and 1894 were supplemented extensively by Dedekind.

where a and b are relatively prime, contains an infinite number of primes. Euler[32] and Legendre[33] made this conjecture and in 1808 Legendre[34] gave a proof that was faulty. In 1837 Dirichlet[35] gave a correct proof. This result generalizes Euclid's theorem on the infinitude of primes in the sequence $1, 2, 3, \ldots$. Dirichlet's analytical proof was long and complicated. Specifically it used what are now called the Dirichlet series, $\sum_{n=1}^{\infty} a_n n^{-z}$, wherein the a_n and z are complex. Dirichlet also proved that the sum of the reciprocals of the primes in the sequence $\{a + nb\}$ diverges. This extends a result of Euler on the usual primes (see below). In 1841[36] Dirichlet proved a theorem on the primes in progressions of complex numbers $a + bi$.

The chief problem involving the introduction of analysis concerned the function $\pi(x)$ which represents the number of primes not exceeding x. Thus $\pi(8)$ is 4 since 2, 3, 5, and 7 are prime and $\pi(11)$ is 5. As x increases the additional primes become scarcer and the problem was, What is the proper analytical expression for $\pi(x)$? Legendre, who had proved that no rational expression would serve, at one time gave up hope that any expression could be found. Then Euler, Legendre, Gauss, and others surmised that

$$(5) \qquad \lim_{x \to \infty} \frac{\pi(x)}{x/\log x} = 1.$$

Gauss used tables of primes (he actually studied all the primes up to 3,000,000) to make conjectures about $\pi(x)$ and inferred[37] that $\pi(x)$ differs little from $\int_2^x dt/\log t$. He knew also that

$$\lim_{x \to \infty} \frac{\int_2^x dt/\log t}{x/\log x} = 1.$$

In 1848 Pafnuti L. Tchebycheff (1821–94), a professor at the University of Petrograd, took up the question of the number of prime numbers less than or equal to x and made a big step forward in this old problem. In a key paper, "Sur les nombres premiers"[38] Tchebycheff proved that

$$A_1 < \frac{\pi(x)}{x/\log x} < A_2,$$

where $0.922 < A_1 < 1$ and $1 < A_2 < 1.105$, but did not prove that the function tends to a limit. This inequality was improved by many mathe-

32. *Opuscula Analytica*, 2, 1783.
33. *Mém. de l'Acad. des Sci., Paris*, 1785, 465–559, pub. 1788.
34. *Théorie des nombres*, 2nd ed., p. 404.
35. *Abh. König. Akad. der Wiss., Berlin*, 1837, 45–81 and 108–10 = *Werke*, 1, 307–42.
36. *Abh. König. Akad. der Wiss., Berlin*, 1841, 141–61 = *Werke*, 2, 509–32.
37. *Werke*, 2, 444–47.
38. *Mém. Acad. Sci. St. Peters.*, 7, 1854, 15–33; also *Jour. de Math.*, (1), 17, 1852, 366–90 = *Œuvres*, 1, 51–70.

maticians including Sylvester, who among others doubted in 1881 that the function had a limit. In his work Tchebycheff used what we now call the Riemann zeta function,

$$\zeta(z) = \sum_{n=1}^{\infty} \frac{1}{n^z}$$

though he used it for real values of z. (This series is a special case of Dirichlet's series.) Incidentally he also proved in the same paper that for $n > 3$ there is always at least one prime between n and $2n - 2$.

The zeta function for real z appears in a work of Euler[39] where he introduced

$$\zeta(s) = \sum_{n=1}^{\infty} \frac{1}{n^s} = \prod_{n=1}^{\infty} \left(1 - \frac{1}{p_n^s}\right)^{-1}.$$

Here the p_n's are prime numbers. Euler used the function to prove that the sum of the reciprocals of the prime numbers diverges. For s an even positive integer Euler knew the value of $\zeta(s)$ (Chap. 20, sec. 4). Then in a paper read in 1749[40] Euler asserted that for real s

$$\zeta(1 - s) = 2(2\pi)^{-s} \cos \frac{\pi s}{2} \, \Gamma(s)\zeta(s).$$

He verified the equation to the point where, he said, there was no doubt about it. This relation was established by Riemann in the 1859 paper referred to below. Riemann, using the zeta function for complex z, attempted to prove the prime number theorem, that is, (5) above.[41] He pointed out that to further the investigation one would have to know the complex zeros of $\zeta(z)$. Actually, $\zeta(z)$ fails to converge for $x \leq 1$ when $z = x + iy$, but the values of ζ in the half-plane $x \leq 1$ are defined by analytic continuation. He expressed the hypothesis that all the zeros in the strip $0 \leq x \leq 1$ lie on the line $x = 1/2$. This hypothesis is still unproven.[42]

In 1896 Hadamard,[43] by applying the theory of entire functions (of a complex variable), which he investigated for the purpose of proving the prime number theorem, and by proving the crucial fact that $\zeta(z) \neq 0$ for $x = 1$, was able finally to prove the prime number theorem. Charles-Jean de la Vallée Poussin (1866–1962) obtained the result for the zeta function

39. *Comm. Acad. Sci. Petrop.*, 9, 1737, 160–88, pub. 1744 = *Opera*, (1), 14, 216–44.
40. *Hist. de l'Acad. de Berlin*, 17, 1761, 83–106, pub. 1768 = *Opera*, (1), 15, 70–90.
41. *Monatsber. Berliner Akad.*, 1859, 671–80 = *Werke*, 145–55.
42. In 1914 Godfrey H. Hardy proved (*Comp. Rend.*, 158, 1914, 1012–14 = *Coll. Papers*, 2, 6–9) that an infinity of zeros of $\zeta(z)$ lie on the line $x = \frac{1}{2}$.
43. *Bull. Soc. Math. de France*, 14, 1896, 199–220 = *Œuvres*, 1, 189–210.

and proved the prime number theorem at the same time.[44] This theorem is a central one in analytic number theory.

Bibliography

Bachmann, P.: "Über Gauss' zahlentheoretische Arbeiten," *Nachrichten König. Ges. der Wiss. zu Gött.*, 1911, 455–508; also in Gauss: *Werke*, 10_2, 1–69.

Bell, Eric T.: *The Development of Mathematics*, 2nd ed., McGraw-Hill, 1945, Chaps. 9–10.

Carmichael, Robert D.: "Some Recent Researches in the Theory of Numbers," *Amer. Math. Monthly*, 39, 1932, 139–60.

Dedekind, Richard: *Über die Theorie der ganzen algebraischen Zahlen* (reprint of the eleventh supplement to Dirichlet's *Zahlentheorie*), F. Vieweg und Sohn, 1964.

———: *Gesammelte mathematische Werke*, 3 vols., F. Vieweg und Sohn, 1930–32, Chelsea (reprint), 1968.

———: "Sur la théorie des nombres entiers algébriques," *Bull. des Sci. Math.*, (1), 11, 1876, 278–88; (2), 1, 1877, 17–41, 69–92, 144–64, 207–48 = *Ges. math. Werke*, 3, 263–96.

Dickson, Leonard E.: *History of the Theory of Numbers*, 3 vols., Chelsea (reprint), 1951.

———: *Studies in the Theory of Numbers* (1930), Chelsea (reprint), 1962.

———: "Fermat's Last Theorem and the Origin and Nature of the Theory of Algebraic Numbers," *Annals. of Math.*, (2), 18, 1917, 161–87.

——— et al.: *Algebraic Numbers, Report of Committee on Algebraic Numbers*, National Research Council, 1923 and 1928; Chelsea (reprint), 1967.

Dirichlet, P. G. L.: *Werke* (1889–97); Chelsea (reprint), 1969, 2 vols.

Dirichlet, P. G. L., and R. Dedekind: *Vorlesungen über Zahlentheorie*, 4th ed., 1894 (contains Dedekind's Supplement); Chelsea (reprint), 1968.

Gauss, C. F.: *Disquisitiones Arithmeticae*, trans. A. A. Clarke, Yale University Press, 1965.

Hasse, H.: "Bericht über neuere Untersuchungen und Probleme aus der Theorie der algebraischen Zahlkörper," *Jahres. der Deut. Math.-Verein.*, 35, 1926, 1–55 and 36, 1927, 233–311.

Hilbert, David: "Die Theorie der algebraischen Zahlkörper," *Jahres. der Deut. Math.-Verein.*, 4, 1897, 175–546 = *Gesammelte Abhandlungen*, 1, 63–363.

Klein, Felix: *Vorlesungen über die Entwicklung der Mathematik im 19. Jahrhundert*, Chelsea (reprint), 1950, Vol. 1.

Kronecker, Leopold: *Werke*, 5 vols. (1895–1931), Chelsea (reprint), 1968. See especially, Vol. 2, pp. 1–10 on the law of quadratic reciprocity.

———: *Grundzüge einer arithmetischen Theorie der algebraischen Grössen*, G. Reimer, 1882 = *Jour. für Math.*, 92, 1881/82, 1–122 = *Werke*, 2, 237–388.

Landau, Edmund: *Handbuch der Lehre von der Verteilung der Primzahlen*, B. G. Teubner, 1909, Vol. 1, pp. 1–55.

Mordell, L. J.: "An Introductory Account of the Arithmetical Theory of Algebraic

44. *Ann. Soc. Sci. Bruxelles*, (1), 20 Part II, 1896, 183–256, 281–397.

Numbers and its Recent Development," *Amer. Math. Soc. Bull.*, 29, 1923, 445–63.

Reichardt, Hans, ed.: *C. F. Gauss, Leben und Werk*, Haude und Spenersche Verlagsbuchhandlung, 1960, pp. 38–91; also B. G. Teubner, 1957.

Scott, J. F.: *A History of Mathematics*, Taylor and Francis, 1958, Chap. 15.

Smith, David E.: *A Source Book in Mathematics*, Dover (reprint), 1959, Vol. 1, 107–48.

Smith, H. J. S.: Collected Mathematical Papers, 2 vols. (1890–94), Chelsea (reprint), 1965. Vol. 1 contains Smith's *Report on the Theory of Numbers*, which is also published separately by Chelsea, 1965.

Vandiver, H. S.: "Fermat's Last Theorem," *Amer. Math. Monthly*, 53, 1946, 555–78.

35
The Revival of Projective Geometry

The doctrines of pure geometry often, and in many questions, give a simple and natural way to penetrate to the origin of truths, to lay bare the mysterious chain which unites them, and to make them known individually, luminously and completely.   MICHEL CHASLES

1. The Renewal of Interest in Geometry

For over one hundred years after the introduction of analytic geometry by Descartes and Fermat, algebraic and analytic methods dominated geometry almost to the exclusion of synthetic methods. During this period some men, for example the English mathematicians who persisted in trying to found the calculus rigorously on geometry, produced new results synthetically. Geometric methods, elegant and intuitively clear, always captivated some minds. Maclaurin, especially, preferred synthetic geometry to analysis. Pure geometry, then, retained some life even if it was not at the heart of the most vital developments of the seventeenth and eighteenth centuries. In the early nineteenth century several great mathematicians decided that synthetic geometry had been unfairly and unwisely neglected and made a positive effort to revive and extend that approach.

One of the new champions of synthetic methods, Jean-Victor Poncelet, did concede the limitations of the older pure geometry. He says, "While analytic geometry offers by its characteristic method general and uniform means of proceeding to the solution of questions which present themselves . . . while it arrives at results whose generality is without bound, the other [synthetic geometry] proceeds by chance; its way depends completely on the sagacity of those who employ it and its results are almost always limited to the particular figure which one considers." But Poncelet did not believe that synthetic methods were necessarily so limited and he proposed to create new ones which would rival the power of analytic geometry.

Michel Chasles (1793–1880) was another great proponent of geometrical methods. In his *Aperçu historique sur l'origine et le développement des méthodes en géométrie* (1837), a historical study in which Chasles admitted that he ignored the German writers because he did not know the language, he states that the mathematicians of his time and earlier had declared geometry a dead language which in the future would be of no use and influence. Not only does Chasles deny this but he cites Lagrange, who was entirely an analyst, as saying in his sixtieth year,[1] when he had encountered a very difficult problem of celestial mechanics, "Even though analysis may have advantages over the old geometric methods, which one usually, but improperly, calls synthesis, there are nevertheless problems in which the latter appear more advantageous, partly because of their intrinsic clarity and partly because of the elegance and ease of their solutions. There are even some problems for which the algebraic analysis in some measure does not suffice and which, it appears, the synthetic methods alone can master." Lagrange cites as an example the very difficult problem of the attraction of an ellipsoid of revolution exerted on a point (unit mass) on its surface or inside. This problem had been solved purely synthetically by Maclaurin.

Chasles also gives an extract from a letter he received from Lambert Adolphe Quetelet (1796–1874), the Belgian astronomer and statistician. Quetelet says, "It is not proper that most of our young mathematicians value pure geometry so lightly." The young men complain of lack of generality of method, continues Quetelet, but is this the fault of geometry or of those who cultivate geometry, he asks. To counter this lack of generality, Chasles gives two rules to prospective geometers. They should generalize special theorems to obtain the most general result, which should at the same time be simple and natural. Second, they should not be satisfied with the proof of a result if it is not part of a general method or doctrine on which it is naturally dependent. To know when one really has the true basis for a theorem, he says, there is always a principal truth which one will recognize because other theorems will result from a simple transformation or as a ready consequence. The great truths, which are the foundations of knowledge, always have the characteristics of simplicity and intuitiveness.

Other mathematicians attacked analytic methods in harsher language. Carnot wished "to free geometry from the hieroglyphics of analysis." Later in the century Eduard Study (1862–1922) referred to the machine-like process of coordinate geometry as the "clatter of the coordinate mill."

The objections to analytic methods in geometry were based on more than a personal preference or taste. There was, first of all, a genuine question of whether analytic geometry was really geometry since algebra was the essence of the method and results, and the geometric significance of both were hidden. Moreover, as Chasles pointed out, analysis through its formal

1. *Nouv. Mém. de l'Acad. de Berlin,* 1773, 121–48, pub. 1775 = *Œuvres,* 3, 617–58.

processes neglects all the small steps which geometry continually makes. The quick and perhaps penetrating steps of analysis do not reveal the sense of what is accomplished. The connection between the starting point and the final result is not clear. Chasles asks, "Is it then sufficient in a philosophic and basic study of a science to know that something is true if one does not know why it is so and what place it should take in the series of truths to which it belongs?" The geometric method, on the other hand, permits simple and intuitively evident proofs and conclusions.

There was another argument which, first voiced by Descartes, still appealed in the nineteenth century. Geometry was regarded as the truth about space and the real world. Algebra and analysis were not in themselves significant truths even about numbers and functions. They were merely methods of arriving at truths, and artificial at that. This view of algebra and analysis was gradually disappearing. Nevertheless the criticism was still vigorous in the early nineteenth century because the methods of analysis were incomplete and even logically unsound. The geometers rightly questioned the validity of the analytic proofs and credited them with merely suggesting results. The analysts could retort only that the geometric proofs were clumsy and inelegant.

The upshot of the controversy is that the pure geometers reasserted their role in mathematics. As if to revenge themselves on Descartes because his creation of analytic geometry had caused the abandonment of pure geometry, the early nineteenth-century geometers made it their objective to beat Descartes at the game of geometry. The rivalry between analysts and geometers grew so bitter that Steiner, who was a pure geometer, threatened to quit writing for Crelle's *Journal für Mathematik* if Crelle continued to publish the analytical papers of Plücker.

The stimulus to revive synthetic geometry came primarily from one man, Gaspard Monge. We have already discussed his valuable contributions to analytic and differential geometry and his inspiring lectures at the Ecole Polytechnique during the years 1795 to 1809. Monge himself did not intend to do more than bring geometry back into the fold of mathematics as a suggestive approach to and an interpretation of analytic results. He sought only to stress both modes of thought. However, his own work in geometry and his enthusiasm for it inspired in his pupils, Charles Dupin, François-Joseph Servois, Charles-Julien Brianchon, Jean-Baptiste Biot (1774–1862), Lazare-Nicholas-Marguerite Carnot, and Jean-Victor Poncelet, the urge to revitalize pure geometry.

Monge's contribution to pure geometry was his *Traité de géométrie descriptive* (1799). This subject shows how to project orthogonally a three-dimensional object on two planes (a horizontal and a vertical one) so that from this representation one can deduce mathematical properties of the object. The scheme is useful in architecture, the design of fortifications, perspective, carpentry, and stonecutting and was the first to treat the projection of a three-dimensional figure into two two-dimensional ones. The ideas and method of descriptive geometry did not prove to be the avenue to subsequent developments in geometry or, for that matter, to any other part of mathematics.

2. Synthetic Euclidean Geometry

Though the geometers who reacted to Monge's inspiration turned to projective geometry, we shall pause to note some new results in synthetic Euclidean geometry. These results, perhaps minor in significance, nevertheless exhibit new themes and the almost inexhaustible richness of this old subject. Actually hundreds of new theorems were produced, of which we can give just a few examples.

Associated with every triangle ABC are nine particular points, the midpoints of the sides, the feet of the three altitudes, and the midpoints of the segments which join the vertices to the point of intersection of the altitudes. All nine points lie on one circle, called the nine-point circle. This theorem was first published by Gergonne and Poncelet.[2] It is often credited to Karl Wilhelm Feuerbach (1800-34), a high-school teacher, who published his proof in *Eigenschaften einiger merkwürdigen Punkte des geradlinigen Dreiecks* (Properties of Some Distinctive Points of the Rectilinear Triangle, 1822). In this book Feuerbach added another fact about the nine-point circle. An excircle (escribed circle) is one which is tangent to one of the sides and to the extensions of the other two sides. (The center of an escribed circle lies on the bisectors of two exterior angles and the remote interior angle.) Feuerbach's theorem states that the nine-point circle is tangent to the inscribed circle and the three excircles.

In a small book published in 1816, *Über einige Eigenschaften des ebenen geradlinigen Dreiecks* (On Some Properties of Plane Rectilinear Triangles), Crelle showed how to determine a point P inside a triangle such that the lines joining P to the vertices of the triangle and the sides of the triangle make equal angles. That is, $\sphericalangle 1 = \sphericalangle 2 = \sphericalangle 3$ in Figure 35.1. There is also a point P' different from P such that $\sphericalangle P'AC = \sphericalangle P'CB = \sphericalangle P'BA$.

The conic sections, we know, were treated definitively by Apollonius as sections of a cone and then introduced as plane loci in the seventeenth century. In 1822 Germinal Dandelin (1794–1847) proved a very interesting theorem about the conic sections in relation to the cone.[3] His theorem states that if two spheres are inscribed in a circular cone so that they are tangent to a given plane cutting the cone in a conic section, the points of contact of the

2. *Ann. de Math.*, 11, 1820/21, 205–20.
3. *Nouv. Mém. de l'Acad. Roy. des Sci.*, Bruxelles, 2, 1822, 169–202.

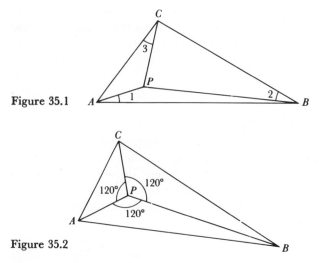

Figure 35.1

Figure 35.2

spheres with the plane are the foci of the conic section, and the intersections of the plane with the planes of the circles along which the spheres touch the cone are the directrices of the conic.

Another interesting theme pursued in the nineteenth century was the solution of maximum and minimum problems by purely geometric methods, that is, without relying upon the calculus of variations. Of the several theorems Jacob Steiner proved by using synthetic methods, the most famous result is the isoperimetric theorem: Of all plane figures with a given perimeter the circle bounds the greatest area. Steiner gave various proofs.[4] Unfortunately, Steiner assumed that there exists a curve that does have maximum area. Dirichlet tried several times to persuade him that his proofs were incomplete on that account but Steiner insisted that this was self-evident. Once, however, he did write (in the first of the 1842 papers):[5] "and the proof is readily made if one assumes that there is a largest figure."

The proof of the existence of a maximizing curve baffled mathematicians for a number of years until Weierstrass in his lectures of the 1870s at Berlin resorted to the calculus of variations.[6] Later Constantin Caratheodory (1873–1950) and Study[7] in a joint paper rigorized Steiner's proofs without employing that calculus. Their proofs (there were two) were direct rather than indirect as in Steiner's method. Hermann Amandus Schwarz, who did great work in partial differential equations and analysis and who

4. *Jour. für Math.*, 18, 1838, 281–96; and 24, 1842, 83–162, 189–250; the 1842 papers are in his *Ges. Werke*, 2, 177–308.
5. *Ges. Werke*, 2, 197.
6. *Werke*, 7, 257–64, 301–2.
7. *Math. Ann.*, 68, 1909, 133–40 = Caratheodory, *Ges. math. Schriften*, 2, 3–11.

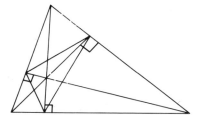

Figure 35.3

served as a professor at several universities including Göttingen and Berlin, gave a rigorous proof for the isoperimetric problem in three dimensions.[8]

Steiner also proved (in the first of the 1842 papers) that of all triangles with a given perimeter, the equilateral has the greatest area. Another of his results[9] states that if A, B, and C are three given points (Fig. 35.2) and if each of the angles of triangle ABC is less than 120°, then the point P for which $PA + PB + PC$ is a minimum is such that each of the angles at P is 120°. If, however, one angle of the triangle, say angle A, is equal to or larger than 120°, then P coincides with A. This result had been proven much earlier by Cavalieri (*Exercitationes Geometricae \Sex*, 1647) but it was undoubtedly unknown to Steiner. Steiner also extended the result to n points.

Schwarz solved the following problem: Given an acute-angled triangle, consider all triangles such that each has its vertices on the three sides of the original triangle; the problem is to find the triangle that has least perimeter. Schwarz proved synthetically[10] that the vertices of this triangle of minimum perimeter are the feet of the altitudes of the given triangle (Fig. 35.3).[11]

A theorem novel for Euclidean geometry was discovered in 1899 by Frank Morley, professor of mathematics at Johns Hopkins University, and proofs were subsequently published by many men.[12] The theorem states that if the angle trisectors are drawn at each vertex of a triangle, adjacent trisectors meet at the vertices of an equilateral triangle (Fig. 35.4). The novelty lies in the involvement of angle trisectors. Up to the middle of the nineteenth century no mathematician would have considered such lines because only those elements and figures that are constructible were regarded as having legitimacy in Euclidean geometry. Constructibility guaranteed

8. *Nachrichten König. Ges. der Wiss. zu Gött.*, 1884, 1–13 = *Ges. math. Abh.*, 2, 327–40.
9. *Monatsber. Berliner Akad.*, 1837, 144 = *Ges. Werke*, 2, 93 and 729–31.
10. Paper unpublished, *Ges. Math. Abh.*, 2, 344–45.
11. Schwarz's proof can be found in Richard Courant and Herbert Robbins, *What Is Mathematics?*, Oxford University Press, 1941, pp. 346–49. A proof using the calculus was given by J. F. de' Toschi di Fagnano (1715–97) in the *Acta Eruditorum*, 1775, p. 297. There were less elegant geometrical proofs before Schwarz's.
12. For a proof and references to published proofs see H. S. M. Coxeter, *Introduction to Geometry*, John Wiley and Sons, 1961, pp. 23–25.

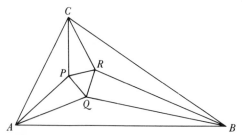

Figure 35.4

existence. However, the conception of what established existence changed as we shall see more clearly when we examine the work on the foundations of Euclidean geometry.

A number of efforts were directed toward reducing the use of straightedge and compass along the lines initiated by Mohr and Mascheroni (Chap. 12, sec. 2). In his *Traité* of 1822 Poncelet showed that all constructions possible with straightedge and compass (except the construction of circular arcs) are possible with straightedge alone provided that we are given a fixed circle and its center. Steiner re-proved the same result more elegantly in a small book, *Die geometrischen Constructionen ausgeführt mittelst der geraden Linie und eines festen Kreises* (The Geometrical Constructions Executed by Means of the Straight Line and a Fixed Circle).[13] Though Steiner intended the book for pedagogical purposes, he does claim in the preface that he will establish the *conjecture* which a French mathematician had expressed.

The brief sampling above of Euclidean theorems established by synthetic methods should not leave the reader with the impression that analytic geometric methods were not also used. In fact, Gergonne gave analytic proofs of many geometric theorems which he published in the journal he founded, the *Annales de Mathématiques*.

3. *The Revival of Synthetic Projective Geometry*

The major area to which Monge and his pupils turned was projective geometry. This subject had had a somewhat vigorous but short-lived burst of activity in the seventeenth century (Chap. 14) but was submerged by the rise of analytic geometry, the calculus, and analysis. As we have already noted, Desargues's major work of 1639 was lost sight of until 1845 and Pascal's major essay on conics (1639) was never recovered. Only La Hire's books, which used some of Desargues's results, were available. What the nineteenth-century men learned from La Hire's books they often incorrectly credited to him. On the whole, however, these geometers were ignorant of Desargues's and Pascal's work and had to re-create it.

13. Published 1833 = *Werke*, 1, 461–522.

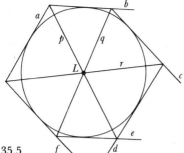

Figure 35.5

The revival of projective geometry was initiated by Lazare N. M. Carnot (1753–1823), a pupil of Monge and father of the distinguished physicist Sadi Carnot. His major work was *Géométrie de position* (1803) and he also contributed the *Essai sur la théorie des transversales* (1806). Monge had espoused the joint use of analysis and pure geometry, but Carnot refused to use analytic methods and started the championship of pure geometry. Many of the ideas we shall shortly discuss more fully are at least suggested in Carnot's work. Thus the principle that Monge called contingent relations and which became known also as the principle of correlativity and more commonly as the principle of continuity is to be found there. To avoid separate figures for various sizes of angles and directions of lines Carnot did not use negative numbers, which he regarded as contradictory, but introduced a complicated scheme called "correspondence of signs."

Among the early nineteenth-century workers in projective geometry we shall just mention François-Joseph Servois and Charles-Julien Brianchon (1785–1864), both of whom made applications of their work to military problems. Though they aided in reconstructing, systematizing, and extending old results, the only new theorem of consequence is Brianchon's famous result,[14] which he proved while still a student at the Ecole. The theorem states that if there are six tangents to a conic (Fig. 35.5), thus forming a circumscribed hexagon, the three lines, each of which joins two opposite vertices, pass through one point. Brianchon derived this theorem by using the pole-polar relationship.

The revival of projective geometry received its main impetus from Poncelet (1788–1867). Poncelet was a pupil of Monge and he also learned much from Carnot. While serving as an officer in Napoleon's campaign against Russia, he was captured and spent the year 1813–14 in a Russian prison at Saratoff. There Poncelet reconstructed without the aid of any books what he had learned from Monge and Carnot and then proceeded to create new results. He later expanded and revised this work and published

14. *Jour de l'Ecole Poly.*, 6, 1806, 297–311.

it as the *Traité des propriétés projectives des figures* (1822). This work was his chief contribution to projective geometry and to the erection of a new discipline. In his later life he was obliged to devote a great deal of time to government service, though he did hold professorships for limited periods.

Poncelet became the most ardent champion of synthetic geometry and even attacked the analysts. He had been friendly with the analyst Joseph-Diez Gergonne (1771–1859) and had published papers in Gergonne's *Annales de Mathématiques*, but his attacks were soon directed to Gergonne also. Poncelet was convinced of the autonomy and importance of pure geometry. Though he admitted the power of analysis he believed that one could give the same power to synthetic geometry. In a paper of 1818, published in Gergonne's *Annales*,[15] he said that the power of analytic methods lay not in the use of algebra but in its generality and this advantage resulted from the fact that the metric properties discovered for a typical figure remain applicable, other than for a possible change of sign, to all related figures which spring from the typical or basic one. This generality could be secured in synthetic geometry by the principle of continuity (which we shall examine shortly).

Poncelet was the first mathematician to appreciate fully that projective geometry was a new branch of mathematics with methods and goals of its own. Whereas the seventeenth-century projective geometers had dealt with specific problems, Poncelet entertained the general problem of seeking all properties of geometrical figures that were common to all sections of any projection of a figure, that is, remain unaltered by projection and section. This is the theme that he and his successors took up. Because distances and angles are altered by projection and section Poncelet selected and developed the theory of involution and of harmonic sets of points but not the concept of cross ratio. Monge had used parallel projection in his work; like Desargues, Pascal, Newton, and Lambert, Poncelet used central projection, that is, projection from a point. This concept Poncelet elevated into a method of approach to geometric problems. Poncelet also considered projective transformation from one space figure to another, of course in purely geometric form. Here he seemed to have lost interest in projective properties and was more concerned with the use of the method in bas-relief and stage design.

His work centers about three ideas. The first is that of homologous figures. Two figures are homologous if one can be derived from the other by one projection and a section, which is called a perspectivity, or by a sequence of projections and sections, that is, a projectivity. In working with homologous figures his plan was to find for a given figure a simpler homologous figure and by studying it find properties which are invariant under projection

---

15. *Ann. de Math.*, 8, 1817/18, 141–55. This paper is reprinted in Poncelet's *Applications d'analyse et de géométrie* (1862–64), 2, 466–76.

and section, and so obtain properties of the more complicated figure. The essence of this method was used by Desargues and Pascal, and Poncelet in his *Traité* praised Desargues's originality in this and other respects.

Poncelet's second leading theme is the principle of continuity. In his *Traité* he phrases it thus: "If one figure is derived from another by a continuous change and the latter is as general as the former, then any property of the first figure can be asserted at once for the second figure." The determination of when both figures are general is not explained. Poncelet's principle also asserts that if a figure should degenerate, as a hexagon does into a pentagon when one side is made to approach zero, any property of the original figure will carry over into an appropriately worded statement for the degenerate figure.

The principle was really not new with Poncelet. In a broad philosophical sense it goes back to Leibniz, who stated in 1687 that when the differences between two cases can be made smaller than any datum in the given, the differences can be made smaller than any given quantity in the result. Since Leibniz's time the principle was recognized and used constantly. Monge began the use of the principle of continuity to establish theorems. He wanted to prove a general theorem but used a special position of the figure to prove it and then maintained that the theorem was true generally, even when some elements in the figures became *imaginary*. Thus to prove a theorem about a line and a surface he would prove it when the line cuts the surface and then maintain that the result holds even when the line no longer cuts the surface and the points of intersection are imaginary. Neither Monge nor Carnot, who also used the principle, gave any justification of it.

Poncelet, who coined the term "principle of continuity," advanced the principle as an absolute truth and applied it boldly in his *Traité*. To "demonstrate" its soundness he takes the theorem on the equality of the products of the segments of intersecting chords of a circle and notes that when the point of intersection moves outside the circle one obtains the equality of the products of the secants and their external segments. Further, when one secant becomes a tangent, the tangent and its external segment become equal and their product continues to equal the product of the other secant and its external segment. All of this was reasonable enough, but Poncelet applied the principle to prove many theorems and, like Monge, extended the principle to make assertions about imaginary figures. (We shall note some examples later.)

The other members of the Paris Academy of Sciences criticized the principle of continuity and regarded it as having only heuristic value. Cauchy, in particular, criticized the principle but unfortunately his criticism was directed at applications made by Poncelet wherein the principle did work. The critics also charged that the confidence which Poncelet and others had in the principle really came from the fact that it could be justified on an algebraic basis. As a matter of fact, the notes Poncelet made in prison show that he did use analysis to test the soundness of the principle. These notes, incidentally, were written up by Poncelet and published by him in two volumes entitled *Applications d'analyse et de géométrie* (1862–64) which is really a revision of his *Traité* of 1822, and in the later work he does use analytic methods. Poncelet admitted that a proof could be based on algebra but he insisted that the principle did not depend on such a proof. However, it is quite certain that Poncelet relied on the algebraic method to see what should be the case and then affirmed the geometric results using the principle as a justification.

Chasles in his *Aperçu* defended Poncelet. Chasles's position was that the algebra is an a posteriori proof of the principle. However, he hedged by pointing out that one must be careful not to carry over from one figure to another any property which depends *essentially* on the elements being real or imaginary. Thus one section of a cone may be a hyperbola, and this has asymptotes. When the section is an ellipse the asymptotes become imaginary. Hence one should not prove a result about the asymptotes alone because these depend upon the particular nature of the section. Also one should not carry the results for a parabola over to the hyperbola because the cutting plane does not have a *general* position in the case of the parabola. Then he discusses the case of two intersecting circles which have a common chord. When the circles no longer intersect, the common chord is imaginary. The fact that the real common chord passes through two *real* points is, he says, an incidental or a contingent property. One must define the chord in some way which does not depend on the fact that it passes through real points when the circles intersect but so that it is a permanent property of the two circles in any position. Thus one can define it as the (real) radical axis, which means that it is a line such that from any point on it the tangents to the two circles are equal, or one can define it by means of the property that any circle drawn about any point of the line as a center cuts the two circles orthogonally.

Chasles, too, insisted that the principle of continuity is suited to treat imaginary elements in geometry. He first explains what one means by the imaginary in geometry. The imaginary elements pertain to a condition or state of a figure in which certain parts are nonexistent, provided these parts are real in another state of the figure. For, he adds, one cannot have any idea of imaginary quantities except by thinking of the related states in which the quantities are real. These latter states are the ones which he called "accidental" and which furnish the key to the imaginary in geometry. To prove results about imaginary elements one has only to take the general condition of the figure in which the elements are real, and then, according to the principle of accidental relations or the principle of continuity, one may conclude that the result holds when the elements are imaginary. "So one sees that the use and the consideration of the imaginary are completely justified." The principle of continuity was accepted during the nineteenth century as intuitively clear and therefore having the status of an axiom. The geometers used it freely and never deemed that it required proof.

Though Poncelet used the principle of continuity to assert results about imaginary points and lines he never gave any general definition of such elements. To introduce some imaginary points he gave a geometrical definition which is complicated and hardly perspicuous. We shall understand these imaginary elements more readily when we discuss them from an algebraic point of view. Despite the lack of clarity in Poncelet's approach he must be credited with introducing the notion of the circular points at infinity, that is, two imaginary points situated on the line at infinity and common to any two circles.[16] He also introduced the imaginary spherical circle which any two spheres have in common. He then proved that two real conics which do not intersect have *two* imaginary common chords, and two conics intersect in four points, real or imaginary.

The third leading idea in Poncelet's work is the notion of pole and polar with respect to a conic. The concept goes back to Apollonius and was used by Desargues (Chap. 14, sec. 3) and others in the seventeenth-century work on projective geometry. Also Euler, Legendre, Monge, Servois, and Brianchon had already used it. But Poncelet gave a general formulation of the transformation from pole to polar and conversely and used it in his *Traité* of 1822 and in his "Mémoire sur la théorie générale des polaires réciproques" presented to the Paris Academy in 1824[17] as a method of establishing many theorems.

One of Poncelet's objectives in studying polar reciprocation with respect to a conic was to establish the principle of duality. The workers in projective geometry had observed that theorems dealing with figures lying in one plane when rephrased by replacing the word "point" by "line" and "line" by "point" not only made sense but proved to be true. The reason for the validity of theorems resulting from such a rephrasing was not clear, and in fact Brianchon questioned the principle. Poncelet thought that the pole and polar relationship was the reason.

However, this relation required the mediation of a conic. Gergonne[18] insisted that the principle was a general one and applied to all statements and theorems except those involving metric properties. Pole and polar were not needed as an intermediary supporting device. He introduced the term "duality" to denote the relationship between the original and new theorem. He also observed that in three-dimensional situations point and plane are dual elements and the line is dual to itself.

To illustrate Gergonne's understanding of the principle of duality let us examine his dualization of Desargues's triangle theorem. We should note

16. *Traité*, 1, 48.
17. *Jour. für Math.*, 4, 1829, 1–71.
18. *Ann. de Math.*, 16, 1825–26, 209–31.

first what the dual of a triangle is. A triangle consists of three points not all on the same line, and the lines joining them. The dual figure consists of three lines not all on the same point, and the points joining them (the points of intersection). The dual figure is again a triangle and so the triangle is called self-dual. Gergonne invented the scheme of writing dual theorems in double columns with the dual alongside the original proposition.

Now let us consider Desargues's theorem, where this time the two triangles and the point O lie in one plane, and let us see what results when we interchange point and line. Gergonne in the 1825–26 paper already referred to wrote this theorem and its dual as follows:

Desargues's Theorem	*Dual of Desargues's Theorem*
If we have two triangles such that lines joining corresponding vertices pass through one point O, then corresponding sides intersect in three points on one straight line.	If we have two triangles such that points which are the joins of corresponding sides lie on one line O, then corresponding vertices are joined by three lines going through one point.

Here the dual theorem is the converse of the original theorem.

Gergonne's formulation of the general principle of duality was somewhat vague and had deficiencies. Though he was convinced it was a universal principle, he could not justify it and Poncelet rightly objected to the deficiencies. Also he disputed with Gergonne over priority of discovery of the principle (which really belongs to Poncelet) and even accused Gergonne of plagiarism. However, Poncelet did rely upon pole and polar and would not grant that Gergonne had taken a step forward in recognizing the wider application of the principle. Later discussions among Poncelet, Gergonne, Möbius, Chasles, and Plücker clarified the principle for all, and Möbius in his *Der barycentrische Calcul* and Plücker, later, gave a good statement of the relationship of the principle of duality to pole and polar: The notion of duality is independent of conics and quadric forms but agrees with pole and polar when the latter can be used. The logical justification of the general principle of duality was not supplied at this time.

The synthetic development of projective geometry was furthered by Jacob Steiner (1796–1863). He is the first of a German school of geometers who took over French ideas, notably Poncelet's, and favored synthetic methods to the extent of even hating analysis. The son of a Swiss farmer, he himself worked on the farm until he was nineteen years old. Though he was largely self-educated he ultimately became a professor at Berlin. In his younger years he was a teacher in Pestalozzi's school and was impressed by the importance of building up the geometric intuition. It was Pestalozzi's principle to get the student to create the mathematics with the lead of the teacher and the use of the Socratic method. Steiner went to extremes. He

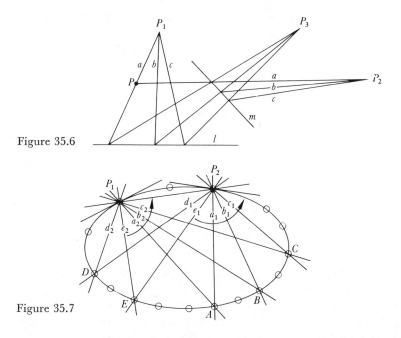

Figure 35.6

Figure 35.7

taught geometry but used no figures, and in training doctoral candidates he darkened the room. In his own later work Steiner took over theorems and proofs published in English and other journals and gave no indication in his own publications that his results had already been established. He had done good original work when younger and sought to maintain his reputation for productivity.

His major work was *Systematische Entwicklung der Abhängigkeit geometrischen Gestalten von einander* (Systematic Development of the Dependence of Geometric Forms on One Another, 1832) and his chief principle was to use projective concepts to build up more complicated structures from simple ones such as points, lines, pencils of lines, planes, and pencils of planes. His results were not especially new but his method was.

To illustrate his principle we shall examine his now standard projective method of defining the conic sections. One starts (Fig. 35.6) with two pencils of lines (families of concurrent lines), say p_1 and p_3, that are perspectively related through a pencil of points on a line l, and the pencils p_3 and p_2 that are perspectively related by means of a pencil of points on another line m. Then the pencils p_1 and p_2 are said to be projectively related. The lines marked a in the pencil with center at P_1 and the pencil with center at P_2 are examples of corresponding lines of the projectivity between the two pencils p_1 and p_2. A conic is now defined as the set of points of intersection of all pairs of corresponding lines of the two projective pencils. Thus P is a point on the conic. Moreover, the conic passes through P_1 and P_2 (Fig. 35.7). In this manner Steiner built up the conics or second degree curves by means of the simpler forms, pencils of lines. However, he did not identify his conics with sections of a cone.

He also built up the ruled quadrics, the hyperboloid of one sheet and the hyperbolic paraboloid, in a similar manner, making projective correspondence the basis of his definitions. Actually his method is not sufficiently general for all of projective geometry.

In his proofs he used cross ratio as a fundamental tool. However, he ignored imaginary elements and referred to them as "the ghosts" or the "shadows of geometry." He also did not use signed quantities though Möbius, whose work we shall soon examine, had already introduced them.

Steiner used the principle of duality from the very outset of his work. Thus he dualized the definition of a conic to obtain a new structure called a line curve. If one starts with two projectively (but not perspectively) related pencils of points then the family of lines (Fig. 35.8) joining the corresponding points of the two pencils is called a line conic. To distinguish such families of lines, which also describe a curve, the usual curve as a locus of points is called a point curve. The tangent lines of a point curve are a line curve and in the case of a conic do constitute the dual curve. Conversely every line conic envelopes a point conic or is the collection of tangents of a point conic.

With Steiner's notion of a dual to a point conic, one can dualize many theorems. Let us take Pascal's theorem and form the dual statement. We shall write the theorem on the left and the new statement on the right.

Pascal's Theorem	*Dual of Pascal's Theorem*
If we take six points, A, B, C, D, E, and F on the point conic, then the lines which join A and B and D and E meet in a point P; the lines which join B and C and E and F meet in a point Q; the lines which join C and D and F and A meet in a point R. The three points P, Q, and R lie on one line l.	If we take six lines a, b, c, d, e, and f on the line conic, then the points which join a and b and d and e are joined by the line p; the points which join b and c and e and f are joined by the line q; the points which join c and d and f and a are joined by the line r. The three lines p, q, and r are on one point L.

Figure 14.12 of Chapter 14 illustrates Pascal's theorem. The dual theorem is the one that Brianchon had discovered by means of the pole and polar relationship (Fig. 35.5 above). Steiner, like Gergonne, did nothing to establish the logical basis for the principle of duality. However, he developed projective geometry systematically by classifying figures and by noting the dual statements as he proceeded. He also studied thoroughly curves and surfaces of the second degree.

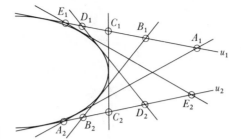

Figure 35.8

Michel Chasles, who devoted his entire life to geometry, followed up on the work of Poncelet and Steiner, though he did not personally know Steiner's work because, as we have noted, Chasles could not read German. Chasles presented his own ideas in his *Traité de géométrie supérieure* (1852) and *Traité des sections coniques* (1865). Since much of Chasles's work either unintentionally duplicated Steiner's or was superseded by more general concepts, we shall note just a few major results that are due to him.

Chasles got the idea of cross ratio from his attempts to understand the lost work of Euclid, *Porisms* (though Steiner and Möbius had already reintroduced it). Desargues, too, had used the concept, but Chasles knew only what La Hire had written about it. Chasles did learn at some time that the idea is also in Pappus because in Note IX of his *Aperçu* (p. 302) he refers to Pappus's use of the idea. One of Chasles's results in this area[19] is that four fixed points of a conic and any fifth point of the conic determine four lines with the same cross ratio.

In 1828 Chasles[20] gave the theorem: Given two sets of collinear points in one-to-one correspondence and such that the cross ratio of any four points on one line equals that of the corresponding points on the other, then the lines joining corresponding points are tangent to a conic that touches the two given lines. This result is the equivalent of Steiner's definition of a line conic, because the cross ratio condition here ensures that the two sets of collinear points are projectively related and the lines joining corresponding points are the lines of Steiner's line conic.

Chasles pointed out that by virtue of the principle of duality lines can be as fundamental as points in the development of plane projective geometry and credits Poncelet and Gergonne with being clear on this point. Chasles also introduced new terminology. The cross ratio he called anharmonic ratio. He introduced the term homography to describe a transformation of a plane into itself or another plane, which carries points into points and lines into lines. This term covers homologous or projectively related figures. He

19. *Correspondance mathématique et physique*, 5, 1829, 6–22.
20. *Correspondance mathématique et physique*, 4, 1828, 363–71.

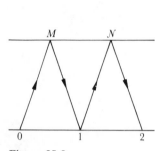

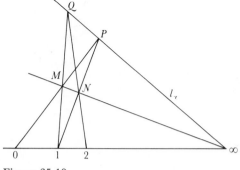

Figure 35.9   Figure 35.10

added the condition that the transformation must preserve cross ratio but this latter fact can be proved. The transformation that carries points into lines and lines into points he called a correlation.

Though he defended pure geometry Chasles thought analytically but presented his proofs and results geometrically. This approach is called the "mixed method" and was used later by others.

By 1850 the general concepts and goals of projective geometry as distinguished from Euclidean geometry were clear; nevertheless the logical relationship of the two geometries was not clarified. The concept of length was used in projective geometry from Desargues to Chasles. In fact the concept of cross ratio was defined in terms of length. Yet length is not a projective concept because it is not invariant under projective transformation. Karl Georg Christian von Staudt (1798–1867), a professor at Erlangen who was interested in foundations, decided to free projective geometry from dependence on length and congruence. The essence of his plan, presented in his *Geometrie der Lage* (Geometry of Position, 1847), was to introduce an analogue of length on a projective basis. His scheme is called "the algebra of throws." One chooses three arbitrary points on a line and assigns to them the symbols 0, 1, ∞. Then by means of a geometric construction (which in itself comes from Möbius)—a "throw"—a symbol is attached to any arbitrary point P.

To see what the construction amounts to in Euclidean geometry suppose we start with the points labelled 0 and 1 on a line (Fig. 35.9). Through the point M on a parallel line draw $0M$ and then draw $1N$ parallel to $0M$. Now draw $1M$ and draw $N2$ parallel to $1M$. It is immediate that $01 = 12$ because opposite sides of a parallelogram are equal. Thus one carries over the length 01 to 12 by a geometric construction.

Now to treat the projective case we start with three points 0, 1, ∞ (Fig. 35.10). The point ∞ lies on l_∞, the line at infinity, but this is just an ordinary line in projective geometry. Now pick a point M and draw a "parallel" to the line 01 through M. This means that the line through M must meet the line 01 at ∞ and so we draw $M\infty$. Now draw $0M$ and prolong it to l_∞. Next we draw through 1 a "parallel" to $0M$. This means the "parallel" through 1 must meet $0M$ on l_∞. We thereby get the line $1P$ and this determines N. Now draw $1M$ and prolong it until it meets l_∞ at Q. The line through N parallel to $1M$ is QN and where it meets 01 we get a point which is labeled 2.

By this type of construction we can attach "rational coordinates" to points on the line 01∞. To assign irrational numbers to points on the line one must introduce an axiom of continuity (Chap. 41). This notion was not well understood at that time and as a consequence von Staudt's work lacked rigor.

Von Staudt's assignment of coordinates to the points of a line did not involve length. His coordinates, though they were the usual number symbols, served as systematic identification symbols for the points. To add or subtract such "numbers" von Staudt could not then use the laws of arithmetic. Instead he gave geometric constructions that defined the operations with these symbols so that, for example, the sum of the numbers 2 and 3 is the number 5. These operations obeyed all the usual laws of numbers. Thus his symbols or coordinates could be treated as ordinary numbers even though they were built up geometrically.

With these labels attached to his points von Staudt could define the cross ratio of four points. If the coordinates of these points are x_1, x_2, x_3, and x_4, then the cross ratio is defined to be

$$\frac{x_1 - x_3}{x_1 - x_4} \bigg/ \frac{x_2 - x_3}{x_2 - x_4}.$$

Thus von Staudt had the fundamental tools to build up projective geometry without depending on the notions of length and congruence.

A harmonic set of four points is one for which the cross ratio is -1. On the basis of harmonic sets von Staudt gave the fundamental definition that two pencils of points are projectively related when under a one-to-one correspondence of their members a harmonic set corresponds to a harmonic set. Four concurrent lines form a harmonic set if the points in which they meet an arbitrary transversal constitute a harmonic set of points. Thus the projectivity of two pencils of lines can also be defined. With these notions von Staudt defined a collineation of the plane into itself as a one-to-one transformation of point to point and line to line and showed that it carries a set of harmonic elements into a set of harmonic elements.

The principal contribution of von Staudt in his *Geometrie der Lage* was to show that projective geometry is indeed more fundamental than Euclidean. Its concepts are logically antecedent. This book and his *Beiträge zur Geometrie der Lage* (Contributions to the Geometry of Position, 1856, 1857, 1860) revealed projective geometry as a subject independent of distance. However, he did use the parallel axiom of Euclidean geometry, which from a logical standpoint was a blemish because parallelism is not a projective invariant. This blemish was removed by Felix Klein.[21]

4. Algebraic Projective Geometry

While the synthetic geometers were developing projective geometry, the algebraic geometers pursued their methods of treating the same subject. The first of the new algebraic ideas was what are now called homogeneous coordinates. One such scheme was created by Augustus Ferdinand Möbius (1790–1868), who like Gauss and Hamilton made his living as an astronomer but devoted considerable time to mathematics. Though Möbius did not take sides in the controversy on synthetic versus algebraic methods his contributions were on the algebraic side.

His scheme for representing the points of a plane by coordinates introduced in his major work, *Der barycentrische Calcul*,[22] was to start with a fixed triangle and to take as the coordinates of any point P in the plane the amount of mass which must be placed at each of the three vertices of the triangle so that P would be the center of gravity of the three masses. When P lies outside the triangle then one or two of the coordinates can be negative. If the three masses are all multiplied by the same constant the point P remains the center of gravity. Hence in Möbius's scheme the coordinates of a point are not unique; only the ratios of the three are determined. The same scheme applied to points in space requires four coordinates. By writing the equations of curves and surfaces in this coordinate system the equations become homogeneous; that is, all terms have the same degree. We shall see examples of the use of homogeneous coordinates shortly.

Möbius distinguished the types of transformations from one plane or space to another. If corresponding figures are equal, the transformation is a congruence and if they are similar, the transformation is a similarity. Next in generality is the transformation that preserves parallelism though not length or shape and this type is called affine (a notion introduced by Euler). The most general transformation that carries lines into lines he called a collineation. Möbius proved in *Der barycentrische Calcul* that every collineation is a projective transformation; that is, it results from a sequence of perspectivities. His proof assumed that the transformation is one-to-one and continuous, but the continuity condition can be replaced by a weaker one. He also gave an analytical representation of the transformation. As Möbius pointed out, one could consider invariant properties of figures under each of the above types of transformations.

21. *Math. Ann.*, 6, 1873, 112–45 = *Ges. Abh.*, 1, 311–43. See also Chap. 38, sec. 3.
22. Published 1827 = *Werke*, 1, 1–388.

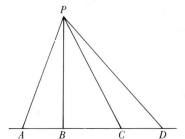

Figure 35.11 A B C D

Möbius introduced signed elements in geometry not only for the line segments but for areas and volumes. Consequently he was able to give a complete treatment of the notion of the signed cross ratio of four points on a line. He also showed that the cross ratio of four lines of a pencil can be expressed in terms of the sines of the angles at the vertex P (Fig. 35.11) by

$$\frac{\sin APB}{\sin APC} \bigg/ \frac{\sin BPD}{\sin CPD}$$

and this ratio is the same as the cross ratio of the four points A, B, C, and D cut out on any transversal to the lines of the pencil. Hence cross ratio is unaltered by projection and section. Möbius had many other ideas which he developed slowly and did not push very far.

The man who gave the algebraic approach to projective geometry its efficacy and vitality is Julius Plücker (1801–68). After serving as a professor of mathematics at several institutions until 1836, he became a professor of mathematics and physics at Bonn, a position he held for the rest of his life. Plücker was primarily a physicist and in fact an experimental physicist in which activity he made many notable discoveries. From 1863 on he again devoted himself to mathematics.

Plücker, too, introduced homogeneous coordinates but in a manner different from Möbius's. His first notion, trilinear coordinates,[23] was also presented in the second volume of his *Analytisch-geometrische Entwickelungen* (1828 and 1831). He starts with a fixed triangle and takes the coordinates of any point P to be the signed perpendicular distances from P to the sides of the triangle; each distance can be multiplied by the same arbitrary constant. Later in the second volume he introduced a special case which amounts to regarding one line of the triangle as the line at infinity. This is equivalent to replacing the usual rectangular Cartesian coordinates x and y by $x = x_1/x_3$ and $y = x_2/x_3$ so that the equations of curves become homogeneous in x_1, x_2, and x_3. The latter notion is the one which became more widely used.

23. *Jour. für Math.*, 5, 1830, 1–36 = *Wiss. Abh.*, 1, 124–58.

By using homogeneous coordinates and Euler's theorem on homogeneous functions, which states that if $f(tx_1, tx_2, tx_3) = t^n f(x_1, x_2, x_3)$ then

$$x_1 \frac{\partial f}{\partial x_1} + x_2 \frac{\partial f}{\partial x_2} + x_3 \frac{\partial f}{\partial x_3} = nf(x_1, x_2, x_3),$$

Plücker was able to give elegant algebraic representations of geometric ideas. Thus if $f(x_1, x_2, x_3) = 0$ is the equation of a conic with (x_1, x_2, x_3) as the coordinates of a point on the conic then

$$\frac{\partial f}{\partial x_1} x_1' + \frac{\partial f}{\partial x_2} x_2' + \frac{\partial f}{\partial x_3} x_3' = 0$$

can be interpreted, when x_1', x_2', x_3' are the running coordinates, as the equation of the tangent at the point (x_1, x_2, x_3) or, when x_1, x_2, x_3 are the running coordinates, as the equation of the polar line of the arbitrary point (x_1', x_2', x_3') with respect to the conic.

Using homogeneous coordinates Plücker gave the algebraic formulation of the infinitely distant line, the circular points at infinity, and other notions. In the homogeneous coordinate system (x_1, x_2, x_3) the equation of the infinitely distant line is $x_3 = 0$. This line is not exceptional in projective geometry, but in our visualization of the geometric elements each normal point of the Euclidean plane is at a finite position given by $x = x_1/x_3$ and $y = x_2/x_3$ and so we are obliged to think of points on $x_3 = 0$ as infinitely distant.

The equation of a circle

$$(x - a)^2 + (y - b)^2 = r^2$$

becomes on the introduction of homogeneous Cartesian coordinates x_1, x_2, and x_3 through $x = x_1/x_3$ and $y = x_2/x_3$

$$(x_1 - ax_3)^2 + (x_2 - bx_3)^2 = r^2 x_3^2.$$

Since the equation of the infinitely distant line is $x_3 = 0$, the intersection of this line with the circle is given by

$$x_1^2 + x_2^2 = 0 \quad \text{and} \quad x_3 = 0,$$

and this is the equation of the circular points at ∞. These circular points have the coordinates $(1, i, 0)$ and $(1, -i, 0)$ or triples proportional to them. Likewise the equation of the spherical circle (*Kugelkreis*) at infinity is

$$x_1^2 + x_2^2 + x_3^2 = 0 \quad \text{and} \quad x_4 = 0.$$

If we write the equation of a straight line in the homogeneous form (we shall use $x, y,$ and z in place of $x_1, x_2,$ and x_3)

$$Ax + By + Cz = 0$$

and now require that the line pass through the points (x_1, y_1, z_1) and $(1, i, 0)$ then the resulting nonhomogeneous equation of the line is

$$x - x_0 + i(y - y_0) = 0$$

wherein $x_0 = x_1/z_1$, and $y_0 = y_1/z_1$. Likewise the equation of the line through (x_1, y_1, z_1) and $(1, -i, 0)$ is

$$x - x_0 - i(y - y_0) = 0.$$

Each of these lines is perpendicular to itself because the slope equals its negative reciprocal. Sophus Lie called them crazy lines; they are now called isotropic lines.

Plücker's efforts to treat duality algebraically led him to a beautiful idea, line coordinates.[24] If the equation of a line in homogeneous coordinates is

$$ux + vy + wz = 0,$$

and if x, y, and z are fixed quantities, u, v, and w or any three numbers proportional to them are the coordinates of a line in the plane. Then just as an equation $f(x_1, x_2, x_3) = 0$ represents a collection of points, so $f(u, v, w) = 0$ represents a collection of lines or a line curve.

With this notion of line coordinates Plücker was able to give an algebraic formulation and proof of the principle of duality. Given any equation $f(r, s, t) = 0$ if we interpret r, s, and t as the homogeneous coordinates x_1, x_2, and x_3 of a point then we have the equation of a point curve, whereas if we interpret them as u, v, and w we have the dual line curve. Any property proved by algebraic processes for the point curve will, because the algebra is the same under either interpretation of the variables, give rise to the dual property of the line curve.

In this second 1830 paper and in Volume 2 of his *Entwickelungen* Plücker also pointed out that the very same curve regarded as a collection of points can also be regarded as the collection of lines tangent to the curve because the tangents determine the shape as much as the points do. The family of tangents is a line curve and has an equation in line coordinates. The degree of this equation is called the class of the curve, whereas the degree of the equation in point coordinates is called the order.

5. *Higher Plane Curves and Surfaces*

The eighteenth-century men had done some work on curves of degree higher than the second (Chap. 23, sec. 3) but the subject was dormant from 1750 to 1825. Plücker took up third and fourth degree curves and used projective concepts freely in this work.

24. *Jour. für Math.*, 6, 1830, 107–46 = *Wiss. Abh.*, 1, 178–219.

In his *System der analytischen Geometrie* (1834) he used a principle, which though helpful was hardly well grounded, to establish canonical forms of curves. To show that the general curve of order (degree) 4, say, could be put into a particular canonical form he argued that if the number of constants were the same in the two forms then one could be converted into the other. Thus he argued that a ternary (three-variable) form of fourth order could always be put in the form

$$C_4 = pqrs + \mu\Omega^2,$$

where p, q, r, and s are linear and Ω is a quadratic form, because both sides contained 14 constants. In his equations μ was real as were the coefficients.

Plücker also took up the number of intersection points of curves, a topic which had also been pursued in the eighteenth century. He used a scheme for representing the family of all curves which go through the intersections of two nth degree curves C_n' and C_n'' which had been introduced by Lamé in a book of 1818. Any curve C_n passing through these intersections can be expressed as

$$C_n = C_n' + \lambda C_n'' = 0$$

where λ is a parameter.

Using this scheme Plücker gave a clear explanation of Cramer's paradox (Chap. 23, sec. 3). A general curve C_n is determined by $n(n + 3)/2$ points because this is the number of essential coefficients in its equation. On the other hand, since two C_n's intersect in n^2 points, through the $n(n + 3)/2$ points in which two C_n's intersect an infinite number of other C_n's pass. The apparent contradiction was explained by Plücker.[25] Any two nth degree curves do indeed meet in n^2 points. However, only $(n/2)(n + 3) - 1$ points are independent. That is, if we take two nth degree curves through the $(n/2)(n + 3) - 1$ points any other nth degree curve through these points will pass through the remaining $(n - 1)(n - 2)/2$ of the n^2 points. Thus when $n = 4$, 13 points are independent. Any two curves through the 13 points determine 16 points but any other curve through the 13 will pass through the remaining three.

Plücker then took up[26] the theory of intersections of an mth degree curve and an nth degree curve. He regarded the latter as fixed and the curves intersecting it as variable. Using the abridged notation C_n for the expression for the nth degree curve and similar notation for the others he wrote

$$C_m = C_m' + A_{m-n}C_n = 0$$

for the case where $m > n$ so that A_{m-n} is a polynomial of degree $m - n$. From this equation Plücker got the correct method of determining the inter-

25. *Annales de Math.*, 19, 1828, 97–106 = *Wiss. Abh.*, 1, 76–82.
26. *Jour. für Math.*, 16, 1837, 47–54.

sections of a C_n with all curves of degree m. Since according to this equation $m - n + 1$ linearly independent curves (the number of coefficients in A_{m-n}) pass through the intersections of C'_m and C_n, Plücker's conclusion was, given $mn - (n - 1)(n - 2)/2$ arbitrary points on C_n, the remaining $(n - 1)(n - 2)/2$ of the mn intersection points with a C_m are determined. The same result was obtained at about the same time by Jacobi.[27]

In his *System* of 1834 and more explicitly in his *Theorie der algebraischen Curven* (1839) Plücker gave what are now called the Plücker formulas which relate the order n and the class k of a curve and the simple singularities. Let d be the number of double points (singular points at which the two tangents are distinct) and r the number of cusps. To the double points there correspond in the line curve double tangents (a double tangent is actually tangent at two distinct points) the number of which is, say, t. To the cusps there correspond osculating tangents (tangents which cross the curve at inflection points) the number of which is, say, w. Then Plücker was able to establish the following dual formulas:

$$k = n(n - 1) - 2d - 3r \qquad n = k(k - 1) - 2t - 3w$$
$$w = 3n(n - 2) - 6d - 8r \qquad r = 3k(k - 2) - 6t - 8w.$$

The number of any one element includes real and imaginary cases.

In the case when $n = 3$, $d = 0$, and $r = 0$, then w, the number of inflection points, is 9. Up to Plücker's time, De Gua and Maclaurin had proved that a line through two inflection points of the general curve of third order goes through a third one, and the fact that a general C_3 had three real inflection points had been assumed from Clairaut's time on. In his *System* of 1834 Plücker proved that a C_3 has either one or three real inflection points, and in the latter case they lie on one line. He also arrived at the more general result which takes complex elements into account. A general C_3 has nine inflection points of which six are imaginary. To derive this result he used his principle of counting constants to show that

$$C_3 = fgh - l^3,$$

where f, g, h, and l are linear forms, and derived the result of De Gua and Maclaurin. He then showed (with incomplete arguments) that the nine inflection points of C_3 lie three on a line so that there are twelve such lines. Ludwig Otto Hesse (1811–74), who served as a professor at several universities, completed Plücker's proof[28] and showed that the twelve lines can be grouped into four triangles.

As an additional example of the discovery of general properties of curves we shall consider the problem of the inflection points of an nth degree

27. *Jour. für Math.*, 15, 1836, 285–308 = *Werke*, 3, 329–54.
28. *Jour. für Math.*, 28, 1844, 97–107 = *Ges. Abh.*, 123–35.

curve $f(x, y) = 0$. Plücker had expressed the usual calculus condition for an inflection point when $y = f(x)$, namely, $d^2y/dx^2 = 0$, in the form appropriate to $f(x, y) = 0$ and obtained an equation of degree $3n - 4$. Since the original curve and the new curve must have $n(3n - 4)$ intersections, it appeared that the original curve has $n(3n - 4)$ points of inflection. Because this number was too large Plücker suggested that the curve of the equation of degree $3n - 4$ has a tangential contact with each of the n infinite branches of the original curve $f = 0$, so that $2n$ of the common points are not inflection points, and so obtained the correct number, $3n(n - 2)$. Hesse clarified this point [29] by using homogeneous coordinates; that is, he replaced x by x_1/x_3, and y by x_2/x_3, and by using Euler's theorem on homogeneous functions he showed that Plücker's equation for the inflection points can be written as

$$H = \begin{vmatrix} f_{11} & f_{12} & f_{13} \\ f_{21} & f_{22} & f_{23} \\ f_{31} & f_{32} & f_{33} \end{vmatrix} = 0,$$

where the subscripts denote partial derivatives. This equation is of degree $3(n - 2)$ and so meets the curve $f(x_1, x_2, x_3) = 0$ of the nth degree in the correct number of inflection points. The determinant itself is called the Hessian of f, a notion introduced by Hesse.[30]

Plücker among others took up the investigation of quartic curves. He was the first to discover (*Theorie der algebraischen Curven*, 1839) that such curves contain 28 double tangents of which eight at most were real. Jacobi then proved [31] that a curve of nth order has in general $n(n - 2)(n^2 - 9)/2$ double tangents.

The work in algebraic geometry also covered figures in space. Though the representation of straight lines in space had already been introduced by Euler and Cauchy, Plücker in his *System der Geometrie des Raumes* (System of Geometry of Space, 1846) introduced a modified representation

$$x = rz + \rho, \qquad y = sz + \sigma$$

in which the four parameters r, ρ, s, and σ fix the line. Now lines can be used to build up all of space, since, for example, planes are no more than collections of lines, and points are intersections of lines. Plücker then said that if lines are regarded as the fundamental element of space, space is four-dimensional because four parameters are needed to cover all of space with lines. The notion of a four-dimensional space of points he rejected as too metaphysical. That the dimension depends on the space-element is the new thought.

The study of figures in space included surfaces of the third and fourth degrees. A ruled surface is generated by a line moving according to some law. The hyperbolic paraboloid (saddle-surface) and hyperboloid of one sheet are examples, as is the helicoid. If a surface of the second degree contains one line it contains an infinity of lines and it is ruled. (It must then be a cone, cylinder, the hyperbolic paraboloid, or the hyperboloid of one sheet.) However, this is not true of cubic surfaces.

As an example of a remarkable property of cubic surfaces there was Cayley's discovery in 1849[32] of the existence of exactly 27 lines on every surface of the third degree. Not all need be real but there are surfaces for which they are all real. Clebsch gave an example in 1871.[33] These lines have special properties. For example, each is cut by ten others. Much further work was devoted to the study of these lines on cubic surfaces.

Among many discoveries concerning surfaces of fourth order one of Kummer's results deserves mention. He had worked with families of lines that represent rays of light, and by considering the associated focal surfaces[34] he was led to introduce a fourth degree surface (and of class four) with 16 double points and 16 double planes as the focal surface of a system of rays of second order. This surface, known as the Kummer surface, embraces as a special case the Fresnel wave surface, which represents the wave front of light propagating in anisotropic media.

The work on synthetic and algebraic projective geometry of the first half of the nineteenth century opened up a brilliant period for geometrical researches of all kinds. The synthetic geometers dominated the period. They attacked each new result to discover some general principle, often not demonstrable geometrically, but from which they nevertheless derived a spate of consequences tied one to the other and to the general principle. Fortunately, algebraic methods were also introduced and, as we shall see, ultimately dominated the field. However, we shall interrupt the history of projective geometry to consider some revolutionary new creations that affected all subsequent work in geometry and, in fact, altered radically the face of mathematics.

Bibliography

Berzolari, Luigi: "Allgemeine Theorie der höheren ebenen algebraischen Kurven," *Encyk. der Math. Wiss.*, B. G. Teubner, 1903–15, III C4, 313–455.

32. *Cambridge and Dublin Math. Jour.*, 4, 1849, 118–32 = *Math. Papers*, 1, 445–56.
33. *Math. Ann.*, 4, 1871, 284–345.
34. *Monatsber. Berliner Akad.*, 1864, 246–60, 495–99.

Boyer, Carl B.: *History of Analytic Geometry*, Scripta Mathematica, 1956, Chaps. 8–9.

Brill, A., and M. Noether: "Die Entwicklung der Theorie der algebraischen Functionen in älterer und neuerer Zeit," *Jahres. der Deut. Math.-Verein.*, 3, 1892/3, 109–566, 287–312 in particular.

Cajori, Florian: *A History of Mathematics*, 2nd ed., Macmillan, 1919, pp. 286–302, 309–14.

Coolidge, Julian L.: *A Treatise on the Circle and the Sphere*, Oxford University Press, 1916.

————: *A History of Geometrical Methods*, Dover (reprint), 1963, Book I, Chap. 5 and Book II, Chap. 2.

————: *A History of the Conic Sections and Quadric Surfaces*, Dover (reprint), 1968.

————: "The Rise and Fall of Projective Geometry," *American Mathematical Monthly*, 41, 1934, 217–28.

Fano, G.: "Gegensatz von synthetischer und analytischer Geometrie in seiner historischen Entwicklung im XIX. Jahrhundert," *Encyk. der Math. Wiss.*, B. G. Teubner 1907–10, III AB4a, 221–88.

Klein, Felix: *Elementary Mathematics from an Advanced Standpoint*, Macmillan, 1939; Dover (reprint), 1945, Geometry, Part 2.

Kötter, Ernst: "Die Entwickelung der synthetischen Geometrie von Monge bis auf Staudt," 1847, *Jahres. der Deut. Math.-Verein.*, Vol. 5, Part II, 1896 (pub. 1901), 1–486.

Möbius, August F.: *Der barycentrische Calcul* (1827), Georg Olms (reprint), 1968. Also in Vol. 1 of *Gesammelte Werke*, pp. 1–388.

————: *Gesammelte Werke*, 4 vols., S. Hirzel, 1885–87; Springer-Verlag (reprint), 1967.

Plücker, Julius: *Gesammelte wissenschaftliche Abhandlungen*, 2 vols., B. G. Teubner, 1895–96.

Schoenflies, A.: "Projektive Geometrie," *Encyk. der Math. Wiss.*, B. G. Teubner, 1907–10, III AB5, 389–480.

Smith, David Eugene: *A Source Book in Mathematics*, Dover (reprint), 1959, Vol. 2, pp. 315–23, 331–45, 670–76.

Steiner, Jacob: *Geometrical Constructions With a Ruler* (a translation of his 1833 book), Scripta Mathematica, 1950.

————: *Gesammelte Werke*, 2 vols., G. Reimer, 1881–82; Chelsea (reprint), 1971.

Zacharias, M.: "Elementargeometrie and elementare nicht-euklidische Geometrie in synthetischer Behandlung," *Encyk. der Math. Wiss.*, B. G. Teubner, 1907–10, III, AB9, 859–1172.

36
Non-Euclidean Geometry

> ... because it seems to be true that many things have, as it were, an epoch in which they are discovered in several places simultaneously, just as the violets appear on all sides in the springtime.　　　　　WOLFGANG BOLYAI

> The character of necessity ascribed to the truths of mathematics and even the peculiar certainty attributed to them is an illusion.　　　　　JOHN STUART MILL

1. *Introduction*

Amidst all the complex technical creations of the nineteenth century the most profound one, non-Euclidean geometry, was technically the simplest. This creation gave rise to important new branches of mathematics but its most significant implication is that it obliged mathematicians to revise radically their understanding of the nature of mathematics and its relation to the physical world. It also gave rise to problems in the foundations of mathematics with which the twentieth century is still struggling. As we shall see, non-Euclidean geometry was the culmination of a long series of efforts in the area of Euclidean geometry. The fruition of this work came in the early nineteenth century during the same decades in which projective geometry was being revived and extended. However, the two domains were not related to each other at this time.

2. *The Status of Euclidean Geometry about 1800*

Though the Greeks had recognized that abstract or mathematical space is distinct from sensory perceptions of space, and Newton emphasized this point,[1] all mathematicians until about 1800 were convinced that Euclidean geometry was the correct idealization of the properties of physical space and of figures in that space. In fact, as we have already noted, there were many attempts to build arithmetic, algebra, and analysis, whose logical foundation

1. *Principia*, Book I, Def. 8, Scholium.

was obscure, on Euclidean geometry and thereby guarantee the truth of these branches too.

Many men actually voiced their absolute trust in the truth of Euclidean geometry. For example, Isaac Barrow, who built his mathematics including his calculus on geometry, lists eight reasons for the certainty of geometry: the clearness of the concepts, the unambiguous definitions, the intuitive assurance and universal truth of its axioms, the clear possibility and easy imaginability of its postulates, the small number of its axioms, the clear conceivability of the mode by which magnitudes are generated, the easy order of the demonstrations, and the avoidance of things not known.

Barrow did raise the question, How are we sure that the geometric principles do apply to nature? His answer was that these are derived from innate reason. Sensed objects are merely the agents which awaken them. Moreover the principles of geometry had been confirmed by constant experience and would continue to be so because the world, designed by God, is immutable. Geometry is then the perfect and certain science.

It is relevant that the philosophers of the late seventeenth and the eighteenth century also raised the question of how we can be sure that the larger body of knowledge that Newtonian science had produced was true. Almost all, notably Hobbes, Locke, and Leibniz, answered that the mathematical laws, like Euclidean geometry, were inherent in the design of the universe. Leibniz did leave some room for doubt when he distinguished between possible and actual worlds. But the only significant exception was David Hume, who in his *Treatise of Human Nature* (1739) denied the existence of laws or necessary sequences of events in the universe. He contended that these sequences were observed to occur and human beings concluded that they always will occur in the same fashion. Science is purely empirical. In particular the laws of Euclidean geometry are not necessary physical truths.

Hume's influence was negated and indeed superseded by Immanuel Kant's. Kant's answer to the question of how we can be sure that Euclidean geometry applies to the physical world, which he gave in his *Critique of Pure Reason* (1781), is a peculiar one. He maintained that our minds supply certain modes of organization—he called them intuitions—of space and time and that experience is absorbed and organized by our minds in accordance with these modes or intuitions. Our minds are so constructed that they compel us to view the external world in only one way. As a consequence certain principles about space are prior to experience; these principles and their logical consequences, which Kant called a priori synthetic truths, are those of Euclidean geometry. The nature of the external world is known to us only in the manner in which our minds oblige us to interpret it. On the grounds just described Kant affirmed, and his contemporaries accepted, that the physical world must be Euclidean. In any case whether one appealed to experience, relied upon innate truths, or accepted Kant's view, the common conclusion was the uniqueness and necessity of Euclidean geometry.

3. The Research on the Parallel Axiom

Though confidence in Euclidean geometry as the correct idealization of physical space remained unshaken from 300 B.C. to about 1800, one concern did occupy the mathematicians during almost all of that long period. The axioms adopted by Euclid were supposed to be self-evident truths about physical space and about figures in that space. However, the parallel axiom in the form stated by Euclid (Chap. IV, sec. 3) was believed to be somewhat too complicated. No one really doubted its truth and yet it lacked the compelling quality of the other axioms. Apparently even Euclid himself did not like his own version of the parallel axiom because he did not call upon it until he had proved all the theorems he could without it.

A related problem which did not bother as many people but which ultimately came to the fore as equally vital is whether one may assume the existence of infinite straight lines in physical space. Euclid was careful to postulate only that one can produce a (finite) straight line as far as necessary so that even the extended straight line was still finite. Also the peculiar wording of Euclid's parallel axiom, that two lines will meet on that side of the transversal where the sum of the interior angles is less than two right angles, was a way of avoiding the outright assertion that there are pairs of lines that will never meet no matter how far they are extended. Nevertheless Euclid did imply the existence of infinite straight lines for, were they finite, they could not be extended as far as necessary in any given context and he *proved* the existence of parallel lines.

The history of non-Euclidean geometry begins with the efforts to eliminate the doubts about Euclid's parallel axiom. From Greek times to about 1800 two approaches were made. One was to replace the parallel axiom by a more self-evident statement. The other was to try to deduce it from the other nine axioms of Euclid; were this possible it would be a theorem and so be beyond doubt. We shall not give a fully detailed account of this work because this history is readily available.[2]

The first major attempt was made by Ptolemy in his tract on the parallel postulate. He tried to prove the assertion by deducing it from the other nine axioms and Euclid's theorems 1 to 28, which do not depend on the parallel axiom. But Ptolemy assumed unconsciously that two straight lines do not enclose a space and that if *AB* and *CD* are parallel (Fig. 36.1) then whatever holds for the interior angles on one side of *FG* must hold on the other.

The fifth-century commentator Proclus was very explicit about his objection to the parallel axiom. He says, "This [postulate] ought even to be struck out of the postulates altogether; for it is a theorem involving many difficulties, which Ptolemy, in a certain book, set himself to solve, and it requires for the demonstration of it a number of definitions as well as theorems, and the converse of it is actually proved by Euclid himself as a

2. See, for example, the reference to Bonola in the bibliography at the end of the chapter.

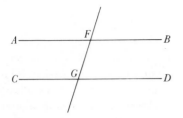

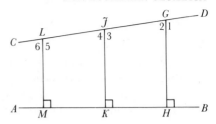

Figure 36.1   Figure 36.2

theorem." Proclus points out that while it is necessary to believe that two lines will tend toward each other on the side of the transversal where the sum of the interior angles is less than two right angles, it is not so clear that these two lines will actually intersect at a finite point. This conclusion is only probable. For, he continues, there are certain curves that approach each other more and more but do not actually meet. Thus, a hyperbola approaches but does not meet its asymptote. Hence might not this be true of the two lines in Euclid's postulate? He then says that up to a certain sum of the interior angles on one side of the transversal the two lines might indeed meet; however, for a value slightly greater but still less than two right angles, the lines might be asymptotic.

Proclus based his own proof of the parallel postulate on an axiom which Aristotle used to prove the universe is finite. The axiom says, "If from one point two straight lines forming an angle be produced indefinitely, the successive distances between the said straight lines [perpendiculars from one onto the other] will ultimately exceed any finite magnitude." Proclus' proof was essentially correct but he substituted one questionable axiom for another.

Nasîr-Eddîn (1201–74), the Persian editor of Euclid, likewise gave a "proof" of Euclid's parallel postulate by assuming that two non-parallel lines approach each other in one direction and diverge in the other. Specifically, if AB and CD (Fig. 36.2) are two lines cut by $GH, JK, LM, \ldots$, if these latter are perpendicular to AB, and if the angles $1, 3, 5, \ldots$ are obtuse while $2, 4, 6, \ldots$ are acute then $GH > JK > LM \ldots$. This fact, Nasîr-Eddîn says, is clearly seen.

Wallis did some work on the parallel axiom in 1663 which he published in 1693.[3] First he reproduced Nasîr-Eddîn's work on the parallel axiom which he had translated for him by a professor of Arabic at Oxford. Incidentally, this was how Nasîr-Eddîn's work on the parallel axiom became known to Europe. Wallis then criticized Nasîr-Eddîn's proof and offered his own proof of Euclid's assertion. His proof rests on the explicit assumption that to each triangle there is a similar one whose sides have any given ratio to the sides of the original one. Wallis believed that this axiom was more evident than

3. *Opera*, 2, 669–78.

the assumption of arbitrarily small subdivision and arbitrarily large extension. In fact, he said, Euclid's axiom that we can construct a circle with
given center and radius presupposes that there is an arbitrarily large radius
at our disposal. Hence one can just as well assume the analogue for
rectilinear figures such as a triangle.

The simplest of the substitute axioms was suggested by Joseph Fenn in
1769, namely, that two intersecting lines cannot both be parallel to a third
straight line. This axiom also appears in Proclus' comments on Proposition 31
of Book I of Euclid's *Elements*. Fenn's statement is entirely equivalent to the
axiom given in 1795 by John Playfair (1748–1819): Through a given point
P not on a line *l*, there is only one line in the plane of *P* and *l* which does not
meet *l*. This is the axiom used in modern books (which for simplicity usually
say there is "one and only one line . . .").

Legendre worked on the problem of the parallel postulate over a period
of about twenty years. His results appeared in books and articles including
the many editions of his *Eléments de géométrie*.[4] In one attack on the problem
he proved the parallel postulate on the assumption that there exist similar
triangles of different sizes; actually his proof was analytical but he assumed
that the unit of length does not matter. Then he gave a proof based on the
assumption that given any three noncollinear points there exists a circle
passing through all three. In still another approach he used all but the parallel
postulate and proved that the sum of the angles of a triangle cannot be
greater than two right angles. He then observed that under these same
assumptions the area is proportional to the defect, that is, two right angles
minus the sum of the angles. He therefore tried to construct a triangle twice
the size of a given triangle so that the defect of the larger one would be twice
that of the given one. Proceeding in this way he hoped to get triangles with
larger and larger defects and thus angle sums approaching zero. This result,
he thought, would be absurd and so the sum of the angles would have to be
180°. This fact in turn would imply the Euclidean parallel axiom. But
Legendre found that the construction reduced to proving that through any
given point within a given angle less than 60° one can always draw a straight
line which meets both sides of the angle. This he could not prove without
Euclid's parallel postulate. In each of the twelve editions (12th ed., 1813) of
Legendre's version of Euclid's *Elements* he added appendices which supposedly
gave proofs of the parallel postulate but each was deficient because it assumed
something implicitly which could not be assumed or assumed an axiom as
questionable as Euclid's.

In the course of his researches Legendre,[5] using the Euclidean axioms
except for the parallel axiom, proved the following significant theorems:
If the sum of the angles of one triangle is two right angles, then it is so in

4. 1st ed., 1794.
5. *Mém. de l'Acad. des Sci., Paris*, 12, 1833, 367–410.

every triangle. Also if the sum is less than two right angles in one triangle it is so in every triangle. Then he gives the proof that if the sum of the angles of any triangle is two right angles, Euclid's parallel postulate holds. This work on the sum of the angles of a triangle was also fruitless because Legendre failed to show (without the aid of the parallel axiom or an equivalent axiom) that the sum of the angles of a triangle cannot be less than two right angles.

The efforts described thus far were mainly attempts to find a more self-evident substitute axiom for Euclid's parallel axiom and many of the proposed axioms did seem intuitively more self-evident. Consequently their creators were satisfied that they had accomplished their objective. However, closer examination showed that these substitute axioms were not really more satisfactory. Some of them made assertions about what happens indefinitely far out in space. Thus, to require that there be a circle through any three points not in a straight line requires larger and larger circles as the three points approach collinearity. On the other hand the substitute axioms that did not involve "infinity" directly, for example, the axiom that there exist two similar but unequal triangles, were seen to be rather complex assumptions and by no means preferable to Euclid's own parallel axiom.

The second group of efforts to solve the problem of the parallel axiom sought to deduce Euclid's assertion from the other nine axioms. The deduction could be direct or indirect. Ptolemy had attempted a direct proof. The indirect method assumes some contradictory assertion in place of Euclid's statement and attempts to deduce a contradiction within this new body of ensuing theorems. For example, since Euclid's parallel axiom is equivalent to the axiom that through a point P not on a line l there is one and only one parallel to l, there are two alternatives to this axiom. One is that there are no parallels to l through P and the other is that there is more than one parallel to l through P. If by adopting each of these in place of the "one parallel" axiom one could show that the new set led to a contradiction, then these alternatives would have to be rejected and the "one parallel" assertion would be *proved*.

The most significant effort of this sort was made by Gerolamo Saccheri (1667–1733), a Jesuit priest and professor at the University of Pavia. He studied the work of Nasîr-Eddîn and Wallis carefully and then adopted his own approach. Saccheri started with the quadrilateral (Fig. 36.3) $ABCD$ in which A and B are right angles and $AC = BD$. It is then easy to prove that $\sphericalangle C = \sphericalangle D$. Now Euclid's parallel axiom is equivalent to the assertion that angles C and D are right angles. Hence Saccheri considered the two possible alternatives:

(1) the hypothesis of the obtuse angle: $\sphericalangle C$ and $\sphericalangle D$ are obtuse;
(2) the hypothesis of the acute angle: $\sphericalangle C$ and $\sphericalangle D$ are acute.

On the basis of the first hypothesis (and the other nine axioms of Euclid)

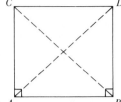

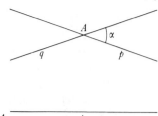

Figure 36.3   Figure 36.4

Saccheri proved that angles C and D must be right angles. Thus with this hypothesis he did deduce a contradiction.

Saccheri next considered the second hypothesis and proved many interesting theorems. He continued until he reached the following theorem: Given any point A and a line b (Fig. 36.4), on the hypothesis of the acute angle there exist in the pencil (family) of lines through A two lines p and q which divide the pencil into two parts. The first of these consists of the lines which intersect b, and the second consists of those lines (lying in angle a) which have a common perpendicular with b somewhere along b. The lines p and q themselves are asymptotic to b. From this result and a lengthy subsequent argument Saccheri deduced that p and b would have a common perpendicular at their common point, which is at infinity. Saccheri believed that this result was at odds with the "nature of the straight line" and therefore concluded that he had arrived at a contradiction.

There remained only the hypothesis that angles C and D of Figure 36.3 are right angles. Saccheri had previously proved that when C and D are right angles, the sum of the angles of any triangle equals 180° and that this fact implies Euclid's parallel postulate. He therefore felt justified in concluding that Euclid was upheld and so published his *Euclides ab Omni Naevo Vindicatus* (Euclid Vindicated from All Faults, 1733). However, since Saccheri did not obtain a contradiction on the basis of the acute angle hypothesis the problem of the parallel axiom was still open.

The efforts to find an acceptable substitute for the Euclidean axiom on parallels or to prove that the Euclidean assertion must be a theorem were so numerous and so futile that in 1759 d'Alembert called the problem of the parallel axiom "the scandal of the elements of geometry."

4. Foreshadowings of Non-Euclidean Geometry

In his dissertation of 1763 Georg S. Klügel (1739–1812), professor of mathematics at the University of Helmstädt, who knew Saccheri's book, made the remarkable observation that the certainty with which men accepted the truth of the Euclidean parallel axiom was based on experience. This observation introduced for the first time the thought that experience rather than self-evidence substantiated the axioms. Klügel expressed doubt that the Euclidean assertion could be proved. He realized that Saccheri had not arrived at a contradiction but merely at results that seemed at variance with experience.

Klügel's paper suggested work on the parallel axiom to Lambert. In his book, *Theorie der Parallellinien* written in 1766 and published in 1786,[6] Lambert, somewhat like Saccheri, considered a quadrilateral with three right angles and considered the possibilities of the fourth angle being right, obtuse, and acute. Lambert did discard the obtuse angle hypothesis because it led to a contradiction. However, unlike Saccheri, Lambert did not conclude that he had obtained a contradiction from the acute angle hypothesis.

The consequences Lambert deduced from both the obtuse and acute angle hypothesis, respectively, even though the former did lead to a contradiction, are noteworthy. His most remarkable result is that under either hypothesis the area of a polygon of n sides is proportional to the difference between the sum of the angles and $2n - 4$ right angles. (Saccheri had this result for a triangle.) He also noted that the obtuse angle hypothesis gave rise to theorems just like those which hold for figures on the surface of a sphere. And he conjectured that the theorems that followed from the acute angle hypothesis would apply to figures on a sphere of imaginary radius. This led him to write a paper[7] on the trigonometric functions of imaginary angles, that is, iA, where A is a real angle and $i = \sqrt{-1}$, which in effect introduced the hyperbolic functions (Chap. 19, sec. 2). We shall see a little more clearly later just what Lambert's observations mean.

Lambert's views on geometry were quite advanced. He realized that any body of hypotheses which did not lead to contradictions offered a possible geometry. Such a geometry would be a valid logical structure even though it might have little to do with real figures. The latter might be suggestive of a particular geometry but do not control the variety of logically developable geometries. Lambert did not reach the more radical conclusion that was introduced somewhat later by Gauss.

Still another forward step was made by Ferdinand Karl Schweikart (1780–1859), a professor of jurisprudence, who devoted spare time to mathematics. He worked on non-Euclidean geometry during the period in which Gauss devoted some thought to the subject but Schweikart arrived at his conclusions independently. He was, however, influenced by Saccheri's and Lambert's work. In a memorandum of 1816 which he sent to Gauss in 1818 for approval, Schweikart actually distinguished two geometries. There is the geometry of Euclid and a geometry based on the assumption that the sum of the angles of a triangle is not two right angles. This latter geometry he

6. *Magazin für reine und angewandte Mathematik*, 1786, 137–64, 325–58.
7. *Hist. de l'Acad. de Berlin*, 24, 1768, 327–54, pub. 1770 = *Opera Mathematica*, 2, 245–69.

called astral geometry, because it might hold in the space of the stars, and its theorems were those which Saccheri and Lambert had established on the basis of the acute angle hypothesis.

Franz Adolf Taurinus (1794–1874), a nephew of Schweikart, took up his uncle's suggestion to study astral geometry. Though he established in his *Geometriae Prima Elementa* (1826) some new results, notably some analytic geometry, he concluded that only Euclid's geometry could be true of physical space but that the astral geometry was *logically consistent*. Taurinus also showed that the formulas which would hold on a sphere of imaginary radius are precisely those that hold in his astral geometry.

The work of Lambert, Schweikart, and Taurinus constitutes advances that warrant recapitulation. All three and other men such as Klügel and Abraham G. Kästner (1719–1800), a professor at Göttingen, were convinced that Euclid's parallel axiom could not be proven, that is, it is independent of Euclid's other axioms. Further, Lambert, Schweikart, and Taurinus were convinced that it is possible to adopt an alternative axiom contradicting Euclid's and build a logically consistent geometry. Lambert made no assertions about the applicability of such a geometry; Taurinus thought it is not applicable to physical space; but Schweikart believed it might apply to the region of the stars. These three men also noted that the geometry on a real sphere has the properties of the geometry based on the obtuse angle hypothesis (if one leaves aside the contradiction which results from the latter) and the geometry on a sphere of imaginary radius has the properties of the geometry based on the acute angle hypothesis. Thus all three recognized the existence of a non-Euclidean geometry but they missed one fundamental point, namely, that Euclidean geometry is not the only geometry that describes the properties of physical space to within the accuracy for which experience can vouch.

5. *The Creation of Non-Euclidean Geometry*

No major branch of mathematics or even a major specific result is the work of one man. At best, some decisive step or proof may be credited to an individual. This cumulative development of mathematics applies especially to non-Euclidean geometry. If one means by the creation of non-Euclidean geometry the recognition that there can be geometries alternative to Euclid's then Klügel and Lambert deserve the credit. If non-Euclidean geometry means the technical development of the consequences of a system of axioms containing an alternative to Euclid's parallel axiom then most credit must be accorded to Saccheri and even he benefited by the work of many men who tried to find a more acceptable substitute axiom for Euclid's. However, the most significant fact about non-Euclidean geometry is that it can be used to describe the properties of physical space as accurately as Euclidean geometry does. The latter is not the necessary geometry of physical space; its physical truth cannot be guaranteed on any a priori grounds. This realization, which did not call for any technical mathematical development because this had already been done, was first achieved by Gauss.

Carl Friedrich Gauss (1777–1855) was the son of a mason in the German city of Brunswick and seemed destined for manual work. But the director of the school at which he received his elementary education was struck by Gauss's intelligence and called him to the attention of Duke Karl Wilhelm. The Duke sent Gauss to a higher school and then in 1795 to the University of Göttingen. Gauss now began to work hard on his ideas. At eighteen he invented the method of least squares and at nineteen he showed that the 17-sided regular polygon is constructible. These successes convinced him that he should turn from philology to mathematics. In 1798 he transferred to the University of Helmstädt and there he was noticed by Johann Friedrich Pfaff who became his teacher and friend. After finishing his doctor's degree Gauss returned to Brunswick where he wrote some of his most famous papers. This work earned for him in 1807 the appointment as professor of astronomy and director of the observatory at Göttingen. Except for one visit to Berlin to attend a scientific meeting he remained at Göttingen for the rest of his life. He did not like to teach and said so. However, he did enjoy social life, was married twice, and raised a family.

Gauss's first major work was his doctoral thesis in which he proved the fundamental theorem of algebra. In 1801 he published the classic *Disquisitiones Arithmeticae*. His mathematical work in differential geometry, the "Disquisitiones Generales circa Superficies Curvas" (General Investigations of Curved Surfaces, 1827), which, incidentally, was the outcome of his interest in surveying, geodesy, and map-making, is a mathematical landmark (Chap. 37, sec. 2). He made many other contributions to algebra, complex functions, and potential theory. In unpublished papers he recorded his innovative work in two major fields: the elliptic functions and non-Euclidean geometry.

His interests in physics were equally broad and he devoted most of his energy to them. When Giuseppe Piazzi (1746–1826) discovered the planet Ceres in 1801 Gauss undertook to determine its path. This was the beginning of his work in astronomy, the activity that absorbed him most and to which he devoted about twenty years. One of his great publications in this area is his *Theoria Motus Corporum Coelestium* (Theory of Motion of the Heavenly Bodies, 1809). Gauss also earned great distinction in his physical research on theoretical and experimental magnetism. Maxwell says in his *Electricity and Magnetism* that Gauss's studies of magnetism reconstructed the whole science, the instruments used, the methods of observation, and the calculation of results. Gauss's papers on terrestrial magnetism are models of physical research and supplied the best method of measuring the earth's magnetic field. His work on astronomy and magnetism opened up a new and brilliant period of alliance between mathematics and physics.

Though Gauss and Wilhelm Weber (1804–91) did not invent the idea of telegraphy, in 1833 they improved on earlier techniques with a practical device which made a needle rotate right and left depending upon the direction of current sent over a wire. Gauss also worked in optics, which had been neglected since Euler's days, and his investigations of 1838–41 gave a totally new basis for the handling of optical problems.

The universality of Gauss's activities is all the more remarkable because his contemporaries had begun to confine themselves to specialized investigations. Despite the fact that he is acknowledged to be the greatest mathematician at least since Newton, Gauss was not so much an innovator as a transitional figure from the eighteenth to the nineteenth century. Although he achieved some new views which did engage other mathematicians he was oriented more to the past than to the future. Felix Klein describes Gauss's position in these words: We could have a tableau of the development of mathematics if we would imagine a chain of high mountains representing the men of the eighteenth century terminated by an imposing summit— Gauss—then a large and rich region filled with new elements of life. Gauss's contemporaries appreciated his genius and by the time of his death in 1855 he was widely venerated and called the "prince of mathematicians."

Gauss published relatively little of his work because he polished whatever he did partly to achieve elegance and partly to achieve for his demonstrations the maximum of conciseness without sacrificing rigor, at least the rigor of his time. In the case of non-Euclidean geometry he published no definitive work. He said in a letter to Bessel of January 27, 1829, that he probably would never publish his findings in this subject because he feared ridicule, or, as he put it, he feared the clamor of the Boeotians, a figurative reference to a dull-witted Greek tribe. Gauss may have been overly cautious, but one must remember that though some mathematicians had been gradually reaching the climax of the work in non-Euclidean geometry the intellectual world at large was still dominated by Kant's teachings. What we do know about Gauss's work in non-Euclidean geometry is gleaned from his letters to friends, two short reviews in the *Göttingische Gelehrte Anzeigen* of 1816 and 1822 and some notes of 1831 found among his papers after his death.[8]

Gauss was fully aware of the vain efforts to establish Euclid's parallel postulate because this was common knowledge in Göttingen and the whole history of these efforts was thoroughly familiar to Gauss's teacher Kästner. Gauss told his friend Schumacher that as far back as 1792 (Gauss was then fifteen) he had already grasped the idea that there could be a logical geometry in which Euclid's parallel postulate did not hold. By 1794 Gauss had found

8. *Werke*, 8, 157–268, contains all of the above and the letter discussed below.

that in his concept of non-Euclidean geometry the area of a quadrangle must be proportional to the difference between 360° and the sum of the angles. However, at this later date and even up to 1799 Gauss was still trying to deduce Euclid's parallel postulate from other more plausible assumptions and he still believed Euclidean geometry to be the geometry of physical space even though he could conceive of other logical non-Euclidean geometries. However, on December 17, 1799, Gauss wrote to his friend the Hungarian mathematician Wolfgang Farkas Bolyai (1775–1856),

> As for me I have already made some progress in my work. However, the path I have chosen does not lead at all to the goal which we seek [deduction of the parallel axiom], and which you assure me you have reached. It seems rather to compel me to doubt the truth of geometry itself. It is true that I have come upon much which by most people would be held to constitute a proof; but in my eyes it proves as good as nothing. For example, if we could show that a rectilinear triangle whose area would be greater than any given area is possible, then I would be ready to prove the whole of [Euclidean] geometry absolutely rigorously.
>
> Most people would certainly let this stand as an axiom; but I, no! It would, indeed, be possible that the area might always remain below a certain limit, however far apart the three angular points of the triangle were taken.

This passage shows that by 1799 Gauss was rather convinced that the parallel axiom cannot be deduced from the remaining Euclidean axioms and began to take more seriously the development of a new and possibly applicable geometry.

From about 1813 on Gauss developed his new geometry which he first called anti-Euclidean geometry, then astral geometry, and finally non-Euclidean geometry. He became convinced that it was logically consistent and rather sure that it might be applicable. In reviews of 1816 and 1822 and in his letter to Bessel of 1829 Gauss reaffirmed that the parallel postulate could not be proved on the basis of the other axioms in Euclid. His letter to Olbers written in 1817[9] is a landmark. In it Gauss says, "I am becoming more and more convinced that the [physical] necessity of our [Euclidean] geometry cannot be proved, at least not by human reason nor for human reason. Perhaps in another life we will be able to obtain insight into the nature of space, which is now unattainable. Until then we must place geometry not in the same class with arithmetic, which is purely a priori, but with mechanics."

To test the applicability of Euclidean geometry and his non-Euclidean geometry Gauss actually measured the sum of the angles of the triangle formed by three mountain peaks, Brocken, Hohehagen, and Inselsberg. The

9. *Werke*, 8, 177.

sides of this triangle were 69, 85, and 197 km. He found[10] that the sum exceeded 180° by 14″85. The experiment proved nothing because the experimental error was much larger than the excess and so the correct sum could have been 180° or even less. As Gauss realized, the triangle was a small one and since in the non-Euclidean geometry the defect is proportional to the area only a large triangle could possibly reveal any significant departure from an angle sum of 180°.

We shall not discuss the specific non-Euclidean theorems that are due to Gauss. He did not write up a full deductive presentation and the theorems he did prove are much like those we shall encounter in the work of Lobatchevsky and Bolyai. These two men are generally credited with the creation of non-Euclidean geometry. Just what is to their credit will be discussed later but they did publish organized presentations of a non-Euclidean geometry on a deductive synthetic basis with the full understanding that this new geometry was logically as legitimate as Euclid's.

Nikolai Ivanovich Lobatchevsky (1793–1856), a Russian, studied at the University of Kazan and from 1827 to 1846 was professor and rector at that university. He presented his views on the foundations of geometry in a paper before the department of mathematics and physics of the University in 1826. However, the paper was never printed and was lost. He gave his approach to non-Euclidean geometry in a series of papers, the first two of which were published in Kazan journals and the third in the *Journal für Mathematik*.[11] The first was entitled "On the Foundations of Geometry" and appeared in 1829–30. The second, entitled "New Foundations of Geometry with a Complete Theory of Parallels" (1835–37), is a better presentation of Lobatchevsky's ideas. He called his new geometry imaginary geometry for reasons which are perhaps already apparent and will be clearer later. In 1840 he published a book in German, *Geometrische Untersuchungen zur Theorie der Parallellinien* (Geometrical Researches on the Theory of Parallels[12]). In this book he laments the slight interest shown in his writings. Though he became blind he dictated a completely new exposition of his geometry and published it in 1855 under the title *Pangéométrie*.

John (János) Bolyai (1802–60), son of Wolfgang Bolyai, was a Hungarian army officer. On non-Euclidean geometry, which he called absolute geometry, he wrote a twenty-six-page paper "The Science of Absolute Space."[13] This was published as an appendix to his father's book *Tentamen Juventutem Studiosam in Elementa Matheseos* (Essay on the Elements of Mathematics for

10. *Werke*, 4, 258.
11. *Jour. für. Math.*, 17, 1837, 295–320.
12. The English translation appears in Bonola. See the bibliography at the end of this chapter.
13. The English translation appears in Bonola. See the bibliography at the end of this chapter.

Studious Youths). Though the two-volume book appeared in 1832–33 and therefore after a publication by Lobatchevsky, Bolyai seems to have worked out his ideas on non-Euclidean geometry by 1825 and was convinced by that time that the new geometry was not self-contradictory. In a letter to his father dated November 23, 1823, John wrote, "I have made such wonderful discoveries that I am myself lost in astonishment." Bolyai's work was so much like Lobatchevsky's that when Bolyai first saw the latter's work in 1835 he thought it was copied from his own 1832–33 publication. On the other hand, Gauss read John Bolyai's article in 1832 and wrote to Wolfgang[14] that he was unable to praise it for to do so would be to praise his own work.

6. The Technical Content of Non-Euclidean Geometry

Gauss, Lobatchevsky, and Bolyai had realized that the Euclidean parallel axiom could not be proved on the basis of the other nine axioms and that some such additional axiom was needed to found Euclidean geometry. Since the parallel axiom was an independent fact it was then at least logically possible to adopt a contradictory statement and develop the consequences of the new set of axioms.

To study the technical content of what these men created, it is just as well to take Lobatchevsky's work because all three did about the same thing. Lobatchevsky gave, as we know, several versions which differ only in details. We shall use his 1835–37 paper as the basis for the account here.

Since, as in Euclid's *Elements*, many theorems can be proved which do not depend at all upon the parallel axiom, such theorems are valid in the new geometry. Lobatchevsky devotes the first six chapters of his paper to the proof of these basic theorems. He assumes at the outset that space is infinite, and he is then able to prove that two straight lines cannot intersect in more than one point and that two perpendiculars to the same line cannot intersect.

In his seventh chapter Lobatchevsky boldly rejects the Euclidean parallel axiom and makes the following assumption: Given a line AB and a point C (Fig. 36.5) then all lines through C fall into two classes with respect to AB, namely, the class of lines which meet AB and the class of lines which do not. To the latter belong two lines p and q which form the boundary between the two classes. These two boundary lines are called parallel lines. More precisely, if C is a point at a perpendicular distance a from the line AB, then there exists an angle[15] $\pi(a)$ such that all lines through C which make with the perpendicular CD an angle less than $\pi(a)$ will intersect AB; all other lines through C do not intersect AB.[16] The two lines which make the

---

14. *Werke*, 8, 220–21.
15. The symbol $\pi(a)$ is standard and so is used here. Actually the π in $\pi(a)$ has nothing to do with the number π.
16. The idea that a specific angle can be associated with a length is due to Lambert.

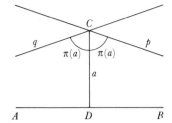

Figure 36.5

angle $\pi(a)$ with AB are the parallels and $\pi(a)$ is called the angle of parallelism. Lines through C other than the parallels and which do not meet AB are called non-intersecting lines, though in Euclid's sense they are parallel to AB and so in this sense Lobatchevsky's geometry contains an infinite number of parallels through C.

If $\pi(a) = \pi/2$ then the Euclidean axiom results. If not, then it follows that $\pi(a)$ increases and approaches $\pi/2$ as a decreases to zero, and $\pi(a)$ decreases and approaches zero as a becomes infinite. The sum of the angles of a triangle is always less than π, decreases as the area of the triangle increases, and approaches π as the area approaches zero. If two triangles are similar then they are congruent.

Now Lobatchevsky turns to the trigonometric part of his geometry. The first step is the determination of $\pi(a)$. The result, if a complete central angle is 2π, is[17]

(1)
$$\tan \frac{\pi(x)}{2} = e^{-x},$$

from which it follows that $\pi(0) = \pi/2$ and $\pi(+\infty) = 0$. The relation (1) is significant in that with each length x it associates a definite angle $\pi(x)$. When $x = 1$, $\tan [\pi(1)/2] = e^{-1}$ so that $\pi(1) = 40°24'$. Thus the unit of length is that length whose angle of parallelism is $40°24'$. This unit of length does not have direct physical significance. It can be physically one inch or one mile. One would choose the physical interpretation which would make the geometry physically applicable.[18]

Then Lobatchevsky deduces formulas connecting sides and angles of the plane triangles of his geometry. In a paper of 1834 he had defined cos x

17. This is a special formulation. In his 1840 work Lobatchevsky gives what amounts to the form usually given in modern texts and which Gauss also has, namely,

(a)
$$\tan \frac{1}{2}\pi(x) = e^{-x/k}$$

where k is a constant, called the space constant. For theoretical purposes the value of k is immaterial. Bolyai also gives the form (a).

18. In the case of the relation $\tan [\pi(x)/2] = e^{-x/k}$, the choice of the value for x which should correspond to, say, $40°24'$ would determine k.

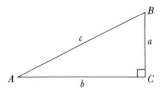

Figure 36.6

and $\sin x$ for real x as the real and imaginary parts of e^{ix}. Lobatchevsky's point was to give a purely analytical foundation for trigonometry and so make it independent of Euclidean geometry. The main trigonometric formulas of his geometry are (Fig. 36.6)

$$\cot \pi(a) = \cot \pi(c) \sin A$$
$$\sin A = \cos B \sin \pi(b)$$
$$\sin \pi(c) = \sin \pi(a) \sin \pi(b).$$

These formulas hold in ordinary spherical trigonometry provided that the sides have imaginary lengths. That is, if one replaces a, b, and c in the usual formulas of spherical trigonometry by ia, ib, and ic one obtains Lobatchevsky's formulas. Since the trigonometric functions of imaginary angles are replaceable by hyperbolic functions one might expect to see these latter functions in Lobatchevsky's formulas. They can be introduced by using the relation $\tan (\pi(x)/2) = e^{-x/k}$. Thus the first of the formulas above can be converted into

$$\sinh \frac{a}{k} = \sinh \frac{c}{k} \sin A.$$

Also whereas in the usual spherical geometry the area of a triangle with angles A, B, and C is $r^2(A + B + C - \pi)$, in the non-Euclidean geometry it is $r^2[\pi - (A + B + C)]$ which amounts to replacing r by ir in the usual formula.

By working with an infinitesimal triangle Lobatchevsky derived in his first paper (1829–30) the formula

$$ds = \sqrt{(dy)^2 + \frac{(dx)^2}{\sin^2 \pi(x)}}$$

for the element of arc on a curve $y = f(x)$ at the point (x, y). Then the entire circumference of a circle of radius r can be calculated. It is

$$C = \pi(e^r - e^{-r}).$$

The expression for the area of a circle proves to be

$$A = \pi(e^{r/2} - e^{-r/2})^2.$$

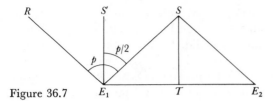

Figure 36.7

He also gives theorems on the area of plane and curved regions and volumes of solids.

The formulas of Euclidean geometry result from the non-Euclidean formulas when the magnitudes are small. Thus if we use the fact that

$$e^r = 1 + r + \frac{r^2}{2!} + \cdots$$

and neglect for small r all but the first two terms then, for example,

$$C = \pi(e^r - e^{-r}) \sim \pi\{1 + r - (1 - r)\} = 2\pi r.$$

In the first paper (1829–30) Lobatchevsky also considered the applicability of his geometry to physical space. The essence of his argument rests on the parallax of stars. Suppose E_1 and E_2 (Fig. 36.7) are the positions of the earth six months apart and S is a star. The parallax p of S is the difference in the directions of E_1S and E_2S measured, say, from the perpendicular E_1S'. If E_1R is the Euclidean parallel to E_2S, then since E_1SE_2 is an isosceles triangle, $\pi/2 - \sphericalangle SE_1E_2$ is half of the change in direction of the star, that is, $p/2$. This angle is 1″.24 for the star Sirius (Lobatchevsky's value). As long as such an angle is not zero, the line from E_1 to the star cannot be the parallel to TS because the line cuts TS. If, however, there were a lower bound to the various parallaxes of all the stars then any line from E_1 making a smaller angle with E_1S' than this lower bound could be taken to be a parallel to TS through E_1 and this geometry would be equally useful so far as stellar measurements are concerned. But then Lobatchevsky showed that the unit of length in his geometry would have to be, physically, more than a half a million times the radius of the earth's path. In other words, Lobatchevsky's geometry could be applicable only in an enormously large triangle.

7. The Claims of Lobatchevsky and Bolyai to Priority

The creation of non-Euclidean geometry is often used as an example of how an idea occurs independently to different people at about the same time. Sometimes this is regarded as pure coincidence and sometimes as evidence of the spirit of the time working its influence in widely separated quarters. The creation of non-Euclidean geometry by Gauss, Lobatchevsky, and Bolyai is not an example of a simultaneous creation nor is the great credit given to Lobatchevsky and Bolyai justified. It is true, as already noted, that they were the first to publish an avowed non-Euclidean geometry and in this act showed more courage than Gauss did. However, the creation of non-Euclidean geometry is hardly their contribution. We have already pointed out that even Gauss was preceded by Lambert, that Schweikart and Taurinus were independent creators, and that Lambert and Taurinus published their work. Moreover the realization that the new geometry may be applicable to physical space is due to Gauss.

Both Lobatchevsky and Bolyai owe much to Gauss. Lobatchevsky's teacher in Kazan was Johann Martin Bartels (1769–1836), a good friend of Gauss. In fact they spent the years 1805 to 1807 together in Brunswick. Subsequently Gauss and Bartels kept in communication with each other. It is extremely unlikely that Bartels did not pass on to Lobatchevsky, who remained at Kazan as a colleague, Gauss's progress in non-Euclidean geometry. In particular Bartels must have known Gauss's doubts as to the truth of Euclidean geometry.

As for John Bolyai, his father Wolfgang was also a close friend of Gauss and a fellow student in Göttingen from 1796 to 1798. Wolfgang and Gauss not only continued to communicate with each other but discussed the specific subject of the parallel axiom, as one of the quotations above indicates. Wolfgang continued to work hard on the problem of the parallel axiom and sent a purported proof to Gauss in 1804. Gauss showed him that the proof was fallacious. By 1817 Gauss was certain not only that the axiom could not be proved but that a logically consistent and physically applicable non-Euclidean geometry could be constructed. Beyond his communication of 1799 to this effect, Gauss transmitted his later thoughts freely to Wolfgang. Wolfgang continued to work on the problem until he published his *Tentamen* of 1832–33. Since he recommended to his son that he take up the problem of the parallel axiom he almost certainly retailed what he knew.

There are contrary views. The mathematician Friedrich Engel (1861–1941) believed that though Lobatchevsky's teacher Bartels was Gauss's friend, Lobatchevsky could hardly have learned more through this connection than that Gauss doubted the physical correctness of the parallel axiom. But this fact in itself was crucial. However, Engel doubted that Lobatchevsky learned even this much from Gauss, for Lobatchevsky had tried from 1816 on to prove the Euclidean parallel axiom; then recognizing the hopelessness of such efforts finally in 1826 created the new geometry. John Bolyai also tried to prove the Euclidean parallel axiom until about 1820 and then turned to the construction of a new geometry. But these continuing efforts to prove the parallel axiom do not imply ignorance of Gauss's thoughts.[19] Since no one, not even Gauss, had shown that Euclid's

19. However, see George Bruce Halsted, *Amer. Math. Monthly*, 6, 1899, 166–72; and 7, 1900, 247–52.

parallel axiom could not be deduced from the other nine axioms both Lobatchevsky and Bolyai may have decided to try their hand at the problem. Having failed, they could appreciate all the more readily the wisdom of Gauss's views on the subject.

As for the technical content contributed by Lobatchevsky and John Bolyai, though they may have created this independently of their predecessors and of each other, Saccheri's and Lambert's work, to say nothing of Schweikart's and Taurinus's, was widely known in Göttingen and was certainly known to Bartels and Wolfgang Bolyai. And when Lobatchevsky refers in his 1835–37 paper to the futility of the efforts over two thousand years to settle the question of the parallel axiom, by inference he admits to the knowledge of the earlier work.

8. The Implications of Non-Euclidean Geometry

We have already stated that the creation of non-Euclidean geometry was the most consequential and revolutionary step in mathematics since Greek times. We shall not treat all the implications of the subject at this time. Instead we shall follow the historical course of events. The impact of the creation and the full realization of its significance were delayed because Gauss did not publish his work on this subject and Lobatchevsky's and Bolyai's work was ignored for about thirty years. Though these men were aware of the importance of their work, the mathematicians generally exhibited their usual reluctance to entertain radical ideas. Also, the key subject in the geometry of the 1830s and 1840s was projective geometry and for this reason too the work on non-Euclidean geometry did not attract the English, French, and German mathematicians. When Gauss's notes and correspondence on non-Euclidean geometry were published after his death in 1855 attention was drawn to the subject. His name gave weight to the ideas and soon thereafter Lobatchevsky's and Bolyai's work was noted by Richard Baltzer (1818–87) in a book of 1866–67. Subsequent developments finally brought mathematicians to the realization of the full import of non-Euclidean geometry.

Gauss did see the most revolutionary implication. The first step in the creation of non-Euclidean geometry was the realization that the parallel axiom could not be proved on the basis of the other nine axioms. It was an independent assertion and so it was possible to adopt a contradictory axiom and develop an entirely new geometry. This Gauss and others did. But Gauss, having realized that Euclidean geometry is not necessarily the geometry of physical space, that is, is not necessarily true, put geometry in the same class with mechanics, and asserted that the quality of truth must be restricted to arithmetic (and its development in analysis). This confidence in arithmetic is in itself curious. Arithmetic at this time had no logical foundation at all. The assurance that arithmetic, algebra, and analysis offered truths about the physical world stemmed entirely from reliance upon experience.

The history of non-Euclidean geometry reveals in a striking manner how much mathematicians are influenced not by the reasoning they perform but by the spirit of the times. Saccheri had rejected the strange theorems of non-Euclidean geometry and concluded that Euclid was vindicated. But one hundred years later Gauss, Lobatchevsky, and Bolyai confidently accepted the new geometry. They believed that their new geometry was logically consistent and hence that this geometry was as valid as Euclid's. But they had no proof of this consistency. Though they proved many theorems and obtained no evident contradictions, the possibility remained open that a contradiction might still be derived. Were this to happen, then the assumption of their parallel axiom would be invalid and, as Saccheri had believed, Euclid's parallel axiom would be a consequence of his other axioms.

Actually Bolyai and Lobatchevsky considered this question of consistency and were partly convinced of it because their trigonometry was the same as for a sphere of imaginary radius and the sphere is part of Euclidean geometry. But Bolyai was not satisfied with this evidence because trigonometry in itself is not a complete mathematical system. Thus despite the absence of any proof of consistency, or of the applicability of the new geometry, which might at least have served as a convincing argument, Gauss, Bolyai, and Lobatchevsky accepted what their predecessors had regarded as absurd. This acceptance was an act of faith. The question of the consistency of non-Euclidean geometry remained open for another forty years.

One more point about the creation of non-Euclidean geometry warrants attention and emphasis. There is a common belief that Gauss, Bolyai, and Lobatchevsky went off into a corner, played with changing the axioms of Euclidean geometry just to satisfy their intellectual curiosity and so created the new geometry. And since this creation has proved to be enormously important for science—a form of non-Euclidean geometry which we have yet to examine has been used in the theory of relativity—many mathematicians have contended that pure intellectual curiosity is sufficient justification for the exploration of any mathematical idea and that the values for science will almost surely ensue as purportedly happened in the case of non-Euclidean geometry. But the history of non-Euclidean geometry does not support this thesis. We have seen that non-Euclidean geometry came about after centuries of work on the parallel axiom. The concern about this axiom stemmed from the fact that it should be, as an axiom, a self-evident truth. Since the axioms of geometry are our basic facts about physical space and vast branches of mathematics and of physical science use the properties of Euclidean geometry the mathematicians wished to be sure that they were relying upon truths. In other words, the problem of the parallel axiom was not only a genuine physical problem but as fundamental a physical problem as there can be.

Bibliography

Bonola, Roberto: *Non-Euclidean Geometry*, Dover (reprint), 1955.

Dunnington, G. W.: *Carl Friedrich Gauss*, Stechert-Hafner, 1960.

Engel, F., and P. Staeckel: *Die Theorie der Parallellinien von Euklid bis auf Gauss*, 2 vols., B. G. Teubner, 1895.

————: *Urkunden zur Geschichte der nichteuklidischen Geometrie*, B. G. Teubner, 1899–1913, 2 vols. The first volume contains the translation from Russian into German of Lobatchevsky's 1829–30 and 1835–37 papers. The second is on the work of the two Bolyais.

Enriques, F.: "Prinzipien der Geometrie," *Encyk. der Math. Wiss.*, B. G. Teubner, 1907–10, III ABI, 1–129.

Gauss, Carl F.: *Werke*, B. G. Teubner, 1900 and 1903, Vol. 8, 157–268; Vol. 9, 297–458.

Heath, Thomas L.: *Euclid's Elements*, Dover (reprint), 1956, Vol. 1, pp. 202–20.

Kagan, V.: *Lobatchevsky and his Contribution to Science*, Foreign Language Pub. House, Moscow, 1957.

Lambert, J. H.: *Opera Mathematica*, 2 vols., Orell Fussli, 1946–48.

Pasch, Moritz, and Max Dehn: *Vorlesungen über neuere Geometrie*, 2nd ed., Julius Springer, 1926, pp. 185–238.

Saccheri, Gerolamo: *Euclides ab Omni Naevo Vindicatus*, English trans. by G. B. Halsted in *Amer. Math. Monthly*, Vols. 1–5, 1894–98; also *Open Court Pub. Co.*, 1920, and Chelsea (reprint), 1970.

Schmidt, Franz, and Paul Staeckel: *Briefwechsel zwischen Carl Friedrich Gauss und Wolfgang Bolyai*, B. G. Teubner, 1899; Georg Olms (reprint), 1970.

Smith, David E.: *A Source Book in Mathematics*, Dover (reprint), 1959, Vol. 2, pp. 351–88.

Sommerville, D. M. Y.: *The Elements of Non-Euclidean Geometry*, Dover (reprint), 1958.

Staeckel, P.: "Gauss als Geometer," *Nachrichten König. Ges. der Wiss. zu Gött.*, 1917, Beiheft, pp. 25–142. Also in Gauss: *Werke*, X_2.

von Walterhausen, W. Sartorius: *Carl Friedrich Gauss*, S. Hirzel, 1856; Springer-Verlag (reprint), 1965.

Zacharias, M.: "Elementargeometrie und elementare nicht-euklidische Geometrie in synthetischer Behandlung," *Encyk. der Math. Wiss.*, B. G. Teubner, 1914–31, III AB9, 859–1172.

37
The Differential Geometry of Gauss and Riemann

> Thou, nature, art my goddess; to thy laws my services are
> bound . . . . CARL F. GAUSS

1. Introduction

We shall now pick up the threads of the development of differential geometry, particularly the theory of surfaces as founded by Euler and extended by Monge. The next great step in this subject was made by Gauss.

Gauss had devoted an immense amount of work to geodesy and map-making starting in 1816. His participation in actual physical surveys, on which he published many papers, stimulated his interest in differential geometry and led to his definitive 1827 paper "Disquisitiones Generales circa Superficies Curvas" (General Investigations of Curved Surfaces).[1] However, beyond contributing this definitive treatment of the differential geometry of surfaces lying in three-dimensional space Gauss advanced the totally new concept that a surface is a space in itself. It was this concept that Riemann generalized, thereby opening up new vistas in non-Euclidean geometry.

2. Gauss's Differential Geometry

Euler had already introduced the idea (Chap. 23, sec. 7) that the coordinates (x, y, z) of any point on a surface can be represented in terms of two parameters u and v; that is, the equations of a surface are given by

$$(1) \qquad x = x(u, v), \qquad y = y(u, v), \qquad z = z(u, v).$$

Gauss's point of departure was to use this parametric representation for the systematic study of surfaces. From these parametric equations we have

$$(2) \quad dx = a\, du + a'\, dv, \qquad dy = b\, du + b'\, dv, \qquad dz = c\, du + c'\, dv$$

1. *Comm. Soc. Gott.*, 6, 1828, 99–146 = *Werke*, 4, 217–58.

wherein $a = x_u$, $a' = x_v$, and so forth. For convenience Gauss introduces the determinants

$$A = \begin{vmatrix} b & c \\ b' & c' \end{vmatrix}, \qquad B = \begin{vmatrix} c & a \\ c' & a' \end{vmatrix}, \qquad C = \begin{vmatrix} a & b \\ a' & b' \end{vmatrix}$$

and the quantity

$$\Delta = \sqrt{A^2 + B^2 + C^2}$$

which he supposes is not identically 0.

The fundamental quantity on any surface is the element of arc length which in (x, y, z) coordinates is

(3) $$ds^2 = dx^2 + dy^2 + dz^2.$$

Gauss now uses the equations (2) to write (3) as

(4) $$ds^2 = E(u, v)\, du^2 + 2F(u, v)\, du\, dv + G(u, v)\, dv^2,$$

where

$$E = a^2 + b^2 + c^2, \qquad F = aa' + bb' + cc', \qquad G = a'^2 + b'^2 + c'^2.$$

The angle between two curves on a surface is another fundamental quantity. A curve on the surface is determined by a relation between u and v, for then x, y, and z become functions of one parameter, u or v, and the equations (1) become the parametric representation of a curve. One says in the language of differentials that at a point (u, v) a curve or the direction of a curve emanating from the point is given by the ratio $du:dv$. If then we have two curves or two directions emanating from (u, v), one given by $du:dv$ and the other by $du':dv'$, and if θ is the angle between these directions, Gauss shows that

(5) $$\cos \theta = \frac{E\, du\, du' + F(du\, dv' + du'\, dv) + G\, dv\, dv'}{\sqrt{E\, du^2 + 2F\, du\, dv + G\, dv^2}\, \sqrt{E\, du'^2 + 2F\, du'\, dv' + G\, dv'^2}}.$$

Gauss undertakes next to study the curvature of a surface. His definition of curvature is a generalization to surfaces of the indicatrix used for space curves by Euler and used for surfaces by Olinde Rodrigues.[2] At each point (x, y, z) on a surface there is a normal with a direction attached. Gauss considers a unit sphere and chooses a radius having the direction of the directed normal on the surface. The choice of radius determines a point (X, Y, Z) on the sphere. If we next consider on the surface any small region surrounding (x, y, z) then there is a corresponding region on the sphere surrounding (X, Y, Z). The curvature of the surface at (x, y, z) is defined as the limit of the ratio of the area of the region on the sphere to the area of the corresponding region on the surface as the two areas shrink to their respective

2. *Corresp. sur l'Ecole Poly.*, 3, 1814–16, 162–82.

points. Gauss evaluates this ratio by noting first that the tangent plane at (X, Y, Z) on the sphere is parallel to the one at (x, y, z) on the surface. Hence the ratio of the two areas is the ratio of their projections on the respective tangent planes. To find this latter ratio Gauss performs an amazing number of differentiations and obtains a result which is still basic, namely, that the (total) curvature K of the surface is

(6)
$$K = \frac{LN - M^2}{EG - F^2}$$

wherein

(7)

$$
L = \begin{vmatrix} x_{uu} & y_{uu} & z_{uu} \\ x_u & y_u & z_u \\ x_v & y_v & z_v \end{vmatrix}, \quad
M = \begin{vmatrix} x_{uv} & y_{uv} & z_{uv} \\ x_u & y_u & z_u \\ x_v & y_v & z_v \end{vmatrix}, \quad
N = \begin{vmatrix} x_{vv} & y_{vv} & z_{vv} \\ x_u & y_u & z_u \\ x_v & y_v & z_v \end{vmatrix}.
$$

Then Gauss shows that his K is the product of the two principal curvatures at (x, y, z), which had been introduced by Euler. The notion of mean curvature, the average of the two principal curvatures, was introduced by Sophie Germain in 1831.[3]

Now Gauss makes an extremely important observation. When the surface is given by the parametric equations (1), the properties of the surface seem to depend on the functions x, y, z. By fixing u, say $u = 3$, and letting v vary, one obtains a curve on the surface. For the various possible fixed values of u one obtains a family of curves. Likewise by fixing v one obtains a family of curves. These two families are the parametric curves on the surface so that each point is given by a pair of numbers (c, d), say, where $u = c$ and $v = d$ are the parametric curves which pass through the point. These coordinates do not necessarily denote distances any more than latitude and longitude do. Let us think of a surface on which the parametric curves have been determined in some way. Then, Gauss affirms, the geometrical properties of the surface are determined solely by the E, F, and G in the expression (4) for ds^2. These functions of u and v are all that matter.

It is certainly the case, as is evident from (4) and (5), that distances and angles on the surface are determined by the E, F, and G. But Gauss's fundamental expression for curvature, (6) above, depends upon the additional quantities L, M, and N. Gauss now proves that

(8) $\quad K = \dfrac{1}{2H} \left\{ \dfrac{\partial}{\partial u} \left[\dfrac{F}{EH} \dfrac{\partial E}{\partial v} - \dfrac{1}{H} \dfrac{\partial G}{\partial u} \right] + \dfrac{\partial}{\partial v} \left[\dfrac{2}{H} \dfrac{\partial F}{\partial u} - \dfrac{1}{H} \dfrac{\partial E}{\partial v} - \dfrac{F}{EH} \dfrac{\partial E}{\partial u} \right] \right\}$

wherein $H = \sqrt{EG - F^2}$ and is equal to Gauss's Δ defined above. Equation (8), called the Gauss characteristic equation, shows that the curvature K and, in particular in view of (6), the quantity $LN - M^2$ depend only upon

3. *Jour. für Math.*, 7, 1831, 1–29.

E, F, and G. Since E, F, and G are functions only of the parametric coordinates on the surface, the curvature too is a function only of the parameters and does not depend at all on whether or how the surface lies in three-space.

Gauss had made the observation that the properties of a surface depend only on E, F, and G. However, many properties other than curvature involve the quantities L, M, and N and not in the combination $LN - M^2$ which appears in equation (6). The analytical substantiation of Gauss's point was made by Gaspare Mainardi (1800–79),[4] and independently by Delfino Codazzi (1824–75),[5] both of whom gave two additional relations in the form of differential equations which together with Gauss's characteristic equation, K having the value in (6), determine L, M, and N in terms of E, F, and G.

Then Ossian Bonnet (1819–92) proved in 1867[6] the theorem that when six functions satisfy the Gauss characteristic equation and the two Mainardi-Codazzi equations, they determine a surface uniquely except as to position and orientation in space. Specifically, if E, F, and G and L, M, and N are given as functions of u and v which satisfy the Gauss characteristic equation and the Mainardi-Codazzi equations and if $EG - F^2 \neq 0$, then there exists a surface given by three functions x, y, and z as functions of u and v which has the first fundamental form

$$E \, du^2 + 2F \, du \, dv + G \, dv^2,$$

and L, M, and N are related to E, F, and G through (7). This surface is uniquely determined except for position in space. (For real surfaces with real coordinates (u, v) we must have $EG - F^2 > 0$, $E > 0$, and $G \geq 0$). Bonnet's theorem is the analogue of the corresponding theorem on curves (Chap. 23, sec. 6).

The fact that the properties of a surface depend only on the E, F, and G has many implications some of which were drawn by Gauss in his 1827 paper. For example, if a surface is bent without stretching or contracting, the coordinate lines u = const. and v = const. will remain the same and so ds will remain the same. Hence all properties of the surface, the curvature in particular, will remain the same. Moreover, if two surfaces can be put into one-to-one correspondence with each other, that is, if $u' = \phi(u, v)$, $v' = \psi(u, v)$, where u' and v' are the coordinates of points on the second surface, and if the element of distance at corresponding points is the same on the two surfaces, that is, if

$$E \, du^2 + 2F \, du \, dv + G \, dv^2 = E' \, du'^2 + 2F' \, du' \, dv' + G' \, dv'^2$$

4. *Giornale dell' Istituto Lombardo*, 9, 1856, 385–98.
5. *Annali di Mat.*, (3), 2, 1868–69, 101–19.
6. *Jour. de l'Ecole Poly.*, 25, 1867, 31–151.

wherein E, F, and G are functions of u and v and E', F', G' are functions of u' and v', then the two surfaces, which are said to be *isometric*, must have the same geometry. In particular, as Gauss pointed out, they must have the same total curvature at corresponding points. This result Gauss called a "theorema egregium," a most excellent theorem.

As a corollary, it follows that to move a part of a surface over to another part (which means preserving distance) a necessary condition is that the surface have constant curvature. Thus, a part of a sphere can be moved without distortion to another but this cannot be done on an ellipsoid. (However, bending can take place in fitting a surface or part of a surface onto another under an isometric mapping.) Though the curvatures at corresponding points are equal, if two surfaces do not have constant curvature they need not necessarily be isometrically related. In 1839[7] Ferdinand Minding (1806–85) proved that if two surfaces do have constant and equal curvature then one can be mapped isometrically onto the other.

Another topic of great importance that Gauss took up in his 1827 paper is that of finding geodesics on surfaces. (The term geodesic was introduced in 1850 by Liouville and was taken from geodesy.) This problem calls for the calculus of variations which Gauss uses. He approaches this subject by working with the x, y, z representation and proves a theorem stated by John Bernoulli that the principal normal of a geodesic curve is normal to the surface. (Thus the principal normal at a point of a latitude circle on a sphere lies in the plane of the circle and is not normal to the sphere whereas the principal normal at a point on a longitude circle is normal to the sphere.) Any relation between u and v determines a curve on the surface and the relation which gives a geodesic is determined by a differential equation. This equation, which Gauss merely says is a second order equation in u and v but does not give explicitly, can be written in many forms. One is

$$(9) \qquad \frac{d^2v}{du^2} = n\left(\frac{dv}{du}\right)^3 + (2m - \nu)\left(\frac{dv}{du}\right)^2 + (l - 2\mu)\frac{dv}{du} - \lambda,$$

where n, m, μ, ν, l, and λ are functions of E, F, and G.

One must be careful about assuming the existence of a unique geodesic between two points on a surface. Two nearby points on a sphere have a unique geodesic joining them, but two diametrically opposite points are joined by an infinity of geodesics. Similarly, two points on the same generator of a circular cylinder are connected by a geodesic along the generator but also by an infinite number of geodesic helices. If there is but one geodesic arc between two points in a region, that arc gives the shortest path between them in that region. The problem of actually determining the geodesics on particular surfaces was taken up by many men.

7. *Jour. für Math.*, 19, 1839, 370–87.

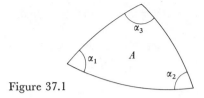

Figure 37.1

In the 1827 article Gauss proved a famous theorem on curvature for a triangle formed by geodesics (Fig. 37.1). Let K be the variable curvature of a surface. $\iint_A K \, dA$ is then the integral of this curvature over the area A. Gauss's theorem applied to the triangle states that

$$\iint_A K \, dA = \alpha_1 + \alpha_2 + \alpha_3 - \pi;$$

that is, the integral of the curvature over a geodesic triangle is equal to the excess of the sum of the angles over 180° or, where the angle sum is less than π, to the defect from 180°. This theorem, Gauss says, ought to be counted as a most elegant theorem. The result generalizes the theorem of Lambert (Chap. 36, sec. 4), which states that the area of a spherical triangle equals the product of its spherical excess and the square of the radius, for in a spherical triangle K is constant and equals $1/R^2$.

One more important piece of work in Gauss's differential geometry must be noted. Lagrange (Chap. 23, sec. 8) had treated the conformal mapping of a surface of revolution into the plane. In 1822 Gauss won a prize offered by the Danish Royal Society of Sciences for a paper on the problem of finding the analytic condition for transforming any surface conformally onto any other surface.[8] His condition, which holds in the neighborhood of corresponding points on the two surfaces, amounts to the fact that a function $P + iQ$, which we shall not specify further, of the parameters T and U by means of which one surface is represented is a function f of $p + iq$ which is the corresponding function of the parameters t and u by which the other surface is represented and $P - iQ$ is $f'(p - iq)$, where f' is f or obtained from f by replacing i by $-i$. The function f depends on the correspondence between the two surfaces, the correspondence being specified by $T = T(t, u)$ and $U(t, u)$. Gauss did not answer the question of whether and in what way a *finite* portion of the surface can be mapped conformally onto the other surface. This problem was taken up by Riemann in his work on complex functions (Chap. 27, sec. 10).

Gauss's work in differential geometry is a landmark in itself. But its implications were far deeper than he himself appreciated. Until this work,

8. *Werke*, 4, 189–216.

surfaces had been studied as figures in three-dimensional Euclidean space. But Gauss showed that the geometry of a surface could be studied by concentrating on the surface itself. If one introduces the u and v coordinates which come from the parametric representation

$$x = x(u, v), \qquad y = y(u, v), \qquad z = z(u, v)$$

of the surface in three-dimensional space and uses the E, F, and G determined thereby then one obtains the Euclidean properties of that surface. However, given these u and v coordinates on the surface and the expression for ds^2 in terms of E, F, and G as functions of u and v, all the properties of the surface follow from this expression. This suggests two vital thoughts. The first is that the surface can be considered as a space in itself because all its properties are determined by the ds^2. One can forget about the fact that the surface lies in a three-dimensional space. What kind of geometry does the surface possess if it is regarded as a space in itself? If one takes the "straight lines" on that surface to be the geodesics, then the geometry is non-Euclidean.

Thus if the surface of the sphere is studied as a space in itself, it has its own geometry and even if the familiar latitude and longitude are used as the coordinates of points, the geometry of that surface is not Euclidean because the "straight lines" or geodesics are arcs of the great circles on the surface. However, the geometry of the spherical surface is Euclidean if it is regarded as a surface in three-dimensional space. The shortest distance between two points on the surface is then the line segment of three-dimensional Euclidean geometry (though it does not lie on the surface). What Gauss's work implied is that there are non-Euclidean geometries at least on surfaces regarded as spaces in themselves. Whether Gauss saw this non-Euclidean interpretation of his geometry of surfaces is not clear.

One can go further. One might think that the proper E, F, and G for a surface is determined by the parametric equations (1). But one could start with the surface, introduce the two families of parametric curves and then pick functions E, F, and G of u and v almost arbitrarily. Then the surface has a geometry determined by these E, F, and G. This geometry is intrinsic to the surface and has no connection with the surrounding space.[8a] Consequently the same surface can have *different* geometries depending on the choice of the functions E, F, and G.

The implications are deeper. If one can pick different sets of E, F, and G and thereby determine different geometries on the same surface, why can't one pick different distance functions in our ordinary three-dimensional space? The common distance function in rectangular coordinates is, of course, $ds^2 = dx^2 + dy^2 + dz^2$ and this is obligatory if one starts with Euclidean geometry because it is just the analytic statement of the Pythagorean theorem. However, given the same rectangular Cartesian coordinates

---

8a. Note that a different parametrization there can be a different E, F, and G and a different curvature. This is what is meant by curvature of space-time in relativity. It is *not* a "curved space."

for the points of space, one might pick a different expression for ds^2 and obtain a quite different geometry for that space, a non-Euclidean geometry. This extension to any space of the ideas Gauss first obtained by studying surfaces was taken up and developed by Riemann.

3. Riemann's Approach to Geometry

The doubts about what we may believe about the geometry of physical space, raised by the work of Gauss, Lobatchevsky, and Bolyai, stimulated one of the major creations of the nineteenth century, Riemannian geometry. The creator was Georg Bernhard Riemann, the deepest philosopher of geometry. Though the details of Lobatchevsky's and Bolyai's work were unknown to Riemann, they were known to Gauss, and Riemann certainly knew Gauss's doubts as to the truth and necessary applicability of Euclidean geometry. Thus in the field of geometry Riemann followed Gauss whereas in function theory he followed Cauchy and Abel. His investigation of geometry was influenced also by the teachings of the psychologist Johann Friedrich Herbart (1776–1841).

Gauss assigned to Riemann the subject of the foundations of geometry as the one on which he should deliver his qualifying lecture, the *Habilitationsvortrag*, for the title of *Privatdozent*. The lecture was delivered in 1854 to the faculty at Göttingen with Gauss present, and was published in 1868 under the title "Über die Hypothesen, welche der Geometrie zu Grunde liegen" (On the Hypotheses Which Lie at the Foundation of Geometry).[9]

In a paper on the conduction of heat, which Riemann wrote in 1861 to compete for a prize offered by the Paris Academy of Sciences and which is often referred to as his *Pariserarbeit*, Riemann found the need to consider further his ideas on geometry and here he gave some technical elaborations of his 1854 paper. This 1861 paper, which did not win the prize, was published posthumously in 1876 in his *Collected Works*.[10] In the second edition of the *Werke* Heinrich Weber in a note explains Riemann's highly compressed material.

The geometry of space offered by Riemann was not just an extension of Gauss's differential geometry. It reconsidered the whole approach to the study of space. Riemann took up the question of just what we may be certain of about physical space. What conditions or facts are presupposed in the very experience of space before we determine by experience the particular axioms that hold in physical space? One of Riemann's objectives was to show that Euclid's particular axioms were empirical rather than, as had been believed, self-evident truths. He adopted the analytical approach because in

9. *Abh. der Ges. der Wiss. zu Gött.*, 13, 1868, 1–20 = *Werke*, 2nd ed., 272–87. An English translation can be found in W. K. Clifford's *Collected Mathematical Papers*. Also in *Nature*, 8, 1873, 14–36, and in D. E. Smith, *A Source Book in Mathematics*, 411–25.
10. *Werke*, 2nd ed., 1892, 391–404.

geometrical proofs we may be misled by our perceptions to assume facts not explicitly recognized. Thus Riemann's idea was that by relying upon analysis we might start with what is surely a priori about space and deduce the necessary consequences. Any other properties of space would then be known to be empirical. Gauss had concerned himself with this very same problem but of this investigation only the essay on curved surfaces was published. Riemann's search for what is a priori led him to study the local behavior of space or, in other words, the differential geometric approach as opposed to the consideration of space as a whole as one finds it in Euclid or in the non-Euclidean geometry of Gauss, Bolyai, and Lobatchevsky. Before examining the details we should be forewarned that Riemann's ideas as expressed in the lecture and in the manuscript of 1854 are vague. One reason is that Riemann adapted it to his audience, the entire faculty at Göttingen. Part of the vagueness stems from the philosophical considerations with which Riemann began his paper.

Guided to a large extent by Gauss's intrinsic geometry of surfaces in Euclidean space, Riemann developed an intrinsic geometry for any space. He preferred to treat n-dimensional geometry even though the three-dimensional case was clearly the important one and he speaks of n-dimensional space as a manifold. A point in a manifold of n dimensions is represented by assigning special values to n variable parameters, $x_1, x_2, \ldots, x_n$, and the aggregate of all such possible points constitutes the n-dimensional manifold itself, just as the aggregate of the points on a surface constitutes the surface itself. The n variable parameters are called coordinates of the manifold. When the x_i's vary continuously, the points range over the manifold.

Because Riemann believed that we know space only locally he started by defining the distance between two generic points whose corresponding coordinates differ only by infinitesimal amounts. He assumes that the square of this distance is

$$(10) \qquad ds^2 = \sum_{i=1}^{n} \sum_{j=1}^{n} g_{ij}\, dx_i\, dx_j,$$

wherein the g_{ij} are functions of the coordinates $x_1, x_2, \ldots, x_n$, $g_{ij} = g_{ji}$, and the right side of (10) is always positive for all possible values of the dx_i. This expression for ds^2 is a generalization of the Euclidean distance formula

$$ds^2 = dx_1^2 + dx_2^2 + \cdots + dx_n^2.$$

He mentions the possibility of assuming for ds the fourth root of a homogeneous function of the fourth degree in the differentials $dx_1, dx_2, \ldots, dx_n$ but did not pursue this possibility. By allowing the g_{ij} to be functions of the coordinates Riemann provided for the possibility that the nature of the space may vary from point to point.

Though Riemann in his paper of 1854 did not set forth explicitly the following definitions he undoubtedly had them in mind because they parallel what Gauss did for surfaces. A curve on a Riemannian manifold is given by the set of n functions

(11) $$x_1 = x_1(t), x_2 = x_2(t), \cdots, x_n = x_n(t).$$

Then the length of a curve between $t = \alpha$ and $t = \beta$ is defined by

(12) $$l = \int_\alpha^\beta ds = \int_\alpha^\beta \frac{ds}{dt} dt = \int_\alpha^\beta \sqrt{\sum_{i,j=1}^n g_{ij} \frac{dx_i}{dt} \frac{dx_j}{dt}} \, dt.$$

The shortest curve between two given points, $t = \alpha$ and $t = \beta$, the geodesic, is then determinable by the method of the calculus of variations. In the notation of that subject it is the curve for which $\delta \int_\alpha^\beta ds = 0$. One must then determine the particular functions of the form (11) which furnish this shortest path between the two points. In terms of the parameter arc length s, the equations of the geodesics prove to be

$$\frac{d^2x_i}{ds^2} + \sum_{\lambda,\mu} \{{}^{\lambda\,\mu}_{\ i}\} \frac{dx_\lambda}{ds} \frac{dx_\mu}{ds} = 0, \qquad i, \lambda, \mu = 1, 2, \cdots, n.$$

This is a system of n second order ordinary differential equations.[11]

The angle θ between two curves meeting at a point $(x_1, x_2, \ldots, x_n)$, one curve determined by the directions dx_i/ds, $i = 1, 2, \ldots, n$, and the other by dx_i'/ds', $i = 1, 2, \ldots, n$, where the primes indicate values belonging to the second direction, is defined by the formula

(13) $$\cos \theta = \sum_{i,i'=1}^n g_{ii'} \frac{dx_i}{ds} \frac{dx_i'}{ds'}.$$

By following the procedures which Gauss used for surfaces, a metrical n-dimensional geometry can be developed with the above definitions as a basis. All the metrical properties are determined by the coefficients g_{ij} in the expression for ds^2.

The second major concept in Riemann's 1854 paper is the notion of curvature of a manifold. Through it Riemann sought to characterize Euclidean space and more generally spaces on which figures may be moved about without change in shape or magnitude. Riemann's notion of curvature for any n-dimensional manifold is a generalization of Gauss's notion of total curvature for surfaces. Like Gauss's notion, the curvature of the manifold is defined in terms of quantities determinable on the manifold itself and there

---

11. For the meaning of the brace symbol see (19) below. Riemann did not give these equations explicitly.

is no need to think of the manifold as lying in some higher-dimensional manifold.

Given a point P in the n-dimensional manifold, Riemann considers a two-dimensional manifold at the point and in the n-dimensional manifold. Such a two-dimensional manifold is formed by a singly infinite set of geodesics through the point and tangent to a plane section of the manifold through the point P. Now a geodesic can be described by the point P and a direction at that point. Let $dx'_1, dx'_2, \ldots, dx'_n$ be the direction of one geodesic and $dx''_1, dx''_2, \ldots, dx''_n$ be the direction of another. Then the ith direction of any one of the singly infinite set of geodesics at P is given by

$$dx_i = \lambda' \, dx'_i + \lambda'' \, dx''_i$$

(subject to the condition $\lambda'^2 + \lambda''^2 + 2\lambda'\lambda'' \cos \theta = 1$ which arises from the condition $\sum g_{ij}(dx_i/ds)(dx_j/ds) = 1$). This set of geodesics forms a two-dimensional manifold which has a Gaussian curvature. Because there is an infinity of such two-dimensional manifolds through P we obtain an infinite number of curvatures at that point in the n-dimensional manifold. But from $n(n-1)/2$ of these measures of curvature, the rest can be deduced. An explicit expression for the measure of curvature can now be derived. This was done by Riemann in his 1861 paper and will be given below. For a manifold which is a surface, Riemann's curvature is exactly Gauss's total curvature. Strictly speaking, Riemann's curvature, like Gauss's, is a property of the metric imposed on the manifold rather than of the manifold itself.

After Riemann had completed his general investigation of n-dimensional geometry and showed how curvature is introduced, he considered more restricted manifolds on which finite spatial forms must be capable of movement without change of size or shape and must be capable of rotation in any direction. This led him to spaces of constant curvature.

When all the measures of curvature at a point are the same and equal to all the measures at any other point, we get what Riemann calls a manifold of constant curvature. On such a manifold it is possible to treat congruent figures. In the 1854 paper Riemann gave the following results but no details: If α be the measure of curvature the formula for the infinitesimal element of distance on a manifold of *constant* curvature becomes (in a suitable coordinate system)

$$(14) \qquad\qquad ds^2 = \frac{\sum\limits_{i=1}^{n} dx_i^2}{(1 + (\alpha/4) \sum x_i^2)^2}.$$

Riemann thought that the curvature α must be positive or zero, so that when $\alpha > 0$ we get a spherical space and when $\alpha = 0$ we get a Euclidean space and conversely. He also believed that if a space is infinitely extended the curvature must be zero. He did, however, suggest that there might be a real surface of constant negative curvature.[12]

To elaborate on Riemann, for $\alpha = a^2 > 0$ and $n = 3$ we get a three-dimensional spherical geometry though we cannot visualize it. The space is finite in extent but boundless. All geodesics in it are of constant length, namely, $2\pi/a$, and return upon themselves. The volume of the space is $2\pi^2/a^3$. For $a^2 > 0$ and $n = 2$ we get the space of the ordinary spherical surface. The geodesics are of course the great circles and are finite. Moreover, any two intersect in two points. Actually it is not clear whether Riemann regarded the geodesics of a surface of constant positive curvature as cutting in one or two points. He probably intended the latter. Felix Klein pointed out later (see the next chapter) that there were two distinct geometries involved.

Riemann also points out a distinction, of which more was made later, between boundlessness [as is the case for the surface of a sphere] and infiniteness of space. Unboundedness, he says, has a greater empirical credibility than any other empirically derived fact such as infinite extent.

Toward the end of his paper Riemann notes that since physical space is a special kind of manifold the geometry of that space cannot be derived only from general notions about manifolds. The properties that distinguish physical space from other triply extended manifolds are to be obtained only from experience. He adds, "It remains to resolve the question of knowing in what measure and up to what point these hypotheses about manifolds are confirmed by experience." In particular the axioms of Euclidean geometry may be only approximately true of physical space. Like Lobatchevsky, Riemann believed that astronomy will decide which geometry fits space. He ends his paper with the prophetic remark: "Either therefore the reality which underlies space must form a discrete manifold or we must seek the ground of its metric relations outside it, in the binding forces which act on it. . . . This leads us into the domain of another science, that of physics, into which the object of our work does not allow us to go today."

This point was developed by William Kingdon Clifford.[13]

> I hold in fact: (1) That small portions of space are of a nature analogous to little hills on a surface which is on the average flat. (2) That this property of being curved or distorted is continually passed on from one portion of space to another after the manner of a wave. (3) That this variation of the curvature of space is really what happens in that phenomenon which we call the motion of matter whether ponderable or ethereal. (4) That in this physical world nothing else takes place but this variation, subject, possibly, to the law of continuity.

12. Such surfaces were already known to Ferdinand Minding (*Jour. für Math.*, 19, 1839, 370–87, pp. 378–80 in particular), including the very one later called the pseudosphere (see Chap. 38, sec. 2). See also Gauss, *Werke*, 8, 265.
13. *Proc. Camb. Phil. Soc.*, 2, 1870, 157–58 = *Math. Papers*, 20–22.

The ordinary laws of Euclidean geometry are not valid for a space whose curvature changes not only from place to place but because of the motion of matter, from time to time. He added that a more exact investigation of physical laws would not be able to ignore these "hills" in space. Thus Riemann and Clifford, unlike most other geometers, felt the need to associate matter with space in order to determine what is true of physical space. This line of thought leads, of course, to the theory of relativity.

In his *Pariserarbeit* (1861) Riemann returned to the question of when a given Riemannian space whose metric is

$$(15) \qquad ds^2 = \sum_{i,j=1}^{n} g_{ij}\, dx_i\, dx_j$$

might be a space of constant curvature or even a Euclidean space. However, he formulated the more general question of when a metric such as (15) can be transformed by the equations

$$(16) \qquad x_i = x_i(y_1, y_2, \ldots, y_n), \qquad i = 1, 2, \ldots, n$$

into a given metric

$$(17) \qquad ds'^2 = \sum_{i,j=1}^{n} h_{ij}\, dy_i\, dy_j$$

with the understanding of course that ds would equal ds' so that the geometries of the two spaces would be the same except for the choice of coordinates. The transformation (16) is not always possible because, as Riemann points out, there are $n(n+1)/2$ independent functions in (15), whereas the transformation introduces only n functions which might be used to convert the g_{ij} into the h_{ij}.

To treat the general question Riemann introduced special quantities p_{ijk} which we shall replace by the more familiar Christoffel symbols with the understanding that

$$p_{ijk} = [^{j\,k}_{i}].$$

The Christoffel symbols, denoted in various ways, are

$$(18) \qquad \Gamma_{\alpha\beta,\lambda} = [^{\alpha\,\beta}_{\lambda}] = [\alpha\beta, \lambda] = \frac{1}{2}\left(\frac{\partial g_{\alpha\lambda}}{\partial x_\beta} + \frac{\partial g_{\beta\lambda}}{\partial x_\alpha} - \frac{\partial g_{\alpha\beta}}{\partial x_\lambda}\right)$$

$$(19) \qquad \Gamma^{\lambda}_{\alpha\beta} = \{^{\alpha\,\beta}_{\lambda}\} = \{\alpha\beta, \lambda\} = \sum_i g^{i\lambda}[^{\alpha\,\beta}_{i}]$$

where $g^{i\lambda}$ is the cofactor divided by g of $g_{i\lambda}$ in the determinant of g. Riemann also introduced what is now known as the Riemann four index symbol

$$(20) \qquad (\mu\lambda, jk) = R_{\lambda\mu, jk} = \frac{\partial \Gamma_{\lambda j,\mu}}{\partial x_k} - \frac{\partial \Gamma_{\lambda k,\mu}}{\partial x_j} + \sum_{i,\alpha} g^{i\alpha}(\Gamma_{\lambda k,\alpha}\Gamma_{\mu j,i} - \Gamma_{\lambda j,\alpha}\Gamma_{\mu k,i}).$$

Then Riemann shows that a necessary condition that ds^2 be transformable to ds'^2 is

$$(21) \qquad (\alpha\delta, \beta\gamma)' = \sum_{r,k,i,h} (rk, ih) \frac{\partial x_r}{\partial y_\alpha} \frac{\partial x_i}{\partial y_\beta} \frac{\partial x_h}{\partial y_\gamma} \frac{\partial x_k}{\partial y_\delta}$$

where the left-hand symbol refers to quantities formed for the ds' metric and (21) holds for all values of α, β, γ, δ, each of which ranges from 1 to n.

And now Riemann turns to the specific question of when a given ds^2 can be transformed to one with constant coefficients. He first derives an explicit expression for the curvature of a manifold. The general definition already given in the 1854 paper makes use of the geodesic lines issuing from a point O of the space. Let d and δ determine two vectors or directions of geodesics emanating from O. (Each direction is specified by the components of the tangent to the geodesic.) Then consider the pencil of geodesic vectors emanating from O and given by $\kappa d + \lambda\delta$ where κ and λ are parameters. If one thinks of d and δ as operating on the $x_i = f_i(t)$ which describe any one curve, then there is a meaning for the second differential $(\kappa d + \lambda\delta)^2 = \kappa^2 d^2 + 2\kappa\lambda \, d\delta + \lambda^2\delta^2$. Riemann then forms

$$(22) \qquad \Omega = \delta\delta \sum g_{ij} \, dx_i \, dx_j - 2 \, d\delta \sum g_{ij} \, dx_i \, dx_j + dd \sum g_{ij} \, dx_i \, dx_j.$$

Here one understands that the d and δ operate formally on the expressions following them (and d and δ commute) so that

$$(23) \qquad \delta\delta \sum g_{ij} \, dx_i \, dx_j = \delta\left[\sum (\delta g_{ij}) \, dx_i \, dx_j + \sum g_{ij}((\delta dx_i) \, dx_j + dx_i \, \delta dx_j)\right]$$

and $\delta g_{ij} = \sum_r (\partial g_{ij}/\partial x_r) \, \delta x_r$. If one calculates Ω one finds that all terms involving third differentials of a function vanish. Only terms involving δx_i, dx_i, $\delta^2 x_i$, δdx_i, and $d^2 x_i$ remain. By calculating these terms and by using the notation

$$p_{ik} = dx_i \, \delta x_k - dx_k \, \delta x_i$$

Riemann obtains

$$(24) \qquad [\Omega] = \sum_{i,k,r,s} (ik, rs) p_{ik} p_{rs}.$$

Now let

$$4\Delta^2 = \sum g_{ij} \, dx_i \, dx_j \cdot \sum g_{ij} \, \delta x_i \, \delta x_j - \left(\sum g_{ij} \, dx_i \, \delta x_j\right)^2.$$

Then the curvature K of a Riemannian manifold is

$$(25) \qquad K = -\frac{[\Omega]}{8\Delta^2}.$$

The overall conclusion is that the necessary and sufficient condition that a given ds^2 can be brought to the form (for $n = 3$)

$$(26) \qquad ds'^2 = c_1\, dx_1^2 + c_2\, dx_2^2 + c_3\, dx_3^2,$$

where the c_i's are constants, is that all the symbols $(\alpha\beta, \gamma\delta)$ be zero. In case the c_i's are all positive the ds' can be reduced to $dy_1^2 + dy_2^2 + dy_3^2$, that is, the space is Euclidean. As we can see from the value of $[\Omega]$, when K is zero, the space is essentially Euclidean.

It is worth noting that Riemann's curvature for an n-dimensional manifold reduces to Gauss's total curvature of a surface. In fact when

$$ds^2 = g_{11}\, dx_1^2 + 2g_{12}\, dx_1\, dx_2 + g_{22}\, dx_2^2,$$

of the 16 symbols $(\alpha\beta, \gamma\delta)$, 12 are zero and for the remaining four we have $(12, 12) = -(12, 21) = -(21, 12) = (21, 21)$. Then Riemann's K reduces to

$$k = \frac{(12, 12)}{g}.$$

By using (20) this expression can be shown to be equal to Gauss's expression for the total curvature of a surface.

4. *The Successors of Riemann*

When Riemann's essay of 1854 was published in 1868, two years after his death, it created intense interest, and many mathematicians hastened to fill in the ideas he sketched and to extend them. The immediate successors of Riemann were Beltrami, Christoffel, and Lipschitz.

Eugenio Beltrami (1835–1900), professor of mathematics at Bologna and other Italian universities, who knew Riemann's 1854 paper but apparently did not know his 1861 paper, took up the matter of proving that the general expression for ds^2 reduces to the form (14) given by Riemann for a space of constant curvature.[14] Beyond this result and proving a few other assertions by Riemann, Beltrami took up the subject of differential invariants, which we shall consider in the next section.

Elwin Bruno Christoffel (1829–1900), who was a professor of mathematics at Zurich and later at Strasbourg, advanced the ideas in both of Riemann's papers. In two key papers[15] Christoffel's major concern was to reconsider and amplify the theme already treated somewhat sketchily by Riemann in his 1861 paper, namely, when one form

$$F = \sum_{i,j} g_{ij}\, dx_i\, dx_j$$

14. *Annali di Mat.*, (2), 2, 1868–69, 232–55 = *Opere Mat.*, 1, 406–29.
15. *Jour. für Math.*, 70, 1869, 46–70 and 241–45 = *Ges. Math. Abh.*, 1, 352 ff., 378 ff.

can be transformed into another

$$F' = \sum_{i,j} g'_{ij} \, dy_i \, dy_j.$$

Christoffel sought necessary and sufficient conditions. It was in this paper, incidentally, that he introduced the Christoffel symbols.

Let us consider first the two-dimensional case where

$$F = a \, dx^2 + 2b \, dx \, dy + c \, dy^2$$

and

$$F' = A \, dX^2 + 2B \, dX \, dY + C \, dY^2$$

and suppose that x and y may be expressed as functions of X and Y so that F becomes F' under the transformation. Of course $dx = (\partial x/\partial X) \, dX + (\partial x/\partial Y) \, dY$. Now when x, y, dx, and dy are replaced in F by their values in X and Y and when one equates coefficients in this new form of F with those of F' one obtains

$$a\left(\frac{\partial x}{\partial X}\right)^2 + 2b \frac{\partial x}{\partial X} \frac{\partial y}{\partial X} + c\left(\frac{\partial y}{\partial X}\right)^2 = A$$

$$a \frac{\partial x}{\partial X} \frac{\partial x}{\partial Y} + b\left(\frac{\partial x}{\partial X} \frac{\partial y}{\partial Y} + \frac{\partial x}{\partial Y} \frac{\partial y}{\partial X}\right) + c \frac{\partial y}{\partial X} \frac{\partial y}{\partial Y} = B$$

$$a\left(\frac{\partial x}{\partial Y}\right)^2 + 2b\left(\frac{\partial x}{\partial Y} \frac{\partial y}{\partial Y}\right) + c\left(\frac{\partial y}{\partial Y}\right)^2 = C.$$

These are three differential equations for x and y as functions of X and Y. If they can be solved then we know how to transform from F to F'. However, there are only two functions involved. There must then be some relations between a, b, and c on the one hand and A, B, and C on the other. By differentiating the three equations above and further algebraic steps the relation proves to be $K = K'$.

For the n-variable case, Christoffel uses the same technique. He starts with

$$F = \sum g_{rs} \, dx_r \, dx_s$$

and

$$F' = \sum g'_{rs} \, dy_r \, dy_s.$$

The transformation is

$$x_i = x_i(y_1, y_2, \cdots, y_n), \qquad i = 1, 2, \cdots, n.$$

He lets $g = |g_{rs}|$. Then if Δ_{rs} is the cofactor of g_{rs} in the determinant let $g^{pq} = \Delta_{pq}/g$. He, like Riemann, introduces independently the four index symbol (without the comma)

$$(gkhi) = \frac{\partial}{\partial x_i} [gh, k] - \frac{\partial}{\partial x_h} [gi, k] + \sum_p (\{gi, p\}[hk, h] - \{gh, p\}[ik, p]).$$

He then deduces $n(n + 1)/2$ partial differential equations for the x_i as functions of the y_i. A typical one is

$$\sum_{r,s} g_{rs} \frac{\partial x_r}{\partial y_\alpha} \frac{\partial x_s}{\partial y_\beta} = g'_{\alpha\beta}.$$

These equations are the necessary and sufficient conditions that a transformation exist for which $F = F'$.

Partly to treat the integrability of this set of equations and partly because Christoffel wishes to consider forms of degree higher than two in the dx_i, he performs a number of differentiations and algebraic steps which show that

$$(27) \qquad (\alpha\delta\beta\gamma)' = \sum_{g,h,i,k} (gkhi) \frac{\partial x_g}{\partial y_\alpha} \frac{\partial x_h}{\partial y_\beta} \frac{\partial x_i}{\partial y_\gamma} \frac{\partial x_k}{\partial y_\delta},$$

where α, β, γ, and δ take all values from 1 to n. There are $n^2(n^2 - 1)/12$ equations of this form. These equations are the necessary and sufficient conditions for the equivalence of two differential forms of fourth order. Indeed let $d^{(1)}x$, $d^{(2)}x$, $d^{(3)}x$, $d^{(4)}x$ be four sets of differentials of x and likewise for the y's. Then if we have the quadrilinear form

$$G_4 = \sum_{g,k,h,i} (gkhi) d^{(1)}x_g d^{(2)}x_k d^{(3)}x_h d^{(4)}x_i$$

the relations (27) are necessary and sufficient that $G_4 = G'_4$, where G'_4 is the analogue of G_4 in the y variables.

This theory can be generalized to μ-ply differential forms. In fact Christoffel introduces

$$(28) \qquad G_\mu = \sum_{i_1,\cdots,i_\mu} (i_1 i_2 \cdots i_\mu) \, \partial x_{i_1} \, \partial x_{i_2} \cdots \partial x_{i_\mu}$$

where the term in parentheses is defined in terms of the g_{rs} much as the four index symbol is and the symbol $\underset{i}{\partial}$ is used to distinguish the differentials of the set of x_i from the set obtained by applying $\underset{j}{\partial}$. He then shows that

$$(29) \qquad (\alpha_1\alpha_2\cdots\alpha_\mu)' = \sum (i_1\cdots i_\mu) \frac{\partial x_{i_1}}{\partial y_{\alpha_1}} \cdots \frac{\partial x_{i_\mu}}{\partial y_{\alpha_\mu}}$$

and obtains necessary and sufficient conditions that G_μ be transformable into G'_μ.

He gives next a general procedure whereby from a μ-ply form G_μ a $(\mu + 1)$-ply form $G_{\mu+1}$ can be derived. The key step is to introduce

$$(30) \quad (i_1 i_2 \cdots i_\mu i) = \frac{\partial}{\partial x_i} (i_1 i_2 \cdots i_\mu)$$

$$- \sum_\lambda [\{i i_1, \lambda\}(\lambda i_2 \cdots i_\mu) + \{i i_2, \lambda\}(i_1 \lambda i_3 \cdots i_\mu) + \cdots].$$

These $(\mu + 1)$-index symbols are the coefficients of the $G_{\mu+1}$ form. The procedure Christoffel uses here is what Ricci and Levi-Civita later called covariant differentiation (Chap. 48).

Whereas Christoffel wrote only one key paper on Riemannian geometry, Rudolph Lipschitz, professor of mathematics at Bonn University, wrote a great number appearing in the *Journal für Mathematik* from 1869 on. Though there are some generalizations of the work of Beltrami and Christoffel, the essential subject matter and results are the same as those of the latter two men. He did produce some new results on subspaces of Riemannian and Euclidean n-dimensional spaces.

The ideas projected by Riemann and developed by his three immediate successors suggested hosts of new problems in both Euclidean and Riemannian differential geometry. In particular the results already obtained in the Euclidean case for three dimensions were generalized to curves, surfaces, and higher-dimensional forms in n dimensions. Of many results we shall cite just one.

In 1886 Friedrich Schur (1856–1932) proved the theorem named after him.[16] In accordance with Riemann's approach to the notion of curvature, Schur speaks of the curvature of an orientation of space. Such an orientation is determined by a pencil of geodesics $\mu\alpha + \lambda\beta$ where α and β are the directions of two geodesics issuing from a point. This pencil forms a surface and has a Gauss curvature which Schur calls the Riemannian curvature of that orientation. His theorem then states that if at each point the Riemannian curvature of a space is independent of the orientation then the Riemannian curvature is constant throughout the space. The manifold is then a space of constant curvature.

5. Invariants of Differential Forms

It was clear from the study of the question of when a given expression for ds^2 can be transformed by a transformation of the form

$$(31) \qquad x_i = x_i(x'_1, x'_2, \cdots, x'_n), \qquad i = 1, 2, \cdots, n$$

16. *Math. Ann.*, 27, 1886, 167–72 and 537–67.

to another such expression with preservation of the value of ds^2 that different coordinate representations can be obtained for the very same manifold. However, the geometrical properties of the manifold must be independent of the particular coordinate system used to represent and study it. Analytically these geometrical properties would be represented by invariants, that is, expressions which retain their form under the change of coordinates and which will consequently have the same value at a given point. The invariants of interest in Riemannian geometry involve not only the fundamental quadratic form, which contains the differentials dx_i and dx_j, but may also contain derivatives of the coefficients and of other functions. They are therefore called differential invariants.

To use the two-dimensional case as an example, if

(32) $$ds^2 = E\,du^2 + 2F\,du\,dv + G\,dv^2$$

is the element of distance for a surface then the Gaussian curvature K is given by formula (8) above. If now the coordinates are changed to

(33) $$u' = f(u, v), \qquad v' = g(u, v)$$

then there is the theorem that if $E\,du^2 + 2F\,du\,dv + G\,dv^2$ transforms into $E'\,du'^2 + 2F'\,du'\,dv' + G'\,dv'^2$ then $K = K'$, where K' is the same expression as in (8) but in the accented variables. Hence the Gaussian curvature of a surface is a scalar invariant. The invariant K is said to be an invariant attached to the form (32) and involves only E, F, and G and their derivatives.

The study of differential invariants was actually initiated in a more limited context by Lamé. He was interested in invariants under transformations from one orthogonal curvilinear coordinate system in three dimensions to another. For rectangular Cartesian coordinates he showed[17] that

(34) $$\Delta_1\phi = \left(\frac{\partial\phi}{\partial x}\right)^2 + \left(\frac{\partial\phi}{\partial y}\right)^2 + \left(\frac{\partial\phi}{\partial z}\right)^2$$

(35) $$\Delta_2\phi = \frac{\partial^2\phi}{\partial x^2} + \frac{\partial^2\phi}{\partial y^2} + \frac{\partial^2\phi}{\partial z^2}$$

are differential invariants (he called them differential parameters). Thus if ϕ is transformed into $\phi'(x', y', z')$ under an orthogonal transformation (rotation of axes) then

$$\left(\frac{\partial\phi}{\partial x}\right)^2 + \left(\frac{\partial\phi}{\partial y}\right)^2 + \left(\frac{\partial\phi}{\partial z}\right)^2 = \left(\frac{\partial\phi'}{\partial x'}\right)^2 + \left(\frac{\partial\phi'}{\partial y'}\right)^2 + \left(\frac{\partial\phi'}{\partial z'}\right)^2$$

at the same point whose coordinates are (x, y, z) in the original system and (x', y', z') in the new coordinate system. The analogous equation holds for $\Delta_2\phi$.

17. *Jour. de l'Ecole Poly.*, 14, 1834, 191–288.

For orthogonal curvilinear coordinates in Euclidean space where ds^2 has the form

(36) $$ds^2 = g_{11}\,du_1^2 + g_{22}\,du_2^2 + g_{33}\,du_3^2$$

Lamé showed (*Leçons sur les coordonnées curvilignes*, 1859, cf. above Chapter 28, sec. 5) that the divergence of the gradient of ϕ, which in rectangular coordinates is given by $\Delta_2\phi$ above, has the invariant form

$$\Delta_2\phi = \frac{1}{\sqrt{g_{11}g_{22}g_{33}}}\left[\frac{\partial}{\partial u_1}\left(\sqrt{\frac{g_{22}g_{33}}{g_{11}}}\frac{\partial\phi}{\partial u_1}\right) + \frac{\partial}{\partial u_2}\left(\sqrt{\frac{g_{33}g_{11}}{g_{22}}}\frac{\partial\phi}{\partial u_2}\right)\right.$$
$$\left. + \frac{\partial}{\partial u_3}\left(\sqrt{\frac{g_{11}g_{22}}{g_{33}}}\frac{\partial\phi}{\partial u_3}\right)\right].$$

Incidentally in this same work Lamé gave conditions on when the ds^2 given by (36) determines a curvilinear coordinate system in Euclidean space and, if it does, how to change to rectangular coordinates.

The investigation of invariants for the theory of surfaces was first made by Beltrami.[18] He gave the two differential invariants

$$\Delta_1\phi = \frac{1}{EG - F^2}\left\{E\left(\frac{\partial\phi}{\partial v}\right)^2 - 2F\frac{\partial\phi}{\partial u}\frac{\partial\phi}{\partial v} + G\left(\frac{\partial\phi}{\partial v}\right)^2\right\}$$

and

$$\Delta_2\phi = \frac{1}{\sqrt{EG - F^2}}\left\{\frac{\partial}{\partial u}\left(\frac{G\phi_u - F\phi_v}{\sqrt{EG - F^2}}\right) + \frac{\partial}{\partial v}\left(\frac{-F\phi_u + E\phi_v}{\sqrt{EG - F^2}}\right)\right\}.$$

These have geometrical meaning. For example in the case of $\Delta_1\phi$, if $\Delta_1\phi = 1$, the curves $\phi(u, v) = $ const. are the orthogonal trajectories of a family of geodesics on the surface.

The search for differential invariants was carried over to quadratic differential forms in n variables. The reason again was that these invariants are independent of particular choices of coordinates; they represent intrinsic properties of the manifold itself. Thus the Riemann curvature is a scalar invariant.

Beltrami, using a method given by Jacobi,[19] succeeded in carrying over to n-dimensional Riemannian spaces the Lamé invariants.[20] Let g as usual be the determinant of the g_{ij} and let g^{ij} be the cofactor divided by g of g_{ij} in g. Then Beltrami showed that Lamé's first invariant becomes

$$\Delta_1(\phi) = \sum_{i,j} g^{ij}\frac{\partial\phi}{\partial x_i}\frac{\partial\phi}{\partial x_j}.$$

18. *Gior. di Mat.*, 2, 1864, 267–82, and succeeding papers in Vols. 2 and 3 = *Opere Mat.*, 1, 107–98.
19. *Jour. für Math.*, 36, 1848, 113–34 = *Werke*, 2, 193–216.
20. *Memorie dell' Accademia delle Scienze dell' Istituto di Bologna*, (2), 8, 1868, 551–90 = *Opere Mat.*, 2, 74–118.

This is the general form for the square of the gradient of ϕ. For the second of Lamé's invariants Beltrami obtained

$$\Delta_2(\phi) = \frac{1}{\sqrt{g}} \sum_i \frac{\partial}{\partial x_i} \left(\sqrt{g} \sum_j g^{ij} \frac{\partial \phi}{\partial x_j} \right).$$

He also introduced the mixed differential invariant

$$\Delta_1(\phi\psi) = \sum_{i,j} g^{ij} \frac{\partial \phi}{\partial x_i} \frac{\partial \psi}{\partial x_j}.$$

This is the general form of the scalar product of the gradients of ϕ and ψ. Of course the form ds^2 is itself an invariant under a change of coordinates. From this, as we found in the previous section, Christoffel derived higher order differential forms, his G_4 and G_μ, which are also invariants. Moreover, he showed how from G_μ one can derive $G_{\mu+1}$, which is also an invariant. The construction of such invariants was also pursued by Lipschitz. The number and variety are extensive. As we shall see this theory of differential invariants was the inspiration for tensor analysis.

Bibliography

Beltrami, Eugenio: *Opere matematiche*, 4 vols., Ulrico Hoepli, 1902–20.
Clifford, William K.: *Mathematical Papers*, Macmillan, 1882; Chelsea (reprint), 1968.
Coolidge, Julian L.: *A History of Geometrical Methods*, Dover (reprint), 1963, pp. 355–87.
Encyklopädie der Mathematischen Wissenschaften, III, Teil 3, various articles, B. G. Teubner, 1902–7.
Gauss, Carl F.: *Werke*, 4, 192–216, 217–58, Königliche Gesellschaft der Wissenschaften zu Göttingen, 1880. A translation, "General Investigations of Curved Surfaces," has been reprinted by Raven Press, 1965.
Helmholtz, Hermann von: "Über die tatsächlichen Grundlagen der Geometrie," *Wissenschaftliche Abhandlungen*, 2, 610–17.
———: "Über die Tatsachen, die der Geometrie zum Grunde liegen," *Nachrichten König. Ges. der Wiss. zu Gött.*, 15, 1868, 193–221; *Wiss. Abh.*, 2, 618–39.
———: "Über den Ursprung Sinn und Bedeutung der geometrischen Sätze"; English translation, "On the Origin and Significance of Geometrical Axioms," in Helmholtz: *Popular Scientific Lectures*, Dover (reprint), 1962, 223–49. Also in James R. Newman: *The World of Mathematics*, Simon and Schuster, 1956, Vol. 1, 647–68.
Jammer, Max: *Concepts of Space*, Harvard University Press, 1954.
Killing, W.: *Die nicht-euklidischen Raumformen in analytischer Behandlung*, B. G. Teubner, 1885.
Klein, F.: *Vorlesungen über die Entwicklung der Mathematik im 19. Jahrhundert*, Chelsea (reprint), 1950, Vol. 1, 6–62; Vol. 2, 147–206.

Pierpont, James: "Some Modern Views of Space," *Amer. Math. Soc. Bull.*, 32, 1926, 225–58.

Riemann, Bernhard: *Gesammelte mathematische Werke*, 2nd ed., Dover (reprint), 1953, pp. 272–87 and 391–404.

Russell, Bertrand: *An Essay on the Foundations of Geometry* (1897), Dover (reprint), 1956.

Smith, David E.: *A Source Book in Mathematics*, Dover (reprint), 1959, Vol. 2, 411–25, 463–75. This contains translations of Riemann's 1854 paper and Gauss's 1822 paper.

Staeckel, P.: "Gauss als Geometer," *Nachrichten König. Ges. der Wiss. zu Gött.*, 1917, Beiheft, 25–140; also in *Werke*, 10_2.

Weatherburn, C. E.: "The Development of Multidimensional Differential Geometry," *Australian and New Zealand Ass'n for the Advancement of Science*, 21, 1933, 12–28.

38
Projective and Metric Geometry

> But it should always be required that a mathematical subject
> not be considered exhausted until it has become intuitively
> evident . . .   FELIX KLEIN

1. Introduction

Prior to and during the work on non-Euclidean geometry, the study of
projective properties was the major geometric activity. Moreover, it was
evident from the work of von Staudt (Chap. 35, sec. 3) that projective
geometry is logically prior to Euclidean geometry because it deals with
qualitative and descriptive properties that enter into the very formation of
geometrical figures and does not use the measures of line segments and
angles. This fact suggested that Euclidean geometry might be some specialization of projective geometry. With the non-Euclidean geometries now at
hand the possibility arose that these, too, at least the ones dealing with spaces
of constant curvature, might be specializations of projective geometry. Hence
the relationship of projective geometry to the non-Euclidean geometries,
which are metric geometries because distance is employed as a fundamental
concept, became a subject of research. The clarification of the relationship
of projective geometry to Euclidean and the non-Euclidean geometries is the
great achievement of the work we are about to examine. Equally vital was the
establishment of the consistency of the basic non-Euclidean geometries.

2. Surfaces as Models of Non-Euclidean Geometry

The non-Euclidean geometries that seemed to be most significant after
Riemann's work were those of spaces of constant curvature. Riemann himself had suggested in his 1854 paper that a space of constant positive curvature
in two dimensions could be realized on a surface of a sphere provided the
geodesic on the sphere was taken to be the "straight line." This non-
Euclidean geometry is now referred to as double elliptic geometry for reasons
which will be clearer later. Prior to Riemann's work, the non-Euclidean
geometry of Gauss, Lobatchevsky, and Bolyai, which Klein later called

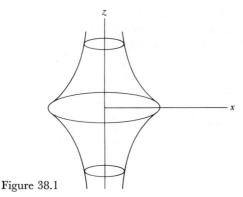

Figure 38.1

hyperbolic geometry, had been introduced as the geometry in a plane in which ordinary (and necessarily infinite) straight lines are the geodesics. The relationship of this geometry to Riemann's varieties was not clear. Riemann and Minding[1] had thought about surfaces of constant negative curvature but neither man related these to hyperbolic geometry.

Independently of Riemann, Beltrami recognized[2] that surfaces of constant curvature are non-Euclidean spaces. He gave a limited representation of hyperbolic geometry on a surface,[3] which showed that the geometry of a restricted portion of the hyperbolic plane holds on a surface of constant negative curvature if the geodesics on this surface are taken to be the straight lines. The lengths and angles on the surface are the lengths and angles of the ordinary Euclidean geometry on the surface. One such surface is known as the pseudosphere (Fig. 38.1). It is generated by revolving a curve called the tractrix about its asymptote. The equation of the tractrix is

$$z = k \log \frac{k + \sqrt{k^2 - x^2}}{x} - \sqrt{k^2 - x^2}$$

and the equation of the surface is

$$z = k \log \frac{\sqrt{k^2 - x^2 - y^2}}{\sqrt{x^2 + y^2}} - \sqrt{k^2 - x^2 - y^2}.$$

The curvature of the surface is $-1/k^2$. Thus the pseudosphere is a model for a limited portion of the plane of Gauss, Lobatchevsky, and Bolyai. On the pseudosphere a figure may be shifted about and just bending will make it conform to the surface, as a plane figure by bending can be fitted to the surface of a circular cylinder.

1. *Jour. für Math.*, 19, 1839, 370–87.
2. *Annali di Mat.*, 7, 1866, 185–204 = *Opere Mat.*, 1, 262–80.
3. *Gior. di Mat.*, 6, 1868, 248–312 = *Opere Mat.*, 1, 374–405.

Beltrami had shown that on a surface of negative constant curvature one can realize a *piece* of the Lobatchevskian plane. However, there is no regular analytic surface of negative constant curvature on which the geometry of the *entire* Lobatchevskian plane is valid. All such surfaces have a singular curve—the tangent plane is not continuous across it—so that a continuation of the surface across this curve will not continue the figures that represent those of Lobatchevsky's geometry. This result is due to Hilbert.[4]

In this connection it is worth noting that Heinrich Liebmann (1874–1939)[5] proved that the sphere is the only closed analytic surface (free of singularities) of constant positive curvature and so the only one that can be used as a Euclidean model for double elliptic geometry.

The development of these models helped the mathematicians to understand and see meaning in the basic non-Euclidean geometries. One must keep in mind that these geometries, in the two-dimensional case, are fundamentally geometries of the plane in which the lines and angles are the usual lines and angles of Euclidean geometry. While hyperbolic geometry had been developed in this fashion, the conclusions still seemed strange to the mathematicians and had been only grudgingly admitted into mathematics. The double elliptic geometry, suggested by Riemann's differential geometric approach, did not even have an axiomatic development as a geometry of the plane. Hence the only meaning mathematicians could see for it was that provided by the geometry on the sphere. A far better understanding of the nature of these geometries was secured through another development that sought to relate Euclidean and projective geometry.

3. Projective and Metric Geometry

Though Poncelet had introduced the distinction between projective and metric properties of figures and had stated in his *Traité* of 1822 that the projective properties were logically more fundamental, it was von Staudt who began to build up projective geometry on a basis independent of length and angle size (Chap. 35, sec. 3). In 1853 Edmond Laguerre (1834–86), a professor at the Collège de France, though primarily interested in what happens to angles under a projective transformation, actually advanced the goal of establishing metric properties of Euclidean geometry on the basis of

4. *Amer. Math. Soc., Trans.*, 2, 1901, 86–99 = *Ges. Abh.*, 2, 437–48. The proof and further historical details can be found in Appendix V of David Hilbert's *Grundlagen der Geometrie*, 7th ed., B. G. Teubner, 1930. The theorem presupposes that the lines of hyperbolic geometry would be the geodesics of the surface and the lengths and angles would be the Euclidean lengths and angles on the surface.
5. *Nachrichten König. Ges. der Wiss. zu Gött.*, 1899, 44–55; *Math. Ann.*, 53, 1900, 81–112; and 54, 1901, 505–17.

projective concepts by supplying a projective basis for the measure of an angle.[6]

To obtain a measure of the angle between two given intersecting lines one can consider two lines through the origin and parallel respectively to the two given lines. Let the two lines through the origin have the equations (in nonhomogeneous coordinates) $y = x \tan \theta$ and $y = x \tan \theta'$. Let $y = ix$ and $y = -ix$ be two (imaginary) lines from the origin to the circular points at infinity, that is, the points $(1, i, 0)$ and $(1, -i, 0)$. Call these four lines u, u', w, and w' respectively. Let ϕ be the angle between u and u'. Then Laguerre's result is that

$$(1) \qquad\qquad \phi = \theta' - \theta = \frac{i}{2} \log (uu', ww'),$$

where (uu', ww') is the cross ratio of the four lines.[7] What is significant about the expression (1) is that it may be taken as the definition of the size of an angle in terms of the projective concept of cross ratio. The logarithm function is, of course, purely quantitative and may be introduced in any geometry.

Independently of Laguerre, Cayley made the next step. He approached geometry from the standpoint of algebra, and in fact was interested in the geometric interpretation of quantics (homogeneous polynomial forms), a subject we shall consider in Chapter 39. Seeking to show that metrical notions can be formulated in projective terms he concentrated on the relation of Euclidean to projective geometry. The work we are about to describe is in his "Sixth Memoir upon Quantics."[8]

Cayley's work proved to be a generalization of Laguerre's idea. The latter had used the circular points at infinity to define angle in the plane. The circular points are really a degenerate conic. In two dimensions Cayley introduced any conic in place of the circular points and in three dimensions he introduced any quadric surface. These figures he called the absolutes. Cayley asserted that all metric properties of figures are none other than projective properties augmented by the absolute or in relation to the absolute. He then showed how this principle led to a new expression for angle and an expression for distance between two points.

He starts with the fact that the points of a plane are represented by homogeneous coordinates. These coordinates are not to be regarded as distances or ratios of distances but as an assumed fundamental notion not

6. *Nouvelles Annales de Mathématiques*, 12, 1853, 57–66 = *Œuvres*, 2, 6–15.
7. The cross ratio is itself a complex number. The coefficient $i/2$ ensures that a right angle has size $\pi/2$. The computation of such cross ratios can be found in texts on projective geometry. See, for example, William C. Graustein: *Introduction to Higher Geometry*, Macmillan, 1933, Chap. 8.
8. *Phil. Trans.*, 149, 1859, 61–91 = *Coll. Math. Papers*, 2, 561–606.

requiring or admitting of explanation. To define distance and size of angle he introduces the quadratic form

$$F(x, x) = \sum_{i,j=1}^{3} a_{ij}x_ix_j, \qquad a_{ij} = a_{ji}$$

and the bilinear form

$$F(x, y) = \sum_{i,j=1}^{3} a_{ij}x_iy_j.$$

The equation $F(x, x) = 0$ defines a conic which is Cayley's absolute. The equation of the absolute in line coordinates is

$$G(u, u) = \sum_{i,j=1}^{3} A^{ij}u_iu_j = 0,$$

where A^{ij} is the cofactor of a_{ij} in the determinant $|a|$ of the coefficients of F.

Cayley now defines the distance δ between two points x and y, where $x = (x_1, x_2, x_3)$ and $y = (y_1, y_2, y_3)$, by the formula

$$(2) \qquad \delta = \text{arc cos} \frac{F(x, y)}{[F(x, x)F(y, y)]^{1/2}}.$$

The angle ϕ between two lines whose line coordinates are $u = (u_1, u_2, u_3)$ and $v = (v_1, v_2, v_3)$ is defined by

$$(3) \qquad \cos \phi = \frac{G(u, v)}{[G(u, u)G(v, v)]^{1/2}}.$$

These general formulas become simple if we take for the absolute the particular conic $x_1^2 + x_2^2 + x_3^2 = 0$. Then if (a_1, a_2, a_3) and (b_1, b_2, b_3) are the homogeneous coordinates of two points, the distance between them is given by

$$(4) \qquad \text{arc cos} \frac{a_1b_1 + a_2b_2 + a_3b_3}{\sqrt{a_1^2 + a_2^2 + a_3^2}\sqrt{b_1^2 + b_2^2 + b_3^2}}$$

and the angle ϕ between two lines whose homogeneous line coordinates are (u_1, u_2, u_3) and (v_1, v_2, v_3) is given by

$$(5) \qquad \cos \phi = \frac{u_1v_1 + u_2v_2 + u_3v_3}{\sqrt{u_1^2 + u_2^2 + u_3^2}\sqrt{v_1^2 + v_2^2 + v_3^2}}.$$

With respect to the expression for distance if we use the shorthand that $xy = x_1y_1 + x_2y_2 + x_3y_3$ then, if $a = (a_1, a_2, a_3)$, b and c are three points on a line,

$$\text{arc cos} \frac{ab}{\sqrt{aa}\sqrt{bb}} + \text{arc cos} \frac{bc}{\sqrt{bb}\sqrt{cc}} = \text{arc cos} \frac{ac}{\sqrt{aa}\sqrt{cc}}.$$

That is, the distances add as they should. By taking the absolute conic to be the circular points at infinity, $(1, i, 0)$ and $(1, -i, 0)$, Cayley showed that his formulas for distance and angle reduce to the usual Euclidean formulas. It will be noted that the expressions for length and angle involve the algebraic expression for the absolute. Generally the analytic expression of any Euclidean metrical property involves the relation of that property to the absolute. Metrical properties are not properties of the figure per se but of the figure in relation to the absolute. This is Cayley's idea of the general projective determination of metrics. The place of the metric concept in projective geometry and the greater generality of the latter were described by Cayley as, "Metrical geometry is part of projective geometry."

Cayley's idea was taken over by Felix Klein (1849–1925) and generalized so as to include the non-Euclidean geometries. Klein, a professor at Göttingen, was one of the leading mathematicians in Germany during the last part of the nineteenth and first part of the twentieth century. During the years 1869–70 he learned the work of Lobatchevsky, Bolyai, von Staudt, and Cayley; however, even in 1871 he did not know Laguerre's result. It seemed to him to be possible to subsume the non-Euclidean geometries, hyperbolic and double elliptic geometry, under projective geometry by exploiting Cayley's idea. He gave a sketch of his thoughts in a paper of 1871 [9] and then developed them in two papers.[10] Klein was the first to recognize that we do not need surfaces to obtain models of non-Euclidean geometries.

To start with, Klein noted that Cayley did not make clear just what he had in mind for the meaning of his coordinates. They were either simply variables with no geometrical interpretation or they were Euclidean distances. But to derive the metric geometries from projective geometry it was necessary to build up the coordinates on a projective basis. Von Staudt had shown (Chap. 35, sec. 3) that it was possible to assign numbers to points by his algebra of throws. But he used the Euclidean parallel axiom. It seemed clear to Klein that this axiom could be dispensed with and in the 1873 paper he shows that this can be done. Hence coordinates and cross ratio of four points, four lines, or four planes can be defined on a purely projective basis.

Klein's major idea was that by specializing the nature of Cayley's absolute quadric surface (if one considers three-dimensional geometry) one could show that the metric, which according to Cayley depended on the nature of the absolute, would yield hyperbolic and double elliptic geometry. When the second degree surface is a real ellipsoid, real elliptic paraboloid, or real hyperboloid of two sheets one gets Lobatchevsky's metric geometry, and when the second degree surface is imaginary one gets Riemann's non-Euclidean geometry (of constant positive curvature). If the absolute is

9. *Nachrichten König. Ges. der Wiss. zu Gött.*, 1871, 419–33 = *Ges. Math. Abh.*, 1, 244–53.
10. *Math. Ann.*, 4, 1871, 573–625; and 6, 1873, 112–45 = *Ges. Math. Abh.*, 1, 254–305, 311–43.

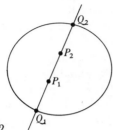

Figure 38.2

taken to be the sphere-circle, whose equation in homogeneous coordinates is $x^2 + y^2 + z^2 = 0$, $t = 0$, then the usual Euclidean metric geometry obtains. Thus the metric geometries become special cases of projective geometry.

To appreciate Klein's ideas let us consider two-dimensional geometry. One chooses a conic in the projective plane; this conic will be the absolute. Its equation is

(6) $$F = \sum_{i,j=1}^{3} a_{ij}x_ix_j = 0$$

in point coordinates and

(7) $$G = \sum_{i,j=1}^{3} A^{ij}u_iu_j = 0$$

in line coordinates. To derive Lobatchevsky's geometry the conic must be real, e.g. in plane homogeneous coordinates $x_1^2 + x_2^2 - x_3^2 = 0$; for Riemann's geometry on a surface of constant positive curvature it is imaginary, for example, $x_1^2 + x_2^2 + x_3^2 = 0$; and for Euclidean geometry, the conic degenerates into two coincident lines represented in homogeneous coordinates by $x_3 = 0$ and on this locus one chooses two imaginary points whose equation is $x_1^2 + x_2^2 = 0$, that is, the circular points at infinity whose homogeneous coordinates are $(1, i, 0)$ and $(1, -i, 0)$. In every case the conic has a real equation.

To be specific let us suppose the conic is the one shown in Figure 38.2. If P_1 and P_2 are two points of a line, this line meets the absolute in two points (real or imaginary). Then the distance is taken to be

(8) $$d = c \log (P_1P_2, Q_1Q_2),$$

where the quantity in parentheses denotes the cross ratio of the four points and c is a constant. This cross ratio can be expressed in terms of the co-

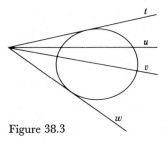

Figure 38.3

ordinates of the points. Moreover, if there are three points P_1, P_2, P_3 on the line then it can readily be shown that

$$(P_1P_2, Q_1Q_2) \cdot (P_2P_3, Q_1Q_2) = (P_1P_3, Q_1Q_2)$$

so that $P_1P_2 + P_2P_3 = P_1P_3$.

Likewise if u and v are two lines (Fig. 38.3) one considers the tangents t and w from their point of intersection to the absolute (the tangents may be imaginary lines); then the angle between u and v is defined to be

$$\phi = c' \log (uv, tw),$$

where again c' is a constant and the quantity in parentheses denotes the cross ratio of the four lines.

To express the values of d and ϕ analytically and to show their dependence upon the choice of the absolute, let the equation of the absolute be given by F and G above. By definition

$$F_{xy} = \sum_{i,j=1}^{3} a_{ij} x_i y_j.$$

One can now show that if $x = (x_1, x_2, x_3)$ and $y = (y_1, y_2, y_3)$ are the coordinates of P_1 and P_2 then

$$d = c \log \frac{F_{xy} + \sqrt{F_{xy}^2 - F_{xx}F_{yy}}}{F_{xy} - \sqrt{F_{xy}^2 - F_{xx}F_{yy}}}.$$

Likewise, if (u_1, u_2, u_3) and (v_1, v_2, v_3) are the coordinates of the two lines, then one can show, using G, that

$$\phi = c' \log \frac{G_{uv} + \sqrt{G_{uv}^2 - G_{uu}G_{vv}}}{G_{uv} - \sqrt{G_{uv}^2 - G_{uu}G_{vv}}}.$$

The constant c' is generally taken to be $i/2$ so as to make ϕ real and a complete central angle 2π.

Klein used the logarithmic expressions above for angle and distance and showed how the metric geometries can be derived from projective geometry.

Thus if one starts with projective geometry then by his choice of the absolute and by using the above expressions for distance and angle one can get the Euclidean, hyperbolic, and elliptic geometries as special cases. The nature of the metric geometry is fixed by the choice of the absolute. Incidentally, Klein's expressions for distance and angle can be shown to be equal to Cayley's.

If one makes a projective (i.e. linear) transformation of the projective plane into itself which transforms the absolute into itself (though points on the absolute go into other points) then because cross ratio is unaltered by a linear transformation, distance and angle will be unaltered. These particular linear transformations which leave the absolute fixed are the rigid motions or congruence transformations of the particular metric geometry determined by the absolute. A general projective transformation will not leave the absolute invariant. Thus projective geometry proper is more general in the transformations it allows.

Another contribution of Klein to non-Euclidean geometry was the observation, which he says he first made in 1871[11] but published in 1874,[12] that there are two kinds of elliptic geometry. In the double elliptic geometry two points do not always determine a unique straight line. This is evident from the spherical model when the two points are diametrically opposite. In the second elliptic geometry, called single elliptic, two points always determine a unique straight line. When looked at from the standpoint of differential geometry, the differential form ds^2 of a surface of constant positive curvature is (in homogeneous coordinates)

$$ds^2 = \frac{dx_1^2 + dx_2^2 + dx_3^2}{\{1 + (a^2/4)(x_1^2 + x_2^2 + x_3^2)\}^2}.$$

In both cases $a^2 > 0$. However, in the first type the geodesics are curves of finite length $2\pi/a$ or if R is the radius, $2\pi R$, and are closed (return upon themselves). In the second the geodesics are of length π/a or πR and are still closed.

A model of a surface which has the properties of single elliptic geometry and which is due to Klein,[13] is provided by a hemisphere including the boundary. However, one must identify any two points on the boundary which are diametrically opposite. The great circular arcs on the hemisphere are the "straight lines" or geodesics of this geometry and the ordinary angles on the surface are the angles of the geometry. Single elliptic geometry (sometimes called elliptic in which case double elliptic geometry is called spherical) is then also realized on a space of constant positive curvature.

11. *Math. Ann.*, 4, 1871, 604. See also, *Math. Ann.*, 6, 1873, 125; and *Math. Ann.*, 37, 1890, 554–57.
12. *Math. Ann.*, 7, 1874, 549–57; 9, 1876, 476–82 = *Ges. math. Abh.*, 2, 63–77.
13. See the references in footnote 11.

One cannot actually unite the pairs of points which are identified in this model at least in three-dimensional space. The surface would have to cross itself and points that coincide on the intersection would have to be regarded as distinct.

We can now see why Klein introduced the terminology hyperbolic for Lobatchevsky's geometry, elliptic for the case of Riemann's geometry on a surface of constant positive curvature, and parabolic for Euclidean geometry. This terminology was suggested by the fact that the ordinary hyperbola meets the line at infinity in two points and correspondingly in hyperbolic geometry each line meets the absolute in two real points. The ordinary ellipse has no real points in common with the line at infinity and in elliptic geometry, likewise, each line has no real points in common with the absolute. The ordinary parabola has only one real point in common with the line at infinity and in Euclidean geometry (as extended in projective geometry) each line has one real point in common with the absolute.

The import which gradually emerged from Klein's contributions was that projective geometry is really logically independent of Euclidean geometry. Moreover, the non-Euclidean and Euclidean geometries were also seen to be special cases or subgeometries of projective geometry. Actually the strictly logical or rigorous work on the axiomatic foundations of projective geometry and its relations to the subgeometries remained to be done (Chap. 42). But by making apparent the basic role of projective geometry Klein paved the way for an axiomatic development which could start with projective geometry and derive the several metric geometries from it.

4. Models and the Consistency Problem

By the early 1870s several basic non-Euclidean geometries, the hyperbolic and the two elliptic geometries, had been introduced and intensively studied. The fundamental question which had yet to be answered in order to make these geometries legitimate branches of mathematics was whether they were consistent. All of the work done by Gauss, Lobatchevsky, Bolyai, Riemann, Cayley, and Klein might still have proved to be nonsense if contradictions were inherent in these geometries.

Actually the proof of the consistency of two-dimensional double elliptic geometry was at hand, and possibly Riemann appreciated this fact though he made no explicit statement. Beltrami[14] had pointed out that Riemann's two-dimensional geometry of constant positive curvature is realized on a sphere. This model makes possible the proof of the consistency of two-dimensional double elliptic geometry. The axioms (which were not explicit at this time) and the theorems of this geometry are all applicable to

14. *Annali di Mat.*, (2), 2, 1868–69, 232–55 = *Opere Matematiche*, 1, 406–29.

the geometry of the surface of the sphere provided that line in the double elliptic geometry is interpreted as great circle on the surface of the sphere. If there should be contradictory theorems in this double elliptic geometry then there would be contradictory theorems about the geometry of the surface of the sphere. Now the sphere is part of Euclidean geometry. Hence if Euclidean geometry is consistent, then double elliptic geometry must also be so. To the mathematicians of the 1870s the consistency of Euclidean geometry was hardly open to question because, apart from the views of a few men such as Gauss, Bolyai, Lobatchevsky, and Riemann, Euclidean geometry was still the necessary geometry of the physical world and it was inconceivable that there could be contradictory properties in the geometry of the physical world. However, it is important, especially in the light of later developments, to realize that this proof of the consistency of double elliptic geometry depends upon the consistency of Euclidean geometry.

The method of proving the consistency of double elliptic geometry could not be used for single elliptic geometry or for hyperbolic geometry. The hemispherical model of single elliptic geometry cannot be realized in three-dimensional Euclidean geometry though it can be in four-dimensional Euclidean geometry. If one were willing to believe in the consistency of the latter, then one might accept the consistency of single elliptic geometry. However, though n-dimensional geometry had already been considered by Grassmann, Riemann, and others, it is doubtful that any mathematician of the 1870s would have been willing to affirm the consistency of four-dimensional Euclidean geometry.

The case for the consistency of hyperbolic geometry could not be made on any such grounds. Beltrami had given the pseudospherical interpretation, which is a surface in Euclidean space, but this serves as a model for only a limited region of hyperbolic geometry and so could not be used to establish the consistency of the entire geometry. Lobatchevsky and Bolyai had considered this problem (Chap. 36, sec. 8) but had not been able to settle it. As a matter of fact, though Bolyai proudly published his non-Euclidean geometry, there is evidence that he doubted its consistency because in papers found after his death he continued to try to prove the Euclidean parallel axiom.

The consistency of the hyperbolic and single elliptic geometries was established by new models. The model for hyperbolic geometry is due to Beltrami.[15] However, the distance function used in this model is due to Klein and the model is often ascribed to him. Let us consider the two-dimensional case.

Within the Euclidean plane (which is part of the projective plane) one selects a real conic which one may as well take to be a circle (Fig. 38.4).

15. *Annali di Mat.*, 7, 1866, 185–204 = *Opere Mat.*, 1, 262–80; *Gior. di Mat.*, 6, 1868, 284–312 = *Opere Mat.*, 1, 374–405.

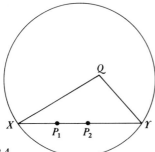

Figure 38.4

According to this representation of hyperbolic geometry the points of the geometry are the points interior to this circle. A line of this geometry is a chord of the circle, say the chord XY (but not including X and Y). If we take any point Q not on XY then we can find any number of lines through Q which do not meet XY. Two of these lines, namely, QX and QY, separate the lines through Q into two classes, those lines which cut XY and those which do not. In other words, the parallel axiom of hyperbolic geometry is satisfied by the points and lines (chords) interior to the circle. Further, let the size of the angle formed by two lines a and b be

$$\sphericalangle(a, b) = \frac{1}{2i} \log (ab, mn),$$

where m and n are the conjugate imaginary tangents from the vertex of the angle to the circle and (ab, mn) is the cross ratio of the four lines a, b, m, and n. The constant $1/2i$ ensures that a right angle has the measure $\pi/2$. The definition of distance between two points is given by formula (8), that is, $d = c \log (P_1P_2, XY)$ with c usually taken to be $k/2$. According to this formula as P_1 or P_2 approaches X or Y, the distance P_1P_2 becomes infinite. Hence in terms of this distance, a chord is an infinite line of hyperbolic geometry.

Thus with the projective definitions of distance and angle size, the points, chords, angles, and other figures interior to the circle satisfy the axioms of hyperbolic geometry. Then the theorems of hyperbolic geometry also apply to these figures inside the circle. In this model the axioms and theorems of hyperbolic geometry are really assertions about special figures and concepts (e.g. distance defined in the manner of hyperbolic geometry) of *Euclidean* geometry. Since the axioms and theorems in question apply to these figures and concepts, regarded as belonging to Euclidean geometry, all of the assertions of hyperbolic geometry are theorems of Euclidean geometry. If, then, there were a contradiction in hyperbolic geometry, this contradiction would be a contradiction within Euclidean geometry. But *if Euclidean geometry is consistent*, then hyperbolic geometry must also be. Thus the

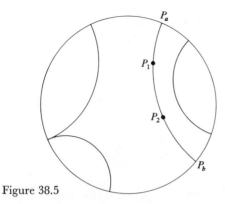

Figure 38.5

consistency of hyperbolic geometry is reduced to the consistency of Euclidean geometry.

The fact that hyperbolic geometry is consistent implies that the Euclidean parallel axiom is independent of the other Euclidean axioms. If this were not the case, that is, if the Euclidean parallel axiom were derivable from the other axioms, it would also be a theorem of hyperbolic geometry for, aside from the parallel axiom, the other axioms of Euclidean geometry are the same as those of hyperbolic geometry. But this theorem would contradict the parallel axiom of hyperbolic geometry and hyperbolic geometry would be inconsistent. The consistency of two-dimensional single elliptic geometry can be shown in the same manner as for hyperbolic geometry because this elliptic geometry is also realized within the projective plane and with the projective definition of distance.

Independently and in connection with his work on automorphic functions Poincaré[16] gave another model which also establishes the consistency of hyperbolic geometry. One form in which this Poincaré model for hyperbolic plane geometry[17] can be expressed takes the absolute to be a circle (Fig. 38.5). Within the absolute the straight lines of the geometry are arcs of circles which cut the absolute orthogonally and straight lines through the center of the absolute. The length of any segment P_1P_2 is given by $\log (P_1P_2, P_aP_b)$, where $(P_1P_2, P_aP_b) = (P_1P_b/P_2P_b)/(P_1P_a/P_2P_a)$, P_a and P_b are the points in which the arc through P_1 and P_2 cuts the absolute, and the lengths P_1P_b, P_2P_b, etc. are the chords. The angle between two intersecting "lines" of this model is the normal Euclidean angle between the two arcs. Two circular arcs which are tangent at a point on the absolute are parallel

16. *Acta Math.*, 1, 1882, 1–62 = *Œuvres*, 2, 108–68; see p. 8 and p. 52 of the paper.
17. This form, attributed to Poincaré, is close to one he gave in the *Bull. Soc. Math. de France*, 15, 1887, 203–16 = *Œuvres*, 11, 79–91. The model described here seems to have been given first by Joseph Wellstein (1869–1919) in H. Weber and J. Wellstein, *Enzyklopädie der Elementar-Mathematik*, 2, 1905, 39–81.

"lines." Since in this model, too, the axioms and theorems of hyperbolic geometry are special theorems of Euclidean geometry, the argument given above apropos of the Beltrami model may be applied here to establish the consistency of hyperbolic geometry. The higher-dimensional analogues of the above models are also valid.

5. Geometry from the Transformation Viewpoint

Klein's success in subsuming the various metric geometries under projective geometry led him to seek to characterize the various geometries not just on the basis of nonmetric and metric properties and the distinctions among the metrics but from the broader standpoint of what these geometries and other geometries which had already appeared on the scene sought to accomplish. He gave this characterization in a speech of 1872, "Vergleichende Betrachtungen über neuere geometrische Forschungen" (A Comparative Review of Recent Researches in Geometry),[18] on the occasion of his admission to the faculty of the University of Erlangen, and the views expressed in it are known as the Erlanger Programm.

Klein's basic idea is that each geometry can be characterized by a group of transformations and that a geometry is really concerned with invariants under this group of transformations. Moreover a subgeometry of a geometry is the collection of invariants under a subgroup of transformations of the original group. Under this definition all theorems of a geometry corresponding to a given group continue to be theorems in the geometry of the subgroup.

Though Klein in his paper does not give the analytical formulations of the groups of transformations he discusses we shall give some for the sake of explicitness. According to his notion of a geometry projective geometry, say in two dimensions, is the study of invariants under the group of transformations from the points of one plane to those of another or to points of the same plane (collineations). Each transformation is of the form

$$
\begin{aligned}
x_1' &= a_{11}x_1 + a_{12}x_2 + a_{13}x_3 \\
x_2' &= a_{21}x_1 + a_{22}x_2 + a_{23}x_3 \\
x_3' &= a_{31}x_1 + a_{32}x_2 + a_{33}x_3
\end{aligned}
$$

(9)

wherein homogeneous coordinates are presupposed, the a_{ij} are real numbers, and the determinant of the coefficients must not be zero. In nonhomogeneous coordinates the transformations are represented by

$$
x' = \frac{a_{11}x + a_{12}y + a_{13}}{a_{31}x + a_{32}y + a_{33}}
$$

(10)

$$
y' = \frac{a_{21}x + a_{22}y + a_{23}}{a_{31}x + a_{32}y + a_{33}}
$$

18. *Math. Ann.*, 43, 1893, 63–100 = *Ges. Math. Abh.*, 1, 460–97. An English translation can be found in the *N.Y. Math. Soc. Bull.*, 2, 1893, 215–49.

and again the determinant of the a_{ij} must not be zero. The invariants under the projective group are, for example, linearity, collinearity, cross ratio, harmonic sets, and the property of being a conic section.

One subgroup of the projective group is the collection of affine transformations.[19] This subgroup is defined as follows: Let any line l_∞ in the projective plane be fixed. The points of l_∞ are called ideal points or points at infinity and l_∞ is called the line at infinity. Other points and lines of the projective plane are called ordinary points and these are the usual points of the Euclidean plane. The affine group of collineations is that subgroup of the projective group which leaves l_∞ invariant (though not necessarily pointwise) and affine geometry is the set of properties and relations invariant under the affine group. Algebraically, in two dimensions and in homogeneous coordinates, the affine transformations are represented by equations (9) above but in which $a_{31} = a_{32} = 0$ and with the same determinant condition. In nonhomogeneous coordinates affine transformations are given by

$$x' = a_{11}x + a_{12}y + a_{13} \qquad \begin{vmatrix} a_{11} & a_{12} \\ a_{21} & a_{22} \end{vmatrix} \neq 0.$$
$$y' = a_{21}x + a_{22}y + a_{23}$$

Under an affine transformation, straight lines go into straight lines, and parallel straight lines into parallel lines. However, lengths and angle sizes are altered. Affine geometry was first noted by Euler and then by Möbius in his *Der barycentrische Calcul*. It is useful in the study of the mechanics of deformations.

The group of any metric geometry is the same as the affine group except that the determinant above must have the value $+1$ or -1. The first of the metric geometries is Euclidean geometry. To define the group of this geometry one starts with l_∞ and supposes there is a fixed involution on l_∞. One requires that this involution has no real double points but has the circular points at ∞ as (imaginary) double points. We now consider all projective transformations which not only leave l_∞ fixed but carry any point of the involution into its corresponding point of the involution, which implies that each circular point goes into itself. Algebraically these transformations of the Euclidean group are represented in nonhomogeneous (two-dimensional) coordinates by

$$x' = \rho(x \cos \theta - y \sin \theta + \alpha), \qquad \rho = \pm 1.$$
$$y' = \rho(x \sin \theta + y \cos \theta + \beta)$$

The invariants are length, size of angle, and size and shape of any figure.

Euclidean geometry as the term is used in this classification is the set of invariants under this class of transformations. The transformations are rotations, translations, and reflections. To obtain the invariants associated with similar figures, the subgroup of the affine group known as the parabolic

19. Klein did not single out this subgroup.

metric group is introduced. This group is defined as the class of projective transformations which leaves the involution on l_∞ invariant, and this means that each pair of corresponding points goes into some pair of corresponding points. In nonhomogeneous coordinates the transformations of the parabolic metric group are of the form

$$x' = ax - by + c$$
$$y' = bex + aey + d$$

wherein $a^2 + b^2 \neq 0$, and $e^2 = 1$. These transformations preserve angle size.

To characterize hyperbolic metric geometry we return to projective geometry and consider an arbitrary, real, nondegenerate conic (the absolute) in the projective plane. The subgroup of the projective group which leaves this conic invariant (though not necessarily pointwise) is called the hyperbolic metric group and the corresponding geometry is hyperbolic metric geometry. The invariants are those associated with congruence.

Single elliptic geometry is the geometry corresponding to the subgroup of projective transformations which leaves a definite imaginary ellipse (the absolute) of the projective plane invariant. The plane of elliptic geometry is the real projective plane and the invariants are those associated with congruence.

Even double elliptic geometry can be encompassed in this transformation viewpoint, but one must start with three-dimensional projective transformations to characterize the two-dimensional metric geometry. The subgroup of transformations consists of those three-dimensional projective transformations which transform a definite sphere (surface) S of the finite portion of space into itself. The spherical surface S is the "plane" of double elliptic geometry. Again the invariants are associated with congruence.

In four metric geometries, that is, Euclidean, hyperbolic, and the two elliptic geometries, the transformations which are permitted in the corresponding subgroup are what are usually called rigid motions and these are the only geometries which permit rigid motions.

Klein introduced a number of intermediate classifications which we shall not repeat here. The scheme below shows the relationships of the principal geometries.

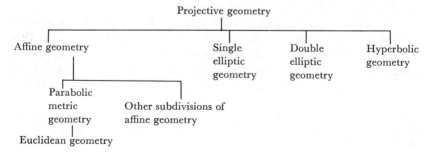

Klein also considered more general geometries than projective. At this time (1872) algebraic geometry was gradually being distinguished as a separate discipline and he characterized this geometry by introducing the transformations which in three dimensions and in nonhomogeneous coordinates read

$$x' = \phi(x, y, z), \qquad y' = \psi(x, y, z), \qquad z' = \chi(x, y, z).$$

He required that the functions ϕ, ψ, and χ be rational and single-valued and that it must be possible to solve for x, y, and z in terms of single-valued rational functions of x', y', z'. These transformations are called Cremona transformations and the invariants under them are the subject matter of algebraic geometry (Chap. 39).

Klein also projected the study of invariants under one-to-one continuous transformations with continuous inverses. This is the class now called homeomorphisms and the study of the invariants under such transformations is the subject matter of topology (Chap. 50). Though Riemann had also considered what are now recognized to be topological problems in his work with Riemann surfaces, the projection of topology as a major geometry was a bold step in 1872.

It has been possible since Klein's days to make further additions and specializations of Klein's classification. But not all of geometry can be incorporated in Klein's scheme. Algebraic geometry today and differential geometry do not come under this scheme.[20] Though Klein's view of geometry did not prove to be all-embracing, it did afford a systematic method of classifying and studying a good portion of geometry and suggested numerous research problems. His "definition" of geometry guided geometrical thinking for about fifty years. Moreover, his emphasis on invariants under transformations carried beyond mathematics to mechanics and mathematical physics generally. The physical problem of invariance under transformation or the problem of expressing physical laws in a manner independent of the coordinate system became important in physical thinking after the invariance of Maxwell's equations under Lorentz transformations (a four-dimensional subgroup of affine geometry) was noted. This line of thinking led to the special theory of relativity.

We shall merely mention here further studies on the classification of geometries which at least in their time attracted considerable attention. Helmholtz and Sophus Lie (1842–99) sought to characterize geometries in which rigid motions are possible. Helmholtz's basic paper, "Über die Thatsachen, die der Geometrie zum Grunde liegen" (On the Facts Which

20. Klein does speak, in the case of differential geometry, of the group of transformations which leave the expression for ds^2 invariant. This leads to differential invariants (*Ges. math. Abh.*, 1, 487).

Underlie Geometry),[21] showed that if the motions of rigid bodies are to be possible in a space then Riemann's expression for ds in a space of constant curvature is the only one possible. Lie attacked the same problem by what is called the theory of continuous transformation groups, a theory he had already introduced in the study of ordinary differential equations, and he characterized the spaces in which rigid motions are possible by means of the kinds of groups of transformations which these spaces permit.[22]

6. *The Reality of Non-Euclidean Geometry*

The interest in the classical synthetic non-Euclidean geometries and in projective geometry declined after the work of Klein and Lie partly because the essence of these structures was so clearly exposed by the transformation viewpoint. The feeling of mathematicians so far as the discovery of additional theorems is concerned was that the mine had been exhausted. The rigorization of the foundations remained to be accomplished and this was an active area for quite a few years after 1880 (Chap. 42).

Another reason for the loss of interest in the non-Euclidean geometries was their seeming lack of relevance to the physical world. It is curious that the first workers in the field, Gauss, Lobatchevsky, and Bolyai did think that non-Euclidean geometry might prove applicable when further work in astronomy had been done. But none of the mathematicians who worked in the later period believed that these basic non-Euclidean geometries would be physically significant. Cayley, Klein, and Poincaré, though they considered this matter, affirmed that we would not ever need to improve on or abandon Euclidean geometry. Beltrami's pseudosphere model had made non-Euclidean geometry real in a mathematical sense (though not physically) because it gave a readily visualizable interpretation of Lobatchevsky's geometry but at the expense of changing the line from the ruler's edge to geodesics on the pseudosphere. Similarly the Beltrami-Klein and Poincaré models made sense of non-Euclidean geometry by changing the concepts either of line, distance, or angle-measure, or of all three, and by picturing them in Euclidean space. But the thought that physical space under the usual interpretation of straight line or even under some other interpretation could be non-Euclidean was dismissed. In fact, most mathematicians regarded non-Euclidean geometry as a logical curiosity.

Cayley was a staunch supporter of Euclidean space and accepted the non-Euclidean geometries only so far as they could be realized in Euclidean space by the use of new distance formulas. In 1883 in his presidential address to the British Association for the Advancement of Science[23] he said that

21. *Nachrichten König. Ges. der Wiss. zu Gött.*, 15, 1868, 193–221 = *Wiss. Abh.*, 2, 618–39.
22. *Theorie der Transformationsgruppen*, 3, 437–543, 1893.
23. *Collected Math. Papers*, 11, 429–59.

non-Euclidean spaces were a priori a mistaken idea, but non-Euclidean geometries were acceptable because they resulted merely from a change in the distance function in Euclidean space. He did not grant independent existence to the non-Euclidean geometries but regarded them as a class of special Euclidean structures or as a way of representing projective relations in Euclidean geometry. It was his view that

> Euclid's twelfth [tenth] axiom in Playfair's form of it does not need demonstration but is part of our notion of space, of the physical space of our own experience—the space, that is, which we become acquainted with by experience, but which is the representation lying at the foundation of all external experience.
>
> Riemann's view may be said to be that, having *in intellectu* a more general notion of space (in fact a notion of non-Euclidean space), we learn by experience that space (the physical space of experience) is, if not exactly, at least to the highest degree of approximation, Euclidean space.

Klein regarded Euclidean space as the necessary fundamental space. The other geometries were merely Euclidean with new distance functions. The non-Euclidean geometries were in effect subordinated to Euclidean geometry.

Poincaré's judgment was more liberal. Science should always try to use Euclidean geometry and vary the laws of physics where necessary. Euclidean geometry may not be true but it is most convenient. One geometry cannot be more true than another; it can only be more convenient. Man creates geometry and then adapts the physical laws to it to make the geometry and laws fit the world. Poincaré insisted[24] that even if the angle sum of a triangle should prove to be greater than 180° it would be better to assume that Euclidean geometry describes physical space and that light travels along curves because Euclidean geometry is simpler. Of course events proved he was wrong. It is not the simplicity of the geometry alone that counts for science but the simplicity of the entire scientific theory. Clearly the mathematicians of the nineteenth century were still tied to tradition in their notions about what makes physical sense. The advent of the theory of relativity forced a drastic change in the attitude toward non-Euclidean geometry.

The delusion of mathematicians that what they are working on at the moment is the most important conceivable subject is illustrated again by their attitude toward projective geometry. The work we have examined in this chapter does indeed show that projective geometry is fundamental to many geometries. However, it does not embrace the evidently vital Riemannian geometry and the growing body of algebraic geometry. Neverthe-

24. *Bull. Soc. Math. de France*, 15, 1887, 203–16 = *Œuvres*, 11, 79–91. He expressed this view again in an article "Les Géométries non-euclidiennes" in the *Revue Générale des Sciences*, 2, 1891, #23. An English translation is in *Nature*, 45, 1892, 404–7. See also his *Science and Hypothesis*, Chapter 3, in *The Foundations of Science*, The Science Press, 1946.

less, Cayley affirmed in his 1859 paper (sec. 3) that, "Projective geometry is all geometry and reciprocally."[25] Bertrand Russell in his *Essay on the Foundations of Geometry* (1897) also believed that projective geometry was necessarily the a priori form of any geometry of physical space. Hermann Hankel, despite the attention he gave to history,[26] did not hesitate to say in 1869 that projective geometry is the royal road to all mathematics. Our examination of the developments already recorded shows clearly that mathematicians can readily be carried away by their enthusiasms.

Bibliography

Beltrami, Eugenio: *Opere matematiche*, Ulrico Hoepli, 1902, Vol. 1.

Bonola, Roberto: *Non-Euclidean Geometry*, Dover (reprint), 1955, pp. 129–264.

Coolidge, Julian L.: *A History of Geometrical Methods*, Dover (reprint), 1963, pp. 68–87.

Klein, Felix: *Gesammelte mathematische Abhandlungen*, Julius Springer, 1921–23, Vols. 1 and 2.

Pasch, Moritz, and Max Dehn: *Vorlesungen über neuere Geometrie*, 2nd ed., Julius Springer, 1926, pp. 185–239.

Pierpont, James: "Non-Euclidean Geometry. A Retrospect," *Amer. Math. Soc. Bull.*, 36, 1930, 66–76.

Russell, Bertrand: *An Essay on the Foundations of Geometry* (1897), Dover (reprint), 1956.

25. Cayley used the term "descriptive geometry" for projective geometry.
26. *Die Entwicklung der Mathematik in den letzten Jahrhunderten* (The Development of Mathematics in the Last Few Centuries), 1869; 2nd ed., 1884.

39
Algebraic Geometry

> In these days the angel of topology and the devil of abstract
> algebra fight for the soul of each individual mathematical
> domain.   HERMANN WEYL

1. Background

While non-Euclidean and Riemannian geometry were being created, the
projective geometers were pursuing their theme. As we have seen, the two
areas were linked by the work of Cayley and Klein. After the algebraic
method became widely used in projective geometry the problem of recognizing what properties of geometrical figures are independent of the coordinate representation commanded attention and this prompted the study
of algebraic invariants.

The projective properties of geometrical figures are those that are
invariant under linear transformations of the figures. While working on these
properties the mathematicians occasionally allowed themselves to consider
higher-degree transformations and to seek those properties of curves and
surfaces that are invariant under these latter transformations. The class of
transformations, which soon superseded linear transformations as the
favorite interest, is called birational because these are expressed algebraically
as rational functions of the coordinates and the inverse transformations are
also rational functions of their coordinates. The concentration on birational
transformations undoubtedly resulted from the fact that Riemann had used
them in his work on Abelian integrals and functions, and in fact, as we shall
see, the first big steps in the study of the birational transformation of curves
were guided by what Riemann had done. These two subjects formed the
content of algebraic geometry in the latter part of the nineteenth century.

The term algebraic geometry is an unfortunate one because originally
it referred to all the work from the time of Fermat and Descartes in which
algebra had been applied to geometry; in the latter part of the nineteenth
century it was applied to the study of algebraic invariants and birational
transformations. In the twentieth century it refers to the last-mentioned
field.

924

2. The Theory of Algebraic Invariants

As we have already noted, the determination of the geometric properties of figures that are represented and studied through coordinate representation calls for the discernment of those algebraic expressions which remain invariant under change of coordinates. Alternatively viewed, the projective transformation of one figure into another by means of a linear transformation preserves some properties of the figure. The algebraic invariants represent these invariant geometrical properties.

The subject of algebraic invariants had previously arisen in number theory (Chap. 34, sec. 5) and particularly in the study of how binary quadratic forms

$$(1) \qquad f = ax^2 + 2bxy + cy^2$$

transform when x and y are transformed by the linear transformation T, namely,

$$(2) \qquad x = \alpha x' + \beta y', \qquad y = \gamma x' + \delta y',$$

where $\alpha\delta - \beta\gamma = r$. Application of T to f produces

$$(3) \qquad f' = a'x'^2 + 2b'x'y' + c'y'^2.$$

In number theory the quantities a, b, c, α, β, γ, and δ are integers and $r = 1$. However, it is true generally that the discriminant D of f satisfies the relation

$$(4) \qquad D' = r^2 D.$$

The linear transformations of projective geometry are more general because the coefficients of the forms and the transformations are not restricted to integers. The term algebraic invariants is used to distinguish those arising under these more general linear transformations from the modular invariants of number theory and, for that matter, from the differential invariants of Riemannian geometry.

To discuss the history of algebraic invariant theory we need some definitions. The nth degree form in one variable

$$f(x) = a_0 x^n + a_1 x^{n-1} + \cdots + a_n$$

becomes in homogeneous coordinates the binary form

$$(5) \qquad f(x_1, x_2) = a_0 x_1^n + a_1 x_1^{n-1} x_2 + \cdots + a_n x_2^n.$$

In three variables the forms are called ternary; in four variables, quaternary; etc. The definitions below apply to forms in n variables.

Suppose we subject the binary form to a transformation T of the form (2). Under T the form $f(x_1, x_2)$ is transformed into the form

$$F(X_1, X_2) = A_0 X_1^n + A_1 X_1^{n-1} X_2 + \cdots + A_n X_2^n.$$

The coefficients of F will differ from those of f and the roots of $F = 0$ will differ from the roots of $f = 0$. Any function I of the coefficients of f which satisfies the relationship

$$I(A_0, A_1, \cdots, A_n) = r^w I(a_0, a_1, \cdots, a_n)$$

is called an invariant of f. If $w = 0$, the invariant is called an absolute invariant of f. The degree of the invariant is the degree in the coefficients and the weight is w. The discriminant of a binary form is an invariant, as (4) illustrates. In this case the degree is 2 and the weight is 2. The significance of the discriminant of any polynomial equation $f(x) = 0$ is that its vanishing is the condition that $f(x) = 0$ have equal roots or, geometrically, that the locus of $f(x) = 0$, which is a series of points, has two coincident points. This property is clearly independent of the coordinate system.

If two (or more) binary forms

$$f_1 = a_0 x_1^m + \cdots + a_m x_2^m$$
$$f_2 = b_0 x_1^n + \cdots + b_n x_2^n$$

are transformed by T into

$$F_1 = A_0 X_1^m + \cdots + A_m X_2^m$$
$$F_2 = B_0 X_1^n + \cdots + B_n X_2^n$$

then any function I of the coefficients which satisfies the relationship

(6) $I(A_0, \cdots, A_m, B_0, \cdots, B_n) = r^w I(a_0, \cdots, a_m, b_0, \cdots, b_n)$

is said to be a joint or simultaneous invariant of the two forms. Thus the linear forms $a_1 x_1 + b_1 x_2$ and $a_2 x_1 + b_2 x_2$ have as a simultaneous invariant the resultant $a_1 b_2 - a_2 b_1$ of the two forms. Geometrically the vanishing of the resultant means that the two forms represent the same point (in homogeneous coordinates). Two quadratic forms $f_1 = a_1 x_1^2 + 2b_1 x_1 x_2 + c_1 x_2^2$ and $f_2 = a_2 x_1^2 + 2b_2 x_1 x_2 + c_2 x_2^2$ possess a simultaneous invariant

$$D_{12} = a_1 c_2 - 2b_1 b_2 + a_2 c_1$$

whose vanishing expresses the fact that f_1 and f_2 represent harmonic pairs of points.

Beyond invariants of a form or system of forms there are covariants. Any function C of the coefficients *and* variables of f which is an invariant under T except for a power of the modulus (determinant) of T is called a covariant of f. Thus, for binary forms, a covariant satisfies the relation

$$C(A_0, A_1, \cdots, A_n, X_1, X_2) = r^w C(a_0, a_1, \cdots, a_n, x_1, x_2).$$

The definitions of absolute and simultaneous covariants are analogous to those for invariants. The degree of a covariant in the coefficients is called its degree and the degree in its variables is called its order. Invariants are thus covariants of order zero. However, sometimes the word invariant is used to mean an invariant in the narrower sense or a covariant.

A covariant of f represents some figure which is not only related to f but projectively related. Thus the Jacobian of two quadratic binary forms $f(x_1, x_2)$ and $\phi(x_1, x_2)$, namely,

$$\begin{vmatrix} \dfrac{\partial f}{\partial x_1} & \dfrac{\partial f}{\partial x_2} \\[2mm] \dfrac{\partial \phi}{\partial x_1} & \dfrac{\partial \phi}{\partial x_2} \end{vmatrix}$$

is a simultaneous covariant of weight 1 of the two forms. Geometrically, the Jacobian set equal to zero represents a pair of points which is harmonic to each of the original pairs represented by f and ϕ and the harmonic property is projective.

The Hessian of a binary form introduced by Hesse,[1]

$$\begin{vmatrix} \dfrac{\partial^2 f}{\partial x_1^2} & \dfrac{\partial^2 f}{\partial x_1\, \partial x_2} \\[2mm] \dfrac{\partial^2 f}{\partial x_1\, \partial x_2} & \dfrac{\partial^2 f}{\partial x_2^2} \end{vmatrix}$$

is a covariant of weight 2. Its geometric meaning is too involved to warrant space here (Cf. Chap. 35, sec. 5). The concept of the Hessian and its covariance applies to any form in n variables.

The work on algebraic invariants was started in 1841 by George Boole (1815–64) whose results[2] were limited. What is more relevant is that Cayley was attracted to the subject by Boole's work and he interested Sylvester in the subject. They were joined by George Salmon (1819–1904), who was a professor of mathematics at Trinity College in Dublin from 1840 to 1866 and then became a professor of divinity at that institution. These three men did so much work on invariants that in one of his letters Hermite dubbed them the invariant trinity.

In 1841 Cayley began to publish mathematical articles on the algebraic side of projective geometry. The 1841 paper of Boole suggested to Cayley the computation of invariants of nth degree homogeneous functions. He called the invariants derivatives and then hyperdeterminants; the term invariant is due to Sylvester.[3] Cayley, employing ideas of Hesse and Eisenstein on determinants, developed a technique for generating his "derivatives." Then he published ten papers on quantics in the *Philosophical Transactions* from 1854

1. *Jour. für Math.*, 28, 1844, 68–96 = *Ges. Abh.*, 89–122.
2. *Cambridge Mathematical Journal*, 3, 1841, 1–20; and 3, 1842, 106–19.
3. *Coll. Math. Papers*, I, 273.

to 1878.[4] Quantics was the term he adopted for homogeneous polynomials in 2, 3, or more variables. Cayley became so much interested in invariants that he investigated them for their own sake. He also invented a symbolic method of treating invariants.

In the particular case of the binary quartic form

$$f = ax_1^4 + 4bx_1^3x_2 + 6cx_1^2x_2^2 + 4dx_1x_2^3 + ex_2^4$$

Cayley showed that the Hessian H and the Jacobian of f and H are covariants and that

$$g_2 = ae - 4bd + 3c^2$$

and

$$g_3 = \begin{vmatrix} a & b & c \\ b & c & d \\ c & d & e \end{vmatrix}$$

are invariants. To these results Sylvester and Salmon added many more.

Another contributor, Ferdinand Eisenstein, who was more concerned with the theory of numbers, had already found for the binary cubic form[5]

$$f = ax_1^3 + 3bx_1^2x_2 + 3cx_1x_2^2 + dx_2^3$$

that the simplest covariant of the second degree is its Hessian H and the simplest invariant is

$$3b^2c^2 + 6abcd - 4b^3d - 4ac^3 - a^2d^2,$$

which is the determinant of the quadratic Hessian as well as the discriminant of f. Also the Jacobian of f and H is another covariant of order three. Then Siegfried Heinrich Aronhold (1819–84), who began work on invariants in 1849, contributed invariants for ternary cubic forms.[6]

The first major problem that confronted the founders of invariant theory was the discovery of particular invariants. This was the direction of the work from about 1840 to 1870. As we can see, many such functions can be constructed because some invariants such as the Jacobian and the Hessian are themselves forms that have invariants and because some invariants taken together with the original form give a new system of forms that have simultaneous invariants. Dozens of major mathematicians including the few we have already mentioned computed particular invariants.

The continued calculation of invariants led to the major problem of invariant theory, which was raised after many special or particular invariants were found; this was to find a complete system of invariants. What

4. Coll. Math. Papers, 2, 4, 6, 7, 10.
5. Jour. für Math., 27, 1844, 89–106, 319–21.
6. Jour. für Math., 55, 1858, 97–191; and 62, 1863, 281–345.

this means is to find for a form of a given number of variables and degree the smallest possible number of rational integral invariants and covariants such that any other rational integral invariant or covariant could be expressed as a rational integral function with numerical coefficients of this complete set. Cayley showed that the invariants and covariants found by Eisenstein for the binary cubic form and the ones he obtained for the binary quartic form are a complete system for the respective cases.[7] This left open the question of a complete system for other forms.

The existence of a finite complete system or basis for binary forms of any given degree was first established by Paul Gordan (1837–1912), who devoted most of his life to the subject. His result[8] is that to each binary form $f(x_1, x_2)$ there belongs a finite complete system of rational integral invariants and covariants. Gordan had the aid of theorems due to Clebsch and the result is known as the Clebsch-Gordan theorem. The proof is long and difficult. Gordan also proved[9] that any finite *system* of binary forms has a finite complete system of invariants and covariants. Gordan's proofs showed how to compute the complete systems.

Various limited extensions of Gordan's results were obtained during the next twenty years. Gordan himself gave the complete system for the ternary quadratic form,[10] for the ternary cubic form,[11] and for a system of two and three ternary quadratics.[12] For the special ternary quartic $x_1^3 x_2 + x_2^3 x_3 + x_3^3 x_1$ Gordan gave a complete system of 54 ground forms.[13]

In 1886 Franz Mertens (1840–1927)[14] re-proved Gordan's theorem for binary systems by an inductive method. He assumed the theorem to be true for any given set of binary forms and then proved it must still be true when the degree of one of the forms is increased by one. He did not exhibit explicitly the finite set of independent invariants and covariants but he proved that it existed. The simplest case, a linear form, was the starting point of the induction and such a form has only powers of itself as covariants.

Hilbert, after writing a doctoral thesis in 1885 on invariants,[15] in 1888[16] also re-proved Gordan's theorem that any given system of binary forms has a finite complete system of invariants and covariants. His proof was a modification of Mertens's. Both proofs were far simpler than Gordan's. But Hilbert's proof also did not present a process for finding the complete system.

7. *Phil. Trans.*, 146, 1856, 101–26 = *Coll. Math. Papers*, 2, 250–75.
8. *Jour. für Math.*, 69, 1868, 323–54.
9. *Math. Ann.*, 2, 1870, 227–80.
10. R. Clebsch and F. Lindemann, *Vorlesungen über Geometrie*, I, 1876, p. 291.
11. *Math. Ann.*, 1, 1869, 56–89, 90–128.
12. Clebsch-Lindemann, p. 288.
13. *Math. Ann.*, 17, 1880, 217–33.
14. *Jour. für Math.*, 100, 1887, 223–30.
15. *Math. Ann.*, 30, 1887, 15–29 = *Ges. Abh.*, 2, 102–16.
16. *Math. Ann.*, 33, 1889, 223–26 = *Ges. Abh.*, 2, 162–64.

In 1888 Hilbert astonished the mathematical community by announcing a totally new approach to the problem of showing that any form of given degree and given number of variables, and any given system of forms in any given number of variables, have a finite complete system of independent rational integral invariants and covariants.[17] The basic idea of this new approach was to forget about invariants for the moment and consider the question: If an infinite system of rational integral expressions in a finite number of variables be given, under what conditions does a finite number of these expressions, a basis, exist in terms of which all the others are expressible as linear combinations with rational integral functions of the same variables as coefficients? The answer is, Always. More specifically, Hilbert's basis theorem, which precedes the result on invariants, goes as follows: By an algebraic form we understand a rational integral homogeneous function in n variables with coefficients in some definite domain of rationality (field). Given a collection of infinitely many forms of any degrees in the n variables, then there is a finite number (a basis) $F_1, F_2, \ldots, F_m$ such that any form F of the collection can be written as

$$F = A_1F_1 + A_2F_2 + \cdots + A_mF_m$$

where $A_1, A_2, \ldots, A_m$ are suitable forms in the n variables (not necessarily in the infinite system) with coefficients in the same domain as the coefficients of the infinite system.

In the application of this theorem to invariants and covariants, Hilbert's result states that for any form or system of forms there is a finite number of rational integral invariants and covariants by means of which every other rational integral invariant or covariant can be expressed as a linear combination of the ones in the finite set. This finite collection of invariants and covariants is the complete invariant system.

Hilbert's existence proof was so much simpler than Gordan's laborious calculation of a basis that Gordan could not help exclaiming, "This is not mathematics; it is theology." However, he reconsidered the matter and said later, "I have convinced myself that theology also has its advantages." In fact he himself simplified Hilbert's existence proof.[18]

In the 1880s and '90s the theory of invariants was seen to have unified many areas of mathematics. This theory was the "modern algebra" of the period. Sylvester said in 1864:[19] "As all roads lead to Rome so I find in my own case at least that all algebraic inquiries, sooner or later, end at the Capitol of modern algebra over whose shining portal is inscribed the Theory of Invariants." Soon the theory became an end in itself, independent of its

17. *Math. Ann.*, 36, 1890, 473–534 = *Ges. Abh.*, 2, 199–257, and in succeeding papers until 1893.
18. *Nachrichten König. Ges. der Wiss. zu Gött.*, 1899, 240–42.
19. *Phil. Trans.*, 154, 1864, 579–666 = *Coll. Math. Papers*, 2, 376–479, p. 380.

origins in number theory and projective geometry. The workers in algebraic invariants persisted in proving every kind of algebraic identity whether or not it had geometrical significance. Maxwell, when a student at Cambridge, said that some of the men there saw the whole universe in terms of quintics and quantics.

On the other hand the physicists of the late nineteenth century took no notice of the subject. Indeed Tait once remarked of Cayley, "Is it not a shame that such an outstanding man puts his abilities to such entirely useless questions?" Nevertheless the subject did make its impact on physics, indirectly and directly, largely through the work in differential invariants.

Despite the enormous enthusiasm for invariant theory in the second half of the nineteenth century, the subject as conceived and pursued during that period lost its attraction. Mathematicians say Hilbert killed invariant theory because he had disposed of all the problems. Hilbert did write to Minkowski in 1893 that he would no longer work in the subject, and said in a paper of 1893 that the most important general goals of the theory were attained. However, this was far from the case. Hilbert's theorem did not show how to compute invariants for any given form or system of forms and so could not provide a single significant invariant. The search for specific invariants having geometrical or physical significance was still important. Even the calculation of a basis for forms of a given degree and number of variables might prove valuable.

What "killed" invariant theory in the nineteenth-century sense of the subject is the usual collection of factors that killed many other activities that were over-enthusiastically pursued. Mathematicians follow leaders. Hilbert's pronouncement, and the fact that he himself abandoned the subject, exerted great influence on others. Also, the calculation of significant specific invariants had become more difficult after the more readily attainable results were achieved.

The computation of algebraic invariants did not end with Hilbert's work. Emmy Noether (1882–1935), a student of Gordan, did a doctoral thesis in 1907 "On Complete Systems of Invariants for Ternary Biquadratic Forms."[20] She also gave a complete system of covariant forms for a ternary quartic, 331 in all. In 1910 she extended Gordan's result to n variables.[21]

The subsequent history of algebraic invariant theory belongs to modern abstract algebra. The methodology of Hilbert brought to the fore the abstract theory of modules, rings, and fields. In this language Hilbert proved that every modular system (an ideal in the class of polynomials in n variables) has a basis consisting of a finite number of polynomials, or every ideal in a polynomial domain of n variables possesses a finite basis provided that in the domain of the coefficients of the polynomials every ideal has a finite basis.

20. *Jour. für Math.*, 134, 1908, 23–90.
21. *Jour. für Math.*, 139, 1911, 118–54.

From 1911 to 1919 Emmy Noether produced many papers on finite bases for various cases using Hilbert's technique and her own. In the subsequent twentieth-century development the abstract algebraic viewpoint dominated. As Eduard Study complained in his text on invariant theory, there was lack of concern for specific problems and only abstract methods were pursued.

3. The Concept of Birational Transformations

We saw in Chapter 35 that, especially during the third and fourth decades of the nineteenth century, the work in projective geometry turned to higher-degree curves. However, before this work had gone very far there was a change in the nature of the study. The projective viewpoint means linear transformations in homogeneous coordinates. Gradually transformations of the second and higher degrees came into play and the emphasis turned to birational transformations. Such a transformation, for the case of two non-homogeneous coordinates, is of the form

$$x' = \phi(x, y), \qquad y' = \psi(x, y)$$

where ϕ and ψ are rational functions in x and y and moreover x and y can be expressed as rational functions of x' and y'. In homogeneous coordinates x_1, x_2, and x_3 the transformations are of the form

$$x_i' = F_i(x_1, x_2, x_3), \qquad i = 1, 2, 3$$

and the inverse is

$$x_i = G_i(x_1', x_2', x_3'), \qquad i = 1, 2, 3,$$

where F_i and G_i are homogeneous polynomials of degree n in their respective variables. The correspondence is one-to-one except that each of a finite number of points may correspond to a curve.

As an illustration of a birational transformation we have inversion with respect to a circle. Geometrically this transformation (Fig. 39.1) is from M to M' or M' to M by means of the defining equation

$$OM \cdot OM' = r^2,$$

where r is the radius of the circle. Algebraically if we set up a coordinate system at O the Pythagorean theorem leads to

$$(7) \qquad x' = \frac{r^2 x}{x^2 + y^2}, \qquad y' = \frac{r^2 y}{x^2 + y^2},$$

where M is (x, y) and M' is (x', y'). Under this transformation circles transform into circles or straight lines and conversely. Inversion is a transformation that carries the entire plane into itself and such birational transformations

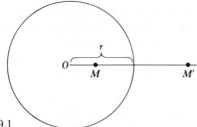

Figure 39.1

are called Cremona transformations. Another example of a Cremona transformation in three (homogeneous) variables is the quadratic transformation

$$(8) \qquad x_1' = x_2 x_3, \qquad x_2' = x_3 x_1, \qquad x_3' = x_1 x_2$$

whose inverse is

$$x_1 = x_2' x_3', \qquad x_2 = x_3' x_1', \qquad x_3 = x_1' x_2'.$$

The term birational transformation is also used in a more general sense, namely, wherein the transformation from the points of one curve into those of another is birational but the transformation need not be birational in the entire plane. Thus (in nonhomogeneous coordinates) the transformation

$$(9) \qquad\qquad X = x^2, \qquad Y = y$$

is not one-to-one in the entire plane but does take any curve C to the right of the y-axis into another in a one-to-one correspondence.

The inversion transformation was the first of the birational transformations to appear. It was used in limited situations by Poncelet in his *Traité* of 1822 (¶370) and then by Plücker, Steiner, Quetelet, and Ludwig Immanuel Magnus (1790–1861). It was studied extensively by Möbius[22] and its use in physics was recognized by Lord Kelvin,[23] and by Liouville,[24] who called it the transformation by reciprocal radii.

In 1854 Luigi Cremona (1830–1903), who served as a professor of mathematics at several Italian universities, introduced the general birational transformation (of the entire plane into itself) and wrote important papers on it.[25] Max Noether (1844–1921), the father of Emmy Noether, then proved the fundamental result[26] that a plane Cremona transformation can be built

22. *Theorie der Kreisverwandschaft* (Theory of Inversion), *Abh. König. Säch. Ges. der Wiss.*, 2 1855, 529–65 = *Werke*, 2, 243–345.
23. *Jour. de Math.*, 10, 1845, 364–67.
24. *Jour. de Math.*, 12, 1847, 265–90.
25. *Gior. di Mat.*, 1, 1863, 305–11 = *Opere*, 1, 54–61; and 3, 1865, 269–80, 363–76 = *Opere*, 2, 193–218.
26. *Math. Ann.*, 3, 1871, 165–227, p. 167 in particular.

up from a sequence of quadratic and linear transformations. Jacob Rosanes (1842–1922) found this result independently[27] and also proved that all one-to-one algebraic transformations of the plane must be Cremona transformations. The proofs of Noether and Rosanes were completed by Guido Castelnuovo (1865–1952).[28]

4. The Function-Theoretic Approach to Algebraic Geometry

Though the nature of the birational transformation was clear, the development of the subject of algebraic geometry as the study of invariants under such transformations was, at least in the nineteenth century, unsatisfactory. Several approaches were used; the results were disconnected and fragmentary; most proofs were incomplete; and very few major theorems were obtained. The variety of approaches has resulted in marked differences in the languages used. The goals of the subject were also vague. Though invariance under birational transformations has been the leading theme, the subject covers the search for properties of curves, surfaces, and higher-dimensional structures. In view of these factors there are not many central results. We shall give a few samples of what was done.

The first of the approaches was made by Clebsch. (Rudolf Friedrich) Alfred Clebsch (1833–72) studied under Hesse in Königsberg from 1850 to 1854. In his early work he was interested in mathematical physics and from 1858 to 1863 was professor of theoretical mechanics at Karlsruhe and then professor of mathematics at Giessin and Göttingen. He worked on problems left by Jacobi in the calculus of variations and the theory of differential equations. In 1862 he published the *Lehrbuch der Elasticität*. However, his chief work was in algebraic invariants and algebraic geometry.

Clebsch had worked on the projective properties of curves and surfaces of third and fourth degrees up to about 1860. He met Paul Gordan in 1863 and learned about Riemann's work in complex function theory. Clebsch then brought this theory to bear on the theory of curves.[29] This approach is called transcendental. Though Clebsch made the connection between complex functions and algebraic curves, he admitted in a letter to Gustav Roch that he could not understand Riemann's work on Abelian functions nor Roch's contributions in his dissertation.

Clebsch reinterpreted the complex function theory in the following manner: The function $f(w, z) = 0$, wherein z and w are complex variables, calls geometrically for a Riemann surface for z and a plane or a portion of a plane for w or, if one prefers, for a Riemann surface to each point of which a pair of values of z and w is attached. By considering only the real parts of z

27. *Jour. für Math.*, 73, 1871, 97–110.
28. *Atti Accad. Torino*, 36, 1901, 861–74.
29. *Jour. für Math.*, 63, 1864, 189–243.

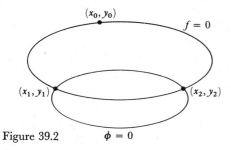

Figure 39.2 $\phi = 0$

and w the equation $f(w, z) = 0$ represents a curve in the real Cartesian plane. z and w may still have complex values satisfying $f(w, z) = 0$ but these are not plotted. This view of real curves with complex points was already familiar from the work in projective geometry. To the theory of birational transformations of the surface corresponds a theory of birational transformations of the plane curve. Under the reinterpretation just described the branch points of the Riemann surface correspond to those points of the curve where a line $x = $ const. meets the curve in two or more consecutive points, that is, is either tangent to the curve or passes through a cusp. A double point of the curve corresponds to a point on the surface where two sheets just touch without any further connection. Higher multiple points on curves also correspond to other peculiarities of Riemann surfaces.

In the subsequent account we shall utilize the following definitions (Cf. Chap. 23, sec. 3): A multiple point (singular point) P of order $k > 1$ of an nth degree plane curve is a point such that a generic line through P cuts the curve in $n - k$ points. The multiple point is ordinary if the k tangents at P are distinct. In counting intersections of an nth degree and an mth degree curve one must take into account the multiplicity of the multiple points on each curve. If it is h on the curve C^n and k on C^m and if the tangents at P of C^n are distinct from those of C^m, then the point of intersection has multiplicity hk. A curve C' is said to be adjoint to a curve C when the multiple points of C are ordinary or cusps and if C' has a point of multiplicity of order $k - 1$ at every multiple point of C of order k.

Clebsch[30] first restated Abel's theorem (Chap. 27, sec. 7) on integrals of the first kind in terms of curves. Abel considered a fixed rational function $R(x, y)$ where x and y are related by any algebraic curve $f(x, y) = 0$, so that y is a function of x. Suppose (Fig. 39.2) $f = 0$ is cut by another algebraic curve

$$\phi(x, y, a_1, a_2, \cdots, a_k) = 0,$$

where the a_i's are the coefficients in $\phi = 0$. Let the intersections of $\phi = 0$ with $f = 0$ be $(x_1, y_1), (x_2, y_2), \cdots, (x_m, y_m)$. (The number m of these is the

30. *Jour. für Math.*, 63, 1864, 189–243.

product of the degrees of f and ϕ.) Given a point (x_0, y_0) on $f = 0$, where y_0 belongs to one branch of $f = 0$, then we can consider the sum

$$I = \sum_{i=1}^{m} \int_{x_0,y_0}^{x_i,y_i} R(x, y) \, dx.$$

The upper limits x_i, y_i all lie on $\phi = 0$ and the integral I is a function of the upper limits. Then there is a characteristic number p of these limits which have to be algebraic functions of the others. This number p depends only on f. Moreover, I can be expressed as the sum of these p integrals and rational and logarithmic functions of the x_i, y_i, $i = 1, 2, \ldots, m$. Further, if the curve $\phi = 0$ is varied by varying the parameters a_1 to a_k then the x_i will also vary and I becomes a function of the a_i through the x_i. The function I of the a_i will be rational in the a_i or at worst involve logarithmic functions of the a_i.

Clebsch also carried over to curves Riemann's concept of Abelian integrals on Riemann surfaces, that is, integrals of the form $\int g(x, y) \, dx$, where g is a rational function and $f(x, y) = 0$. To illustrate the integrals of the first kind consider a plane fourth degree curve C_4 without double points. Here $p = 3$ and there are the three everywhere finite integrals

$$u_1 = \int \frac{x \, dx}{f_y}, \qquad u_2 = \int \frac{y \, dx}{f_y}, \qquad u_3 = \int \frac{dx}{f_y}.$$

What applies to the C_4 carries over to arbitrary algebraic curves $f(x, y) = 0$ of the nth order. In place of the three everywhere finite integrals there are now p such integrals, (where p is the genus of $f = 0$). Each has $2p$ periodicity modules (Chap. 27, sec. 8). The integrals are of the form

$$\int \frac{\phi(x, y)}{\partial f/\partial y} \, dx$$

where ϕ is a polynomial (an adjoint) of precisely the degree $n - 3$ which vanishes at the double points and cusps of $f = 0$.

Clebsch's next contribution [31] was to introduce the notion of genus as a concept for classifying curves. If the curve has d double points then the genus $p = (1/2)(n - 1)(n - 2) - d$. Previously there was the notion of the deficiency of a curve (Chap. 23, sec. 3), that is, the maximum possible number of double points a curve of degree n could possess, namely, $(n - 1)(n - 2)/2$, minus the number it actually does possess. Clebsch showed [32] that for curves with only ordinary multiple points (the tangents are all distinct) the genus is the same as the deficiency and the genus is an invariant under birational transformation of the entire plane into itself.[33]

31. *Jour. für Math.*, 64, 1865, 43–65.
32. *Jour. für Math.*, 64, 1865, 98–100.
33. If the multiple (singular) points are of order r_i then the genus p of a curve C is $(n - 1)(n - 2)/2 - (1/2) \sum r_i(r_i - 1)$, where the summation extends over all multiple points. The genus is a more refined concept.

Clebsch's notion of genus is related to Riemann's connectivity of a Riemann surface. The Riemann surface corresponding to a curve of genus p has connectivity $2p + 1$.

The notion of genus can be used to establish significant theorems about curves. Jacob Lüroth (1844–1910) showed[34] that a curve of genus 0 can be birationally transformed into a straight line. When the genus is 1, Clebsch showed that a curve can be birationally transformed into a third degree curve.

In addition to classifying curves by genus, Clebsch, following Riemann, introduced classes within each genus. Riemann had considered[35] the birational transformation of his surfaces. Thus if $f(w, z) = 0$ is the equation of the surface and if

$$w_1 = R_1(w, z), \qquad z_1 = R_2(w, z)$$

are rational functions and if the inverse transformation is rational then $f(w, z)$ can be transformed to $F(w_1, z_1) = 0$. Two algebraic equations $F(w, z) = 0$ (or their surfaces) can be transformed birationally into one another only if both have the same p value. (The number of sheets need not be preserved.) For Riemann no further proof was needed. It was guaranteed by the intuition.

Riemann (in the 1857 paper) regarded all equations (or the surfaces) which are birationally transformable into each other as belonging to the same class. They have the same genus p. However, there are different classes with the same p value (because the branch-points may differ). The most general class of genus p is characterized by $3p - 3$ (complex) constants (coefficients in the equation) when $p > 1$, by one constant when $p = 1$ and by zero constants when $p = 0$. In the case of elliptic functions $p = 1$ and there is one constant. The trigonometric functions, for which $p = 0$, do not have any arbitrary constant. The number of constants was called by Riemann the class modulus. The constants are invariant under birational transformation. Clebsch likewise put all those curves which are derivable from a given one by a one-to-one birational transformation into one class. Those of one class necessarily have the same genus but there may be different classes with the same genus.

5. The Uniformization Problem

Clebsch then turned his attention to what is called the uniformization problem for curves. Let us note first just what this problem amounts to. Given the equation

$$(10) \qquad\qquad w^2 + z^2 = 1$$

34. *Math. Ann.*, 9, 1876, 163–65.
35. *Jour. für Math.*, 54, 1857, 115–55 = *Werke*, 2nd ed., 88–142.

we can represent it in the parametric form

$$(11) \qquad\qquad z = \sin t, \qquad w = \cos t$$

or in the parametric form

$$(12) \qquad\qquad z = \frac{2t}{1 + t^2}, \qquad w = \frac{1 - t^2}{1 + t^2}.$$

Thus even though (10) defines w as a multiple-valued function of z, we can represent z and w as single-valued or uniform functions of t. The parametric equations (11) or (12) are said to uniformize the algebraic equation (10).

For an equation $f(w, z) = 0$ of genus 0 Clebsch[36] showed that each of the variables can be expressed as a rational function of a single parameter. These rational functions are uniformizing functions. When $f = 0$ is interpreted as a curve it is then called unicursal. Conversely if the variables w and z of $f = 0$ are rationally expressible in terms of an arbitrary parameter then $f = 0$ is of genus 0.

When $p = 1$ then Clebsch showed in the same year[37] that w and z can be expressed as rational functions of the parameters ξ and η where η^2 is a polynomial of either the third or fourth degree in ξ. Then $f(w, z) = 0$ or the corresponding curve is called bicursal, a term introduced by Cayley.[38] It is also called elliptic because the equation $(dw/dz)^2 = \eta^2$ leads to elliptic integrals. We can as well say that w and z are expressible as single-valued doubly periodic functions of a single parameter α or as rational functions of $\wp(\alpha)$ where $\wp(\alpha)$ is Weierstrass's function. Clebsch's result on the uniformization of curves of genus 1 by means of elliptic functions of a parameter made it possible to establish for such curves remarkable properties about points of inflection, osculatory conics, tangents from a point to a curve, and other results, many of which had been demonstrated earlier but with great difficulty.

For equations $f(w, z) = 0$ of genus 2, Alexander von Brill (1842–1935) showed[39] that the variables w and z are expressible as rational functions of ξ and η where η^2 is now a polynomial of the fifth or sixth degree in ξ.

Thus functions of genus 0, 1, and 2 can be uniformized. For function $f(w, z) = 0$ of genus greater than 2 the thought was to employ more general functions, namely, automorphic functions. In 1882 Klein[40] gave a general uniformization theorem but the proof was not complete. In 1883 Poincaré announced[41] his general uniformization theorem but he too had no complete

36. *Jour. für Math.*, 64, 1865, 43–65.
37. *Jour. für Math.*, 64, 1865, 210–70.
38. *Proc. Lon. Math. Soc.*, 4, 1871–73, 347–52 = *Coll. Math. Papers*, 8, 181–87.
39. *Jour. für Math.*, 65, 1866, 269–83.
40. *Math. Ann.*, 21, 1883, 141–218 = *Ges. math. Abh.*, 3, 630–710.
41. *Bull. Soc. Math. de France*, 11, 1883, 112–25 = *Œuvres*, 4, 57–69.

proof. Both Klein and Poincaré continued to work hard to prove this theorem but no decisive result was obtained for twenty-five years. In 1907 Poincaré[42] and Paul Koebe (1882–1945)[43] independently gave a proof of this uniformization theorem. Koebe then extended the result in many directions. With the theorem on uniformization now rigorously established an improved treatment of algebraic functions and their integrals has become possible.

6. The Algebraic-Geometric Approach

A new direction of work in algebraic geometry begins with the collaboration of Clebsch and Gordan during the years 1865–70. Clebsch was not satisfied merely to show the significance of Riemann's work for curves. He sought now to establish the theory of Abelian integrals on the basis of the algebraic theory of curves. In 1865 he and Gordan joined forces in this work and produced their *Theorie der Abelschen Funktionen* (1866). One must appreciate that at this time Weierstrass's more rigorous theory of Abelian integrals was not known and Riemann's foundation—his proof of existence based on Dirichlet's principle—was not only strange but not well established. Also at this time there was considerable enthusiasm for the theory of invariants of algebraic forms (or curves) and for projective methods as the first stage, so to speak, of the treatment of birational transformations.

Although the work of Clebsch and Gordan was a contribution to algebraic geometry, it did not establish a purely algebraic theory of Riemann's theory of Abelian integrals. They did use algebraic and geometric methods as opposed to Riemann's function-theoretic methods but they also used basic results of function theory and the function-theoretic methods of Weierstrass. In addition they took some results about rational functions and the intersection point theorem as given. Their contribution amounted to starting from some function-theoretic results and, using algebraic methods, obtaining new results previously established by function-theoretic methods. Rational transformations were the essence of the algebraic method.

They gave the first *algebraic* proof for the invariance of the genus p of an algebraic curve under rational transformations, using as a definition of p the degree and number of singularities of $f = 0$. Then, using the fact that p is the number of linearly independent integrals of the first kind on $f(x_1, x_2, x_3) = 0$ and that these integrals are everywhere finite, they showed that the transformation

$$\rho x_i = \psi_i(y_1, y_2, y_3), \qquad i = 1, 2, 3,$$

42. *Acta Math.*, 31, 1908, 1–63 = *Œuvres*, 4, 70–139.
43. *Math. Ann.*, 67, 1909, 145–224.

transforms an integral of the first kind into an integral of the first kind so that p is invariant. They also gave new proofs of Abel's theorem (by using function-theoretic ideas and methods).

Their work was not rigorous. In particular they, too, in the Plücker tradition counted arbitrary constants to determine the number of intersection points of a C_m with a C_n. Special kinds of double points were not investigated. The significance of the Clebsch-Gordan work for the theory of algebraic functions was to express clearly in algebraic form such results as Abel's theorem and to use it in the study of Abelian integrals. They put the algebraic part of the theory of Abelian integrals and functions more into the foreground and in particular established the theory of transformations on its own foundations.

Clebsch and Gordan had raised many problems and left many gaps. The problems lay in the direction of new algebraic investigations for a purely algebraic theory of algebraic functions. The work on the algebraic approach was continued by Alexander von Brill and Max Noether from 1871 on; their key paper was published in 1874.[44] Brill and Noether based their theory on a celebrated residual theorem (*Restsatz*) which in their hands took the place of Abel's theorem. They also gave an algebraic proof of the Riemann-Roch theorem on the number of constants which appear in algebraic functions $F(w, z)$ which become infinite nowhere except in m prescribed points of a C_n. According to this theorem the most general algebraic function which fulfills this condition has the form

$$F = C_1 F_1 + C_2 F_2 + \cdots + C_\mu F_\mu + C_{\mu+1}$$

where

$$\mu = m - p + \tau,$$

τ is the number of linearly independent functions ϕ (of degree $n - 3$) which vanish in the m prescribed points, and p is the genus of the C_n. Thus if the C_n is a C_4 without double points, then $p = 3$ and the ϕ's are straight lines. For this case when

$$m = 1, \text{ then } \tau = 2 \quad \text{and } \mu = 1 - 3 + 2 = 0;$$
$$m = 2, \text{ then } \tau = 1, \quad \text{and } \mu = 2 - 3 + 1 = 0;$$
$$m = 3, \text{ then } \tau = 1 \text{ or } 0 \text{ and } \mu = 1 \text{ or } 0.$$

When $\mu = 0$ there is no algebraic function which becomes infinite in the given points. When $m = 3$, there is one and only one such function provided the three given points be on a straight line. If the three points do lie on a line $v = 0$, this line cuts the C_4 in a fourth point. We choose a line $u = 0$ through this point and then $F_1 = u/v$.

44. "Über die algebraischen Funktionen und ihre Anwendung in der Geometrie," *Math. Ann.*, 7, 1874, 269–310.

This work replaces Riemann's determination of the most general algebraic function having given points at which it becomes infinite. Also the Brill-Noether result transcends the projective viewpoint in that it deals with the geometry of points on the curve C_n given by $f = 0$, whose mutual relations are not altered by a one-to-one birational transformation. Thus for the first time the theorems on points of intersection of curves were established algebraically. The counting of constants as a method was dispensed with.

The work in algebraic geometry continued with the detailed investigation of algebraic space curves by Noether[45] and Halphen.[46] Any space curve C can be projected birationally into a plane curve C_1. All such C_1 coming from C have the same genus. The genus of C is therefore defined to be that of any such C_1 and the genus of C is invariant under birational transformation of the space.

The topic which has received the greatest attention over the years is the study of singularities of plane algebraic curves. Up to 1871, the theory of algebraic functions considered from the algebraic viewpoint had limited itself to curves which had distinct or separated double points and at worst only cusps (*Rückkehrpunkte*). Curves with more complicated singularities were believed to be treatable as limiting cases of curves with double points. But the actual limiting procedure was vague and lacked rigor and unity. The culmination of the work on singularities is two famous transformation theorems. The first states that every plane irreducible algebraic curve can be transformed by a Cremona transformation to one having no singular points other than multiple points with distinct tangents. The second asserts that by a transformation birational only on the curve every plane irreducible algebraic curve can be transformed into another having only double points with distinct tangents. The reduction of curves to these simpler forms facilitates the application of many of the methodologies of algebraic geometry.

However, the numerous proofs of these theorems, especially the second one, have been incomplete or at least criticized by mathematicians (other than the author). There really are two cases of the second theorem, real curves in the projective plane and curves in the complex function theory sense where x and y each run over a complex plane. Noether[47] in 1871 used a sequence of quadratic transformations which are one-to-one in the entire plane to prove the first theorem. He is generally credited with the proof but actually he merely indicated a proof which was perfected and modified by many writers.[48] Kronecker, using analysis and algebra, developed a method for proving the second theorem. He communicated this method verbally to

45. *Jour. für Math.*, 93, 1882, 271–318.
46. *Jour. de l'Ecole Poly.*, Cahier 52, 1882, 1–200 = *Œuvres*, 3, 261–455.
47. *Nachrichten König. Ges. der Wiss. zu Gött.*, 1871, 267–78.
48. See also Noether, *Math. Ann.*, 9, 1876, 166–82; and 23, 1884, 311–58.

Riemann and Weierstrass in 1858, lectured on it from 1870 on and published it in 1881.[49] The method used rational transformations, which with the aid of the equation of the given plane curve are one-to-one, and transformed the singular case into the "regular one;" that is, the singular points become just double points with distinct tangents. The result, however, was not stated by Kronecker and is only implicit in his work.

This second theorem to the effect that all multiple points can be reduced to double points by birational transformations on the curve was first explicitly stated and proved by Halphen in 1884.[50] Many other proofs have been given but none is universally accepted.

7. The Arithmetic Approach

In addition to the transcendental approach and the algebraic-geometric approach there is what is called the arithmetic approach to algebraic curves, which is, however, in concept at least, purely algebraic. This approach is really a group of theories which differ greatly in detail but which have in common the construction and analysis of the integrands of the three kinds of Abelian integrals. This approach was developed by Kronecker in his lectures,[51] by Weierstrass in his lectures of 1875–76, and by Dedekind and Heinrich Weber in a joint paper.[52] The approach is fully presented in the text by Kurt Hensel and Georg Landsberg: *Theorie der algebraischen Funktionen einer Variabeln* (1902).

The central idea of this approach comes from the work on algebraic numbers by Kronecker and Dedekind and utilizes an analogy between the algebraic integers of an algebraic number field and the algebraic functions on the Riemann surface of a complex function. In the theory of algebraic numbers one starts with an irreducible polynomial equation $f(x) = 0$ with integral coefficients. The analogue for algebraic geometry is an irreducible polynomial equation $f(\zeta, z) = 0$ whose coefficients of the powers of ζ are polynomials in z (with, say, real coefficients). In number theory one then considers the field $R(x)$ generated by the coefficients of $f(x) = 0$ and one of its roots. In the geometry one considers the field of all $R(\zeta, z)$ which are algebraic and one-valued on the Riemann surface. One then considers in the number theory the integral algebraic numbers. To these there correspond the algebraic functions $G(\zeta, z)$ which are entire, that is, become infinite only at $z = \infty$. The decomposition of the algebraic integers into real prime factors

49. *Jour. für Math.*, 91, 1881, 301–34 = *Werke*, 2, 193–236.
50. Reproduced in an appendix to a French edition (1884) of G. Salmon's *Higher Plane Curves* and in E. Picard's *Traité d'analyse*, 2, 1893, 364 ff. = *Œuvres*, 4, 1–93.
51. *Jour. für Math.*, 91, 1881, 301–34 and 92, 1882, 1–122 = *Werke*, 2, 193–387.
52. "Theorie der algebraischen Funktionen einer Veränderlichen," *Jour. für Math.*, 92, 1882, 181–290 = Dedekind's *Werke*, I, 238–350.

and units respectively corresponds to the decomposition of the $G(\zeta, z)$ into factors such that each vanishes at one point only of the Riemann surface and factors that vanish nowhere, respectively. Where Dedekind introduced ideals in the number theory to discuss divisibility, in the geometric analogue one replaces a factor of a $G(\zeta, z)$ which vanishes at one point of the Riemann surface by the collection of all functions of the field of $R(\zeta, z)$ which vanish at that point. Dedekind and Weber used this arithmetic method to treat the field of algebraic functions and they obtained the classic results.

Hilbert[53] continued what is essentially the algebraic or arithmetic approach to algebraic geometry of Dedekind and Kronecker. One principal theorem, Hilbert's *Nullstellensatz*, states that every algebraic structure (figure) of arbitrary extent in a space of arbitrarily many homogeneous variables $x_1, \ldots, x_n$ can always be represented by a finite number of homogeneous equations

$$F_1 = 0, F_2 = 0, \cdots, F_\mu = 0$$

so that the equation of any other structure containing the original one can be represented by

$$M_1 F_1 + \cdots + M_\mu F_\mu = 0,$$

where the M's are arbitrary homogeneous integral forms whose degree must be so chosen that the left side of the equation is itself homogeneous.

Hilbert following Dedekind called the collection of the $M_i F_i$ a module (the term is now ideal and module now is something more general). One can state Hilbert's result thus: Every algebraic structure of R_n determines the vanishing of a finite module.

8. *The Algebraic Geometry of Surfaces*

Almost from the beginning of work in the algebraic geometry of curves, the theory of surfaces was also investigated. Here too the direction of the work turned to invariants under linear and birational transformations. Like the equation $f(x, y) = 0$, the polynomial equation $f(x, y, z) = 0$ has a double interpretation. If x, y, and z take on real values then the equation represents a two-dimensional surface in three-dimensional space. If, however, these variables take on complex values, then the equation represents a four-dimensional manifold in a six-dimensional space.

The approach to the algebraic geometry of surfaces paralleled that for curves. Clebsch employed function-theoretic methods and introduced[54]

53. "Über die Theorie der algebraischen Formen," *Math. Ann.*, 36, 1890, 473–534 = *Ges. Abh.*, 2, 199–257.
54. *Comp. Rend.*, 67, 1868, 1238–39.

double integrals which play the role of Abelian integrals in the theory of curves. Clebsch noted that for an algebraic surface $f(x, y, z) = 0$ of degree m with isolated multiple points and ordinary multiple lines, certain surfaces of degree $m - 4$ ought to play the role which the adjoint curves of degree $m - 3$ play with respect to a curve of degree m. Given a rational function $R(x, y, z)$ where $x, y,$ and z are related by $f(x, y, z) = 0$, if one seeks the double integrals

$$\int\int R(x, y, z)\, dx\, dy,$$

which always remain finite when the integrals extend over a two-dimensional region of the four-dimensional surface, one finds that they are of the form

$$\int\int \frac{Q(x, y, z)}{f_z}\, dx\, dy$$

where Q is a polynomial of degree $m - 4$. $Q = 0$ is an adjoint surface which passes through the multiple lines of $f = 0$ and has a multiple line of order $k - 1$ at least in every multiple line of f of order k and has a multiple point of order $q - 2$ at least in every isolated multiple point of f of order q. Such an integral is called a double integral of the first kind. The number of linearly independent integrals of this class, which is the number of essential constants in $Q(x, y, z)$, is called the geometrical genus p_g of $f = 0$. If the surface has no multiple lines of points

$$p_g = \frac{(m - 1)(m - 2)(m - 3)}{6}.$$

Max Noether[55] and Hieronymus G. Zeuthen (1839–1920)[56] proved that p_g is an invariant under birational transformations of the surface (not of the whole space).

Up to this point the analogy with the theory of curves is good. The double integrals of the first kind are analogous to the Abelian integrals of the first kind. But now a first difference becomes manifest. It is necessary to calculate the number of essential constants in the polynomials Q of degree $m - 4$ which behave at multiple points of the surface in such a manner that the integral remains finite. But one can find by a precise formula the number of conditions thus involved only for a polynomial of sufficiently large degree N. If one puts into this formula $N = m - 4$, one might find a number different from p_g. Cayley[57] called this new number the numerical (arithmetic) genus p_n of the surface. The most general case is where $p_n = p_g$. When

55. *Math. Ann.*, 2, 1870, 293–316.
56. *Math. Ann.*, 4, 1871, 21–49.
57. *Phil. Trans.*, 159, 1869, 201–29 = *Coll. Math. Papers*, 6, 329–58; and *Math. Ann.*, 3, 1871, 526–29 = *Coll. Math. Papers*, 8, 394–97.

the equality does not hold one has $p_n < p_g$ and the surface is called irregular; otherwise it is called regular. Then Zeuthen[58] and Noether[59] established the invariance of the number p_n when it is not equal to p_g.

Picard[60] developed a theory of double integrals of the second kind. These are the integrals which become infinite in the manner of

$$(13) \qquad \int \int \left(\frac{\partial U}{\partial x} + \frac{\partial V}{\partial y} \right) dx \, dy$$

where U and V are rational functions of x, y, and z and $f(x, y, z) = 0$. The number of different integrals of the second kind, different in the sense that no linear combination of these integrals reduces to the form (13), is finite; this is a birational invariant of the surface $f = 0$. But it is not true here, as in the case of curves, that the number of distinct Abelian integrals of the second kind is $2p$. This new invariant of algebraic surfaces does not appear to be tied to the numerical or the geometrical genus.

Far less has been accomplished for the theory of surfaces than for curves. One reason is that the possible singularities of surfaces are much more complicated. There is the theorem of Picard and Georges Simart proven by Beppo Levi (1875–1928)[61] that any (real) algebraic surface can be birationally transformed into a surface free of singularities which must, however, be in a space of five dimensions. But this theorem does not prove to be too helpful.

In the case of curves the single invariant number, the genus p, is capable of definition in terms of the characteristics of the curve or the connectivity of the Riemann surface. In the case of $f(x, y, z) = 0$ the number of characterizing arithmetical birational invariants is unknown.[62] We shall not attempt to describe further the few limited results for the algebraic geometry of surfaces.

The subject of algebraic geometry now embraces the study of higher-dimensional figures (manifolds or varieties) defined by one or more algebraic equations. Beyond generalization in this direction, another type, namely, the use of more general coefficients in the defining equations, has also been undertaken. These coefficients can be members of an abstract ring or field and the methods of abstract algebra are applied. The several methods of pursuing algebraic geometry as well as the abstract algebraic formulation introduced in the twentieth century have led to sharp differences in language and methods of approach so that one class of workers finds it very difficult to understand another. The emphasis in this century has been on the abstract algebraic approach. It does seem to offer sharp formulations of theorems and

58. *Math. Ann.*, 4, 1871, 21–49.
59. *Math. Ann.*, 8, 1875, 495–533.
60. *Jour. de Math.*, (5), 5, 1899, 5–54, and later papers.
61. *Annali di Mat.*, (2), 26, 1897, 219–53.
62. The manifolds involved cannot be characterized even topologically.

proofs thereby settling much controversy about the meaning and correctness of the older results. However, much of the work seems to have far more bearing on algebra than on geometry.

Bibliography

Baker, H. F.: "On Some Recent Advances in the Theory of Algebraic Surfaces," *Proc. Lon. Math. Soc.*, (2), 12, 1912–13, 1–40.

Berzolari, L.: "Allgemeine Theorie der höheren ebenen algebraischen Kurven," *Encyk. der Math. Wiss.*, B. G. Teubner, 1903–15, III C4, 313–455.

————: "Algebraische Transformationen und Korrespondenzen," *Encyk. der Math. Wiss.*, B. G. Teubner, 1903–15, III, 2, 2nd half B, 1781–2218. Useful for results on higher-dimensional figures.

Bliss, G. A.: "The Reduction of Singularities of Plane Curves by Birational Transformations," *Amer. Math. Soc. Bull.*, 29, 1923, 161–83.

Brill, A., and M. Noether: "Die Entwicklung der Theorie der algebraischen Funktionen," *Jahres. der Deut. Math.-Verein.*, 3, 1892–93, 107–565.

Castelnuovo, G., and F. Enriques: "Die algebraischen Flächen vom Gesichtspunkte der birationalen Transformationen," *Encyk. der Math. Wiss.*, B. G. Teubner, 1903–15, III C6b, 674–768.

————: "Sur quelques récents résultats dans la théorie des surfaces algébriques," *Math. Ann.*, 48, 1897, 241–316.

Cayley, A.: *Collected Mathematical Papers*, Johnson Reprint Corp., 1963, Vols. 2, 4, 6, 7, 10, 1891–96.

Clebsch, R. F. A.: "Versuch einer Darlegung und Würdigung seiner Wissenschaftlichen Leistungen," *Math. Ann.*, 7, 1874, 1–55. An article by friends of Clebsch.

Coolidge, Julian L.: *A History of Geometrical Methods*, Dover (reprint), 1963, pp. 195–230, 278–92.

Cremona, Luigi: *Opere mathematiche*, 3 vols., Ulrico Hoepli, 1914–17.

Hensel, Kurt, and Georg Landsberg: *Theorie der algebraischen Funktionen einer Variabeln* (1902), Chelsea (reprint), 1965, pp. 694–702 in particular.

Hilbert, David: *Gesammelte Abhandlungen*, Julius Springer, 1933, Vol. 2.

Klein, Felix: *Vorlesungen über die Entwicklung der Mathematik im 19 Jahrhundert*, 1, 155–66, 295–319; 2, 2–26, Chelsea (reprint), 1950.

Meyer, Franz W.: "Bericht über den gegenwärtigen Stand der Invariantentheorie," *Jahres. der Deut. Math.-Verein.*, 1, 1890–91, 79–292.

National Research Council: *Selected Topics in Algebraic Geometry*, Chelsea (reprint), 1970.

Noether, Emmy: "Die arithmetische Theorie der algebraischen Funktionen einer Veränderlichen in ihrer Beziehung zu den übrigen Theorien und zu der Zahlentheorie," *Jahres. der Deut. Math.-Verein.*, 28, 1919, 182–203.

40
The Instillation of Rigor in Analysis

> But it would be a serious error to think that one can find
> certainty only in geometrical demonstrations or in the
> testimony of the senses.   A. L. CAUCHY

1. Introduction

By about 1800 the mathematicians began to be concerned about the looseness in the concepts and proofs of the vast branches of analysis. The very concept of a function was not clear; the use of series without regard to convergence and divergence had produced paradoxes and disagreements; the controversy about the representations of functions by trigonometric series had introduced further confusion; and, of course, the fundamental notions of derivative and integral had never been properly defined. All these difficulties finally brought on dissatisfaction with the logical status of analysis.

Abel, in a letter of 1826 to Professor Christoffer Hansteen,[1] complained about "the tremendous obscurity which one unquestionably finds in analysis. It lacks so completely all plan and system that it is peculiar that so many men could have studied it. The worst of it is, it has never been treated stringently. There are very few theorems in advanced analysis which have been demonstrated in a logically tenable manner. Everywhere one finds this miserable way of concluding from the special to the general and it is extremely peculiar that such a procedure has led to so few of the so-called paradoxes."

Several mathematicians resolved to bring order out of chaos. The leaders of what is often called the critical movement decided to rebuild analysis solely on the basis of arithmetical concepts. The beginnings of the movement coincide with the creation of non-Euclidean geometry. An entirely different group, except for Gauss, was involved in the latter activity and it is therefore difficult to trace any direct connection between it and the decision to found analysis on arithmetic. Perhaps the decision was reached because the hope of

1. *Œuvres*, 2, 263–65.

grounding analysis on geometry, which many seventeenth-century men often asserted could be done, was blasted by the increasing complexity of the eighteenth-century developments in analysis. However, Gauss had already expressed his doubts as to the truth of Euclidean geometry as early as 1799, and in 1817 he decided that truth resided only in arithmetic. Moreover, during even the early work by Gauss and others on non-Euclidean geometry, flaws in Euclid's development had already been noted. Hence it is very likely that both factors caused distrust of geometry and prompted the decision to found analysis on arithmetical concepts. This certainly was what the leaders of the critical movement undertook to do.

Rigorous analysis begins with the work of Bolzano, Cauchy, Abel, and Dirichlet and was furthered by Weierstrass. Cauchy and Weierstrass are best known in this connection. Cauchy's basic works on the foundations of analysis are his *Cours d'analyse algébrique*,[2] *Résumé des leçons sur le calcul infinitésimal*,[3] and *Leçons sur le calcul différentiel*.[4] Actually Cauchy's rigor in these works is loose by modern standards. He used phrases such as "approach indefinitely," "as little as one wishes," "last ratios of infinitely small increments," and "a variable approaches its limit." However, if one compares Lagrange's *Théorie des fonctions analytiques*[5] and his *Leçons sur le calcul des fonctions*[6] and the influential book by Lacroix, *Traité du calcul différentiel et du calcul intégral*[7] with the *Cours d'analyse algébrique* of Cauchy one begins to see the striking difference between the mathematics of the eighteenth century and that of the nineteenth. Lagrange, in particular, was purely formal. He operated with symbolic expressions. The underlying concepts of limit, continuity, and so on are not there.

Cauchy is very explicit in his introduction to the 1821 work that he seeks to give rigor to analysis. He points out that the free use for all functions of the properties that hold for algebraic functions, and the use of divergent series are not justified. Though Cauchy's work was but one step in the direction of rigor, he himself believed and states in his *Résumé* that he had brought the ultimate in rigor into analysis. He did give the beginnings of precise proofs of theorems and properly limited assertions at least for the elementary functions. Abel in his paper of 1826 on the binomial series (sec. 5) praised this achievement of Cauchy: "The distinguished work [the *Cours d'analyse*] should be read by everyone who loves rigor in mathematical investigations." Cauchy abandoned the explicit representations of Euler and the power series of Lagrange and introduced new concepts to treat functions.

2. 1821, *Œuvres*, (2) III.
3. 1823, *Œuvres*, (2), IV, 1–261.
4. 1829, *Œuvres*, (2), IV, 265–572.
5. 1797; 2nd ed., 1813 = *Œuvres*, 9.
6. 1801; 2nd ed., 1806 = *Œuvres*, 10.
7. 3 vols., 1st ed., 1797–1800; 2nd ed., 1810–19.

2. Functions and Their Properties

The eighteenth century mathematicians had on the whole believed that a function must have the same analytic expression throughout. During the latter part of the century, largely as a consequence of the controversy over the vibrating-string problem, Euler and Lagrange allowed functions that have different expressions in different domains and used the word continuous where the same expression held and discontinuous at points where the expression changed form (though in the modern sense the entire function could be continuous). While Euler, d'Alembert, and Lagrange had to reconsider the concept of function, they did not arrive at any widely accepted definition nor did they resolve the problem of what functions could be represented by trigonometrical series. However, the gradual expansion in the variety and use of functions forced mathematicians to accept a broader concept.

Gauss in his earlier work meant by a function a closed (finite analytical) expression and when he spoke of the hypergeometric series $F(\alpha, \beta, \gamma, x)$ as a function of α, β, γ, and x he qualified it by the remark, "insofar as one can regard it as a function." Lagrange had already used a broader concept in regarding power series as functions. In the second edition of his *Mécanique analytique* (1811–15) he used the word function for almost any kind of dependence on one or more variables. Even Lacroix in his *Traité* of 1797 had already introduced a broader notion. In the introduction he says, "Every quantity whose value depends on one or several others is called a function of the latter, whether one knows or one does not know by what operations it is necessary to go from the latter to the first quantity." Lacroix gives as an example a root of an equation of the fifth degree as a function of its coefficients.

Fourier's work opened up even more widely the question of what a function is. On the one hand, he insisted that functions need not be representable by any analytic expressions. In his *The Analytical Theory of Heat*[8] he says, "In general the function $f(x)$ represents a succession of values or ordinates each of which is arbitrary. . . . We do not suppose these ordinates to be subject to a common law; they succeed each other in any manner whatever. . . ." Actually he himself treated only functions with a finite number of discontinuities in any finite interval. On the other hand, to a certain extent Fourier was supporting the contention that a function must be representable by an analytic expression, though this expression was a Fourier series. In any case Fourier's work shook the eighteenth-century belief that all functions were at worst extensions of algebraic functions. The algebraic functions and even the elementary transcendental functions were no longer the prototype of functions. Since the properties of algebraic functions could no longer be carried over to all functions, the question then arose as to what one really

8. English translation, p. 430, Dover (reprint), 1955.

means by a function, by continuity, differentiability, integrability, and other properties.

In the positive reconstructions of analysis which many men undertook the real number system was taken for granted. No attempt was made to analyze this structure or to build it up logically. Apparently the mathematicians felt they were on sure ground as far as this area was concerned.

Cauchy begins his 1821 work with the definition of a variable. "One calls a quantity which one considers as having to successively assume many values different from one another a variable." As for the concept of function, "When variable quantities are so joined between themselves that, the value of one of these being given, one may determine the values of all the others, one ordinarily conceives these diverse quantities expressed by means of the one among them, which then takes the name independent variable; and the other quantities expressed by means of the independent variable are those which one calls functions of this variable." Cauchy is also explicit that an infinite series is one way of specifying a function. However, an analytical expression for a function is not required.

In a paper on Fourier series which we shall return to later, "Über die Darstellung ganz willkürlicher Functionen durch Sinus-und Cosinusreihen" (On the Representation of Completely Arbitrary Functions by Sine and Cosine Series),[9] Dirichlet gave the definition of a (single-valued) function which is now most often employed, namely, that y is a function of x when to each value of x in a given interval there corresponds a unique value of y. He added that it does not matter whether throughout this interval y depends upon x according to one law or more or whether the dependence of y on x can be expressed by mathematical operations. In fact in 1829[10] he gave the example of a function of x which has the value c for all rational values of x and the value d for all irrational values of x.

Hankel points out that the best textbooks of at least the first half of the century were at a loss as to what to do about the function concept. Some defined a function essentially in Euler's sense; others required that y vary with x according to some law but did not explain what law meant; some used Dirichlet's definition; and still others gave no definition. But all deduced consequences from their definitions which were not logically implied by the definitions.

The proper distinction between continuity and discontinuity gradually emerged. The careful study of the properties of functions was initiated by Bernhard Bolzano (1781–1848), a priest, philosopher, and mathematician of Bohemia. Bolzano was led to this work by trying to give a purely arithmetical proof of the fundamental theorem of algebra in place of Gauss's first proof (1799) which used geometric ideas. Bolzano had the correct

9. *Repertorium der Physik*, 1, 1837, 152–74 = *Werke*, 1, 135–60.
10. *Jour. für Math.*, 4, 1829, 157–69 = *Werke*, 1, 117–32.

concepts for the establishment of the calculus (except for a theory of real numbers), but his work went unnoticed for half a century. He denied the existence of infinitely small numbers (infinitesimals) and infinitely large numbers, both of which had been used by the eighteenth-century writers. In a book of 1817 whose long title starts with *Rein analytischer Beweis* (see the bibliography) Bolzano gave the proper definition of continuity, namely, $f(x)$ is continuous in an interval if at any x in the interval the difference $f(x + \omega) - f(x)$ can be made as small as one wishes by taking ω sufficiently small. He proves that polynomials are continuous.

Cauchy, too, tackled the notions of limit and continuity. As with Bolzano the limit concept was based on purely arithmetical considerations. In the *Cours* (1821) Cauchy says, "When the successive values attributed to a variable approach indefinitely a fixed value so as to end by differing from it by as little as one wishes, this last is called the limit of all the others. Thus, for example, an irrational number is the limit of diverse fractions which furnish closer and closer approximate values of it." This example was a bit unfortunate because many took it to be a definition of irrational numbers in terms of limit whereas the limit could have no meaning if irrationals were not already present. Cauchy omitted it in his 1823 and 1829 works.

In the preface to his 1821 work Cauchy says that to speak of the continuity of functions he must make known the principal properties of infinitely small quantities. "One says [*Cours*, p. 5] that a variable quantity becomes infinitely small when its numerical value decreases indefinitely in such a way as to converge to the limit 0." Such variables he calls infinitesimals. Thus Cauchy clarifies Leibniz's notion of infinitesimal and frees it of metaphysical ties. Cauchy continues, "One says that a variable quantity becomes infinitely large when its numerical value increases indefinitely in such a manner as to converge to the limit ∞." However, ∞ means not a fixed quantity but something indefinitely large.

Cauchy is now prepared to define continuity of a function. In the *Cours* (pp. 34–35) he says. "Let $f(x)$ be a function of the variable x, and suppose that, for each value of x intermediate between two given limits [bounds], this function constantly assumes a finite and unique value. If, beginning with a value of x contained between these limits, one assigns to the variable x an infinitely small increment α, the function itself will take on as an increment the difference $f(x + \alpha) - f(x)$ which will depend at the same time on the new variable α and on the value of x. This granted, the function $f(x)$ will be, between the two limits assigned to the variable x, a *continuous* function of the variable if, for each value of x intermediate between these two limits, the numerical value of the difference $f(x + \alpha) - f(x)$ decreases indefinitely with that of α. In other words, *the function* f(x) *will remain continuous with respect to* x *between the given limits, if, between these limits, an infinitely small increment of the variable always produces an infinitely small increment of the function itself.*

"We also say that the function $f(x)$ is a continuous function of x in the neighborhood of a particular value assigned to the variable x, as long as it [the function] is continuous between those two limits of x, no matter how close together, which enclose the value in question." He then says that $f(x)$ is discontinuous at x_0 if it is not continuous in every interval around x_0.

In his *Cours* (p. 37) Cauchy asserted that if a function of several variables is continuous in each one separately it is a continuous function of all the variables. This is not correct.

Throughout the nineteenth century the notion of continuity was explored and mathematicians learned more about it, sometimes producing results that astonished them. Darboux gave an example of a function which took on all intermediate values between two given values in passing from $x = a$ to $x = b$ but was not continuous. Thus a basic property of continuous functions is not sufficient to insure continuity.[11]

Weierstrass's work on the rigorization of analysis improved on Bolzano, Abel, and Cauchy. He, too, sought to avoid intuition and to build on arithmetical concepts. Though he did this work during the years 1841–56 when he was a high-school teacher much of it did not become known until 1859 when he began to lecture at the University of Berlin.

Weierstrass attacked the phrase "a variable approaches a limit," which unfortunately suggests time and motion. He interprets a variable simply as a letter standing for any one of a set of values which the letter may be given. Thus motion is eliminated. A continuous variable is one such that if x_0 is any value of the set of values of the variable and δ any positive number there are other values of the variable in the interval $(x_0 - \delta, x_0 + \delta)$.

To remove the vagueness in the phrase "becomes and remains less than any given quantity," which Bolzano and Cauchy used in their definitions of continuity and limit of a function, Weierstrass gave the now accepted definition that $f(x)$ is continuous at $x = x_0$ if given any positive number ε, there exists a δ such that for all x in the interval $|x - x_0| < \delta$, $|f(x) - f(x_0)| < \varepsilon$. A function $f(x)$ has a limit L at $x = x_0$ if the same statement holds but with L replacing $f(x_0)$. A function $f(x)$ is continuous in an interval of x values if it is continuous at each x in the interval.

During the years in which the notion of continuity itself was being refined, the efforts to establish analysis rigorously called for the proof of many theorems about continuous functions which had been accepted intuitively. Bolzano in his 1817 publication sought to prove that if $f(x)$ is negative for $x = a$ and positive for $x = b$, then $f(x)$ has a zero between a and b. He considered the sequence of functions (for fixed x)

(1) $F_1(x), F_2(x), F_3(x), \cdots, F_n(x), \cdots$

11. Consider $y = \sin(1/x)$ for $x \neq 0$ and $y = 0$ for $x = 0$. This function goes through all values from one assumed at a negative value of x to one assumed at a positive value of x. However, it is not continuous at $x = 0$.

and introduced the theorem that if for n large enough we can make the difference $F_{n+r} - F_n$ less than any given quantity, no matter how large r is, then there exists a fixed magnitude X such that the sequence comes closer and closer to X, and indeed as close as one wishes. His determination of the quantity X was somewhat obscure because he did not have a clear theory of the real number system and of irrational numbers in particular on which to build. However, he had the idea of what we now call the Cauchy condition for the convergence of a sequence (see below).

In the course of the proof Bolzano established the existence of a least upper bound for a bounded set of real numbers. His precise statement is: If a property M does not apply to all values of a variable quantity x, but to all those that are smaller than a certain u, there is always a quantity U which is the largest of those of which it can be asserted that all smaller x possess the property M. The essence of Bolzano's proof of this lemma was to divide the bounded interval into two parts and select a particular part containing an infinite number of members of the set. He then repeats the process until he closes down on the number which is the least upper bound of the given set of real numbers. This method was used by Weierstrass in the 1860s, with due credit to Bolzano, to prove what is now called the Weierstrass-Bolzano theorem. It establishes for any bounded infinite set of points the existence of a point such that in every neighborhood of it there are points of the set.

Cauchy had used without proof (in one of his proofs of the existence of roots of a polynomial) the existence of a minimum of a continuous function defined over a closed interval. Weierstrass in his Berlin lectures proved for any continuous function of one or more variables defined over a closed bounded domain the existence of a minimum value and a maximum value of the function.

In work inspired by the ideas of Georg Cantor and Weierstrass, Heine defined uniform continuity for functions of one or several variables[12] and then proved[13] that a function which is continuous on a closed bounded interval of the real numbers is uniformly continuous. Heine's method introduced and used the following theorem: Let a closed interval $[a, b]$ and a countably infinite set Δ of intervals, all in $[a, b]$, be given such that every point x of $a \leq x \leq b$ is an interior point of at least one of the intervals of Δ. (The endpoints a and b are regarded as interior points when a is the left-hand end of an interval and b the right-hand end of another interval.) Then a set consisting of a *finite* number of the intervals of Δ has the same property, namely, every point of the closed interval $[a, b]$ is an interior point of at least one of this finite set of intervals (a and b can be endpoints).

Emile Borel (1871–1956), one of the leading French mathematicians of this century, recognized the importance of being able to select a finite number of covering intervals and first stated it as an independent theorem

12. *Jour. für Math.*, 71, 1870, 353–65.
13. *Jour. für Math.*, 74, 1872, 172–88.

for the case when the original set of intervals Δ is countable.[14] Though many German and French mathematicians refer to this theorem as Borel's, since Heine used the property in his proof of uniform continuity the theorem is also known as the Heine-Borel theorem. The merit of the theorem, as Lebesgue pointed out, is not in the proof of it, which is not difficult, but in perceiving its importance and enunciating it as a distinct theorem. The theorem applies to closed sets in any number of dimensions and is now basic in set theory.

The extension of the Heine-Borel theorem to the case where a finite set of covering intervals can be selected from an uncountably infinite set is usually credited to Lebesgue who claimed to have known the theorem in 1898 and published it in his *Leçons sur l'intégration* (1904). However, it was first published by Pierre Cousin (1867–1933) in 1895.[15]

3. *The Derivative*

D'Alembert was the first to see that Newton had essentially the correct notion of the derivative. D'Alembert says explicitly in the *Encyclopédie* that the derivative must be based on the limit of the ratio of the differences of dependent and independent variables. This version is a reformulation of Newton's prime and ultimate ratio. D'Alembert did not go further because his thoughts were still tied to geometric intuition. His successors of the next fifty years still failed to give a clear definition of the derivative. Even Poisson believed that there are positive numbers that are not zero, but which are smaller than any given number however small.

Bolzano was the first (1817) to define the derivative of $f(x)$ as the quantity $f'(x)$ which the ratio $[f(x + \Delta x) - f(x)]/\Delta x$ approaches indefinitely closely as Δx approaches 0 through positive and negative values. Bolzano emphasized that $f'(x)$ was not a quotient of zeros or a ratio of evanescent quantities but a number which the ratio above approached.

In his *Résumé des leçons*[16] Cauchy defined the derivative in the same manner as Bolzano. He then unified this notion and the Leibnizian differentials by defining dx to be any finite quantity and dy to be $f'(x) \, dx$.[17] In other words, one introduces two quantities dx and dy whose ratio, by definition, is $f'(x)$. Differentials have meaning in terms of the derivative and are merely an auxiliary notion that could be dispensed with logically but are convenient as a way of thinking or writing. Cauchy also pointed out what the differential expressions used throughout the eighteenth century meant in terms of derivatives.

He then clarified the relation between $\Delta y/\Delta x$ and $f'(x)$ through the mean

14. *Ann. de l'Ecole Norm. Sup.*, (3), 12, 1895, 9–55.
15. *Acta Math.*, 19, 1895, 1–61.
16. 1823, *Œuvres*, (2), 4, 22.
17. Lacroix in the first edition of his *Traité* had already defined dy in this manner.

value theorem, that is, $\Delta y = f'(x + \theta \Delta x) \Delta x$, where $0 < \theta < 1$. The theorem itself was known to Lagrange (Chap. 20, sec. 7). Cauchy's proof of the mean value theorem used the continuity of $f'(x)$ in the interval Δx.

Though Bolzano and Cauchy had rigorized (somewhat) the notions of continuity and the derivative, Cauchy and nearly all mathematicians of his era believed and many texts "proved" for the next fifty years that a continuous function must be differentiable (except of course at isolated points such as $x = 0$ for $y = 1/x$). Bolzano did understand the distinction between continuity and differentiability. In his *Funktionenlehre*, which he wrote in 1834 but did not complete and publish,[18] he gave an example of a continuous function which has no finite derivative at any point. Bolzano's example, like his other works, was not noticed.[19] Even if it had been published in 1834 it probably would have made no impression because the curve did not have an analytic representation, and for mathematicians of that period functions were still entities given by analytical expressions.

The example that ultimately drove home the distinction between continuity and differentiability was given by Riemann in the *Habilitationsschrift*, the paper of 1854 he wrote to qualify as a *Privatdozent* at Göttingen, "Über die Darstellbarkeit einer Function durch eine trigonometrische Reihe," (On the Representability of a Function by a Trigonometric Series).[20] (The paper on the foundations of geometry (Chap. 37, sec. 3) was given as a qualifying *lecture*.) Riemann defined the following function. Let (x) denote the difference between x and the nearest integer and let $(x) = 0$ if it is halfway between two integers. Then $-1/2 < (x) < 1/2$. Now $f(x)$ is defined as

$$f(x) = \frac{(x)}{1} + \frac{(2x)}{4} + \frac{(3x)}{9} + \cdots.$$

This series converges for all values of x. However, for $x = p/2n$ where p is an integer prime to $2n$, $f(x)$ is discontinuous and has a jump whose value is $\pi^2/8n^2$. At all other values of x, $f(x)$ is continuous. Moreover, $f(x)$ is discontinuous an infinite number of times in every arbitrarily small interval. Nevertheless, $f(x)$ is integrable (sec. 4). Moreover, $F(x) = \int f(x) \, dx$ is continuous for all x but fails to have a derivative where $f(x)$ is discontinuous. This pathological function did not attract much attention until it was published in 1868.

An even more striking distinction between continuity and differentiability was demonstrated by the Swiss mathematician Charles Cellérier

---

18. *Schriften*, 1, Prague, 1930. It was edited and published by K. Rychlik, Prague, 1930.
19. In 1922 Rychlik proved that the function was nowhere differentiable. See Gerhard Kowalewski, "Über Bolzanos nichtdifferenzierbare stetige Funktion," *Acta Math.*, 44, 1923, 315–19. This article contains a description of Bolzano's function.
20. *Abh. der Ges. der Wiss. zu Gött.*, 13, 1868, 87–132 = *Werke*, 227–64.

(1818–89). In 1860 he gave an example of a function which is continuous but nowhere differentiable, namely,

$$f(x) = \sum_{n=1}^{\infty} a^{-n} \sin a^n x$$

in which a is a large positive integer. This was not published, however, until 1890.[21] The example that attracted the most attention is due to Weierstrass. As far back as 1861 he had affirmed in his lectures that any attempt to prove that differentiability follows from continuity must fail. He then gave the classic example of a continuous nowhere differentiable function in a lecture to the Berlin Academy on July 18, 1872.[22] Weierstrass communicated his example in a letter of 1874 to Du Bois-Reymond and the example was first published by the latter.[23] Weierstrass's function is

$$f(x) = \sum_{n=0}^{\infty} b^n \cos (a^n \pi x)$$

wherein a is an odd integer and b a positive constant less than 1 such that $ab > 1 + (3\pi/2)$. The series is uniformly convergent and so defines a continuous function. The example given by Weierstrass prompted the creation of many more functions that are continuous in an interval or everywhere but fail to be differentiable either on a dense set of points or at any point.[24]

The historical significance of the discovery that continuity does not imply differentiability and that functions can have all sorts of abnormal behavior was great. It made mathematicians all the more fearful of trusting intuition or geometrical thinking.

4. The Integral

Newton's work showed how areas could be found by reversing differentiation. This is of course still the essential method. Leibniz's idea of area and volume as a "sum" of elements such as rectangles or cylinders [the definite integral] was neglected. When the latter concept was employed at all in the eighteenth century it was loosely used.

Cauchy stressed defining the integral as the limit of a sum instead of the inverse of differentiation. There was at least one major reason for the change.

21. *Bull. des Sci. Math.*, (2), 14, 1890, 142–60.
22. *Werke*, 2, 71–74.
23. *Jour. für Math.*, 79, 1875, 21–37.
24. Other examples and references can be found in E. J. Townsend, *Functions of Real Variables*, Henry Holt, 1928, and in E. W. Hobson, *The Theory of Functions of a Real Variable*, 2, Chap. 6, Dover (reprint), 1957.

Fourier, as we know, dealt with discontinuous functions, and the formula for the coefficients of a Fourier series, namely,

$$a_n = \frac{1}{\pi} \int_0^{2\pi} f(x) \cos nx \, dx, \qquad b_n = \frac{1}{\pi} \int_0^{2\pi} f(x) \sin nx \, dx$$

calls for the integrals of such functions. Fourier regarded the integral as a sum (the Leibnizian view) and so had no difficulty in handling even discontinuous $f(x)$. The problem of the analytical meaning of the integral when $f(x)$ is discontinuous had, however, to be considered.

Cauchy's most systematic attack on the definite integral was made in his *Résumé* (1823) wherein he also points out that it is necessary to establish the existence of the definite integral and indirectly of the antiderivative or primitive function before one can use them. He starts with continuous functions.

For continuous $f(x)$ he gives[25] the precise definition of the integral as the limit of a sum. If the interval $[x_0, X]$ is subdivided by the x-values, $x_1, x_2, \ldots, x_{n-1}$, with $x_n = X$, then the integral is

$$\lim_{n \to \infty} \sum_{i=1}^{n} f(\xi_i)(x_i - x_{i-1}),$$

where ξ_i is any value of x in $[x_{i-1}, x_i]$. The definition presupposes that $f(x)$ is continuous over $[x_0, X]$ and that the length of the largest subinterval approaches zero. The definition is arithmetical. Cauchy shows the integral exists no matter how the x_i and ξ_i are chosen. However, his proof was not rigorous because he did not have the notion of uniform continuity. He denotes the limit by the notation proposed by Fourier $\int_{x_0}^{X} f(x) \, dx$ in place of

$$\int f(x) \, dx \begin{bmatrix} x = b \\ x = a \end{bmatrix}$$

often employed by Euler for antidifferentiation.

Cauchy then defines

$$F(x) = \int_{x_0}^{x} f(x) \, dx$$

and shows that $F(x)$ is continuous in $[x_0, X]$. By forming

$$\frac{F(x + h) - F(x)}{h} = \frac{1}{h} \int_{x}^{x+h} f(x) \, dx$$

and using the mean value theorem for integrals Cauchy proves that

$$F'(x) = f(x).$$

25. *Résumé*, 81–84 = *Œuvres*, (2), 4, 122–27.

This is the fundamental theorem of the calculus, and Cauchy's presentation is the first demonstration of it. Then, after showing that all primitives of a given $f(x)$ differ by a constant he defines the indefinite integral as

$$\int f(x)\, dx = \int_a^x f(x)\, dx + C.$$

He points out that

$$\int_a^b f'(x) = f(b) - f(a)$$

presupposes $f'(x)$ continuous. Cauchy then treats the singular (improper) integrals where $f(x)$ becomes infinite at some value of x in the interval of integration or where the interval of integration extends to ∞. For the case where $f(x)$ has a discontinuity at $x = c$ at which value $f(x)$ may be bounded or not Cauchy defines

$$\int_a^b f(x)\, dx = \lim_{\varepsilon_1 \to 0} \int_a^{c-\varepsilon_1} f(x)\, dx + \lim_{\varepsilon_2 \to 0} \int_{c+\varepsilon_2}^b f(x)\, dx$$

when these limits exist. When $\varepsilon_1 = \varepsilon_2$ we get what Cauchy called the principal value.

The notions of area bounded by a curve, length of a curve, volume bounded by surfaces and areas of surfaces had been accepted as intuitively understood, and it had been considered one of the great achievements of the calculus that these quantities could be calculated by means of integrals. But Cauchy, in keeping with his goal of arithmetizing analysis, *defined* these geometric quantities by means of the integrals which had been formulated to calculate them. Cauchy unwittingly imposed a limitation on the concepts he defined because the calculus formulas impose restrictions on the quantities involved. Thus the formula for the length of arc of a curve given by $y = f(x)$ is

$$s = \int_a^b \sqrt{1 + (y')^2}\, dx$$

and this formula presupposes the differentiability of $f(x)$. The question of what are the most general definitions of areas, lengths of curves, and volumes was to be raised later (Chap. 42, sec. 5).

Cauchy had proved the existence of an integral for any continuous integrand. He had also defined the integral when the integrand has jump discontinuities and infinities. But with the growth of analysis the need to consider integrals of more irregularly behaving functions became manifest. The subject of integrability was taken up by Riemann in his paper of 1854 on trigonometric series. He says that it is important at least for mathematics, though not for physical applications, to consider the broader conditions under which the integral formula for the Fourier coefficients holds.

Riemann generalized the integral to cover functions $f(x)$ defined and bounded over an interval $[a, b]$. He breaks up this interval into subintervals[26] $\Delta x_1, \Delta x_2, \ldots, \Delta x_n$ and defines the oscillation of $f(x)$ in Δx_i as the difference between the greatest and least value of $f(x)$ in Δx_i. Then he proves that a necessary and sufficient condition that the sums

$$S = \sum_{i=1}^{n} f(x_i) \, \Delta x_i,$$

where x_i is any value of x in Δx_i, approach a unique limit (that the integral exists) as the maximum Δx_i approaches zero is that the sum of the intervals Δx_i in which the oscillation of $f(x)$ is greater than any given number λ must approach zero with the size of the intervals.

Riemann then points out that this condition on the oscillations allows him to replace continuous functions by functions with isolated discontinuities and also by functions having an everywhere dense set of points of discontinuity. In fact the example he gave of an integrable function with an infinite number of discontinuities in every arbitrarily small interval (sec. 3) was offered to illustrate the generality of his integral concept. Thus Riemann dispensed with continuity and piecewise continuity in the definition of the integral.

In his 1854 paper Riemann with no further remarks gives another necessary and sufficient condition that a bounded function $f(x)$ be integrable on $[a, b]$. It amounts to first setting up what are now called the upper and lower sums

$$S = M_1 \, \Delta x_1 + \cdots + M_n \, \Delta x_n$$
$$s = m_1 \, \Delta x_1 + \cdots + m_n \, \Delta x_n$$

where m_i and M_i are the least and greatest values of $f(x)$ in Δx_i. Then letting $D_i = M_i - m_i$, Riemann states that the integral of $f(x)$ over $[a, b]$ exists if and only if

$$\lim_{\max \Delta x \to 0} \{D_1 \, \Delta x_1 + D_2 \, \Delta x_2 + \cdots + D_n \, \Delta x_n\} = 0$$

for all choices of Δx_i filling out the interval $[a, b]$. Darboux completed this formulation and proved that the condition is necessary and sufficient.[27] There are many values of S each corresponding to a partition of $[a, b]$ into Δx_i. Likewise there are many values of s. Each S is called an upper sum and each s a lower sum. Let the greatest lower bound of the S be J and the least upper bound of the s be I. It follows that $I \leq J$. Darboux's theorem then states that the sums S and s tend respectively to J and I when the number of

26. For brevity we use Δx_i for the subintervals and their lengths.
27. *Ann. de l'Ecole Norm. Sup.*, (2), 4, 1875, 57–112.

Δx_i is increased indefinitely in such a way that the maximum subinterval approaches zero. A bounded function is said to be integrable on $[a, b]$ if $J = I$.

Darboux then shows that a bounded function will be integrable on $[a, b]$ if and only if the discontinuities in $f(x)$ constitute a set of measure zero. By the latter he meant that the points of discontinuity can be enclosed in a finite set of intervals whose total length is arbitrarily small. This very formulation of the integrability condition was given by a number of other men in the same year (1875). The terms upper integral and the notation $\overline{\int_a^b} f(x) \, dx$ for the greatest lower bound J of the S's, and lower integral and the notation $\underline{\int_a^b} f(x) \, dx$ for the least upper bound I of the s's were introduced by Volterra.[28]

Darboux also showed in the 1875 paper that the fundamental theorem of the calculus holds for functions integrable in the extended sense. Bonnet had given a proof of the mean value theorem of the differential calculus which did not use the continuity of $f'(x)$.[29] Darboux using this proof, which is now standard, showed that

$$\int_a^b f'(x) \, dx = f(b) - f(a)$$

when f' is merely integrable in the Riemann-Darboux sense. Darboux's argument was that

$$f(b) - f(a) = \sum_{i=1}^n f(x_i) - f(x_{i-1}),$$

where $a = x_0 < x_1 < x_2 < \cdots < x_n = b$. By the mean value theorem

$$\sum f(x_i) - f(x_{i-1}) = \sum f'(t_i)(x_i - x_{i-1}),$$

where t_i is some value in (x_{i-1}, x_i). Now if the maximum Δx_i, or $x_i - x_{i-1}$, approaches zero then the right side of this last equation approaches $\int_a^b f'(x) \, dx$ and the left side is $f(b) - f(a)$.

One of the favorite activities of the 1870s and the 1880s was to construct functions with various infinite sets of discontinuities that would still be integrable in Riemann's sense. In this connection H. J. S. Smith[30] gave the first example of a function nonintegrable in Riemann's sense but for which the points of discontinuity were "rare." Dirichlet's function (sec. 2) is also nonintegrable in this sense, but it is discontinuous everywhere.

The notion of integration was then extended to unbounded functions and

28. *Gior. di Mat.*, 19, 1881, 333–72.
29. Published in Serret's *Cours de calcul différentiel et intégral*, 1, 1868, 17–19.
30. *Proc. Lon. Math. Soc.*, 6, 1875, 140–53 = *Coll. Papers*, 2, 86–100.

to various improper integrals. The most significant extension was made in the next century by Lebesgue (Chap. 44). However, as far as the elementary calculus was concerned the notion of integral was by 1875 sufficiently broad and rigorously founded.

The theory of double integrals was also tackled. The simpler cases had been treated in the eighteenth century (Chap. 19, sec. 6). In his paper of 1814 (Chap. 27, sec. 4) Cauchy showed that the order of integration in which one evaluates a double integral $\int \int f(x, y) \, dx \, dy$ does matter if the integrand is discontinuous in the domain of integration. Specifically Cauchy pointed out[31] that the repeated integrals

$$\int_0^1 dy \left(\int_0^1 f(x, y) \, dx \right), \ \int_0^1 dx \left(\int_0^1 f(x, y) \, dy \right)$$

need not be equal when f is unbounded.

Karl J. Thomae (1840–1921) extended Riemann's theory of integration to functions of two variables.[32] Then Thomae in 1878[33] gave a simple example of a bounded function for which the second repeated integral exists but the first is meaningless.

In the examples of Cauchy and Thomae the double integral does not exist. But in 1883[34] Du Bois-Reymond showed that even when the double integral exists the two repeated integrals need not. In the case of double integrals too the most significant generalization was made by Lebesgue.

5. Infinite Series

The eighteenth-century mathematicians used series indiscriminately. By the end of the century some doubtful or plainly absurd results from work with infinite series stimulated inquiries into the validity of operations with them. Around 1810 Fourier, Gauss, and Bolzano began the exact handling of infinite series. Bolzano stressed that one must consider convergence and criticized in particular the loose proof of the binomial theorem. Abel was the most outspoken critic of the older uses of series.

In his 1811 paper and his *Analytical Theory of Heat* Fourier gave a satisfactory definition of convergence of an infinite series, though in general he worked freely with divergent series. In the book (p. 196 of the English edition) he describes convergence to mean that as n increases the sum of n terms approaches a fixed value more and more closely and should differ from it only by a quantity which becomes less than any given magnitude. Moreover, he recognized that convergence of a series of functions may obtain only

31. *Mémoire* of 1814; see, in particular, p. 394 of *Œuvres*, (1), 1.
32. *Zeit. für Math. und Phys.*, 21, 1876, 224–27.
33. *Zeit. für Math. und Phys.*, 23, 1878, 67–68.
34. *Jour. für Math.*, 94, 1883, 273–90.

in an interval of x values. He also stressed that a necessary condition for convergence is that the terms approach zero in value. However, the series $1 - 1 + 1 + \cdots$ still fooled him; he took its sum to be $1/2$.

The first important and strictly rigorous investigation of convergence was made by Gauss in his 1812 paper "Disquisitiones Generales Circa Seriem Infinitam" (General Investigations of Infinite Series)[35] wherein he studied the hypergeometric series $F(\alpha, \beta, \gamma, x)$. In most of his work he called a series convergent if the terms from a certain one on decrease to zero. But in his 1812 paper he noted that this is not the correct concept. Because the hypergeometric series can represent many functions for different choices of α, β, and γ it seemed desirable to him to develop an exact criterion for convergence for this series. The criterion is laboriously arrived at but it does settle the question of convergence for the cases it was designed to cover. He showed that the hypergeometric series converges for real and complex x if $|x| < 1$ and diverges if $|x| > 1$. For $x = 1$, the series converges if and only if $\alpha + \beta < \gamma$ and for $x = -1$ the series converges if and only if $\alpha + \beta < \gamma + 1$. The unusual rigor discouraged interest in the paper by mathematicians of the time. Moreover, Gauss was concerned with particular series and did not take up general principles of the convergence of series.

Though Gauss is often mentioned as one of the first to recognize the need to restrict the use of series to their domains of convergence he avoided any decisive position. He was so much concerned to solve concrete problems by numerical calculation that he used Stirling's divergent development of the gamma function. When he did investigate the convergence of the hypergeo- metric series in 1812 he remarked[36] that he did so to please those who favored the rigor of the ancient geometers, but he did not state his own stand on the subject. In the course of his paper[37] he used the development of $\log (2 - 2 \cos x)$ in cosines of multiples of x even though there was no proof of the convergence of this series and there could have been no proof with the techniques available at the time. In his astronomical and geodetic work Gauss, like the eighteenth-century men, followed the practice of using a finite number of terms of an infinite series and neglecting the rest. He stopped including terms when he saw that the succeeding terms were numerically small and of course did not estimate the error.

Poisson too took a peculiar position. He rejected divergent series[38] and even gave examples of how reckoning with divergent series can lead to false results. But he nevertheless made extensive use of divergent series in his representation of arbitrary functions by series of trigonometric and spherical functions.

35. *Comm. Soc. Gott.*, 2, 1813, = *Werke*, 3, 125–62 and 207–29.
36. *Werke*, 3, 129.
37. *Werke*, 3, 156.
38. *Jour. de l'Ecole Poly.*, 19, 1823, 404–509.

Bolzano in his 1817 publication had the correct notion of the condition for the convergence of a sequence, the condition now ascribed to Cauchy. Bolzano also had clear and correct notions about the convergence of series. But, as we have already noted, his work did not become widely known.

Cauchy's work on the convergence of series is the first extensive significant treatment of the subject. In his *Cours d'analyse* Cauchy says, "Let

$$s_n = u_0 + u_1 + u_2 + \cdots + u_{n-1}$$

be the sum of the first n terms [of the infinite series which one considers], n designating a natural number. If, for constantly increasing values of n, the sum s_n approaches indefinitely a certain limit s, the series will be called *convergent*, and the limit in question will be called the *sum* of the series.[39] On the contrary, if while n increases indefinitely, the sum s_n does not approach a fixed limit, the series will be called *divergent* and will have no sum."

After defining convergence and divergence Cauchy states (*Cours*, p. 125) the Cauchy convergence criterion, namely, a sequence $\{S_n\}$ converges to a limit S if and only if $S_{n+r} - S_n$ can be made less in absolute value than any assignable quantity for all r and sufficiently large n. Cauchy proves this condition is necessary but merely remarks that if the condition is fulfilled, the convergence of the sequence is assured. He lacked the knowledge of properties of real numbers to make the proof.

Cauchy then states and proves specific tests for the convergence of series with positive terms. He points out that u_n must approach zero. Another test (*Cours*, 132–35) requires that one find the limit or limits toward which the expression $(u_n)^{1/n}$ tends as n becomes infinite, and designate the greatest of these limits by k. Then the series will be convergent if $k < 1$ and divergent if $k > 1$. He also gives the ratio test which uses $\lim_{n \to \infty} u_{n+1}/u_n$. If this limit is less than 1 the series converges and if greater than 1, the series diverges. Special tests are given if the ratio is 1. There follow comparison tests and a logarithmic test. He proves that the sum $u_n + v_n$ of two convergent series converges to the sum of the separate sums and the analogous result for product. Series with some negative terms, Cauchy shows, converge when the series of absolute values of the terms converge, and he then deduces Leibniz's test for alternating series.

Cauchy also considered the sum of a series

$$\sum u_n(x) = u_1(x) + u_2(x) + u_3(x) + \cdots$$

in which all the terms are continuous, single-valued real functions. The theorems on the convergence of series of constant terms apply here to determine an interval of convergence. He also considers series with complex functions as terms.

39. The correct notion of the limit of a sequence was given by Wallis in 1655 (*Opera*, 1695, 1, 382) but was not taken up.

Lagrange was the first to state Taylor's theorem with a remainder but Cauchy in his 1823 and 1829 texts made the important point that the infinite Taylor series converges to the function from which it is derived if the remainder approaches zero. He gives the example $e^{-x^2} + e^{-1/x^2}$ of a function whose Taylor series does not converge to the function. In his 1823 text he gives the example e^{-1/x^2} of a function which has all derivatives at $x = 0$ but has no Taylor expansion around $x = 0$. Here he contradicts by an example Lagrange's assertion in his *Théorie des fonctions* (Ch. V., Art. 30) that if $f(x)$ has at x_0 all derivatives then it can be expressed as a Taylor series which converges to $f(x)$ for x near x_0. Cauchy also gave[40] an alternative form for the remainder in Taylor's formula.

Cauchy here made some additional missteps with respect to rigor. In his *Cours d'analyse* (pp. 131–32) he states that $F(x)$ is continuous if when $F(x) = \sum_1^\infty u_n(x)$ the series is convergent and the $u_n(x)$ are continuous. In his *Résumé des leçons*[41] he says that if the $u_n(x)$ are continuous and the series converges then one may integrate the series term by term; that is,

$$\int_a^b F \, dx = \sum_1^\infty \int_a^b u_n \, dx.$$

He overlooked the need for uniform convergence. He also asserts for continuous functions[42] that

$$\frac{\partial}{\partial u} \int_a^b f(x, u) \, dx = \int_a^b \frac{\partial f}{\partial u} \, dx.$$

Cauchy's work inspired Abel. Writing to his former teacher Holmboë from Paris in 1826 Abel said[43] Cauchy "is at present the one who knows how mathematics should be treated." In that year[44] Abel investigated the domain of convergence of the binomial series

$$1 + mx + \frac{m(m-1)}{2} x^2 + \frac{m(m-1)(m-2)}{3!} x^3 + \cdots$$

with m and x complex. He expressed astonishment that no one had previously investigated the convergence of this most important series. He proves first that if the series

$$f(\alpha) = v_0 + v_1\alpha + v_2\alpha^2 + \cdots,$$

wherein the v_i's are constants and α is real, converges for a value δ of α then it will converge for every smaller value of α, and $f(\alpha - \beta)$ for β approaching

40. *Exercices de mathématiques*, 1, 1826, 5 = *Œuvres*, (2), 6, 38–42.
41. 1823, *Œuvres*, (2), 4, p. 237.
42. *Exercices de mathématiques*, 2, 1827 = *Œuvres*, (2), 7, 160.
43. *Œuvres*, 2, 259.
44. *Jour. für Math.*, 1, 1826, 311–39 = *Œuvres*, 1, 219–50.

0 will approach $f(\alpha)$ when α is equal to or smaller than δ. The last part says that a convergent *power* series is a continuous function of its argument up to and including δ, for α can be δ.

In this same 1826 paper[45] Abel corrected Cauchy's error on the continuity of the sum of a convergent series of continuous functions. He gave the example of

$$(2) \qquad\qquad \sin x - \frac{\sin 2x}{2} + \frac{\sin 3x}{3} \cdots$$

which is discontinuous when $x = (2n + 1)\pi$ and n is integral, though the individual terms are continuous.[46] Then by using the *idea* of uniform convergence he gave a correct proof that the sum of a uniformly convergent series of continuous functions is continuous in the interior of the interval of convergence. Abel did not isolate the property of uniform convergence of a series.

The notion of uniform convergence of a series $\sum_1^\infty u_n(x)$ requires that given any ε, there exists an N such that for all $n > N$, $|S(x) - \sum_1^n u_n(x)| < \varepsilon$ for all x in some interval. $S(x)$ is of course the sum of the series. This notion was recognized in and for itself by Stokes, a leading mathematical physicist,[47] and independently by Philipp L. Seidel (1821–96).[48] Neither man gave the precise formulation. Rather both showed that if a sum of a series of continuous functions is discontinuous at x_0 then there are values of x near x_0 for which the series converges arbitrarily slowly. Also neither related the need for uniform convergence to the justification of integrating a series term by term. In fact, Stokes accepted[49] Cauchy's use of term-by-term integration. Cauchy ultimately recognized the need for uniform convergence[50] in order to assert the continuity of the sum of a series of continuous functions but even he at that time did not see the error in his use of term-by-term integration of series.

Actually Weierstrass[51] had the notion of uniform convergence as early as 1842. In a theorem that duplicates unknowingly Cauchy's theorem on the existence of power series solutions of a system of first order ordinary differential equations, he affirms that the series converge uniformly and so constitute analytic functions of the complex variable. At about the same time

45. *Œuvres*, 1, 224.
46. The series (2) is the Fourier expansion of $x/2$ in the interval $-\pi < x < \pi$. Hence the series represents the periodic function which is $x/2$ in each 2π interval. Then the series converges to $\pi/2$ when x approaches $(2n + 1)\pi$ from the left and the series converges to $-\pi/2$ when x approaches $(2n + 1)\pi$ from the right.
47. *Trans. Camb. Phil. Soc.*, 8_5, 1848, 533–83 = *Math. and Phys. Papers*, 1, 236–313.
48. *Abh. der Bayer. Akad. der Wiss.*, 1847/49, 379–94.
49. *Papers*, 1, 242, 255, 268, and 283.
50. *Comp. Rend.*, 36, 1853, 454–59 = *Œuvres*, (1), 12, 30–36.
51. *Werke*, 1, 67–85.

Weierstrass used the notion of uniform convergence to give conditions for the integration of a series term by term and conditions for differentiation under the integral sign.

Through Weierstrass's circle of students the importance of uniform convergence was made known. Heine emphasized the notion in a paper on trigonometric series.[52] Heine may have learned of the idea through Georg Cantor who had studied at Berlin and then came to Halle in 1867 where Heine was a professor of mathematics.

During his years as a high-school teacher Weierstrass also discovered that any continuous function over a closed interval of the real axis can be expressed in that interval as an absolutely and uniformly convergent series of polynomials. Weierstrass included also functions of several variables. This result[53] aroused considerable interest and many extensions of this result to the representation of complex functions by a series of polynomials or a series of rational functions were established in the last quarter of the nineteenth century.

It had been assumed that the terms of a series can be rearranged at will. In a paper of 1837[54] Dirichlet proved that in an absolutely convergent series one may group or rearrange terms and not change the sum. He also gave examples to show that the terms of any conditionally convergent series can be rearranged so that the sum is altered. Riemann in a paper written in 1854 (see below) proved that by suitable rearrangement of the terms the sum could be any given number. Many more criteria for the convergence of infinite series were developed by leading mathematicians from the 1830s on throughout the rest of the century.

6. Fourier Series

As we know Fourier's work showed that a wide class of functions can be represented by trigonometric series. The problem of finding precise conditions on the functions which would possess a convergent Fourier series remained open. Efforts by Cauchy and Poisson were fruitless.

Dirichlet took an interest in Fourier series after meeting Fourier in Paris during the years 1822–25. In a basic paper "Sur la convergence des séries trigonométriques"[55] Dirichlet gave the first set of *sufficient* conditions that the Fourier series representing a given $f(x)$ converge and converge to $f(x)$. The proof given by Dirichlet is a refinement of that sketched by Fourier in the concluding sections of his *Analytical Theory of Heat*. Consider $f(x)$ either

52. *Jour. für Math.*, 71, 1870, 353–65.
53. *Sitzungsber. Akad. Wiss. zu Berlin*, 1885, 633–39, 789–905 = *Werke*, 3, 1–37.
54. *Abh. König. Akad. der Wiss.*, Berlin, 1837, 45–81 = *Werke*, 1, 313–342 = *Jour. de Math.*, 4, 1839, 393–422.
55. *Jour. für Math.*, 4, 1829, 157–69 = *Werke*, 1, 117–32.

given periodic with period 2π or given in the interval $[-\pi, \pi]$ and defined to be periodic in each interval of length 2π to the left and right of $[-\pi, \pi]$. Dirichlet's conditions are:

(a) $f(x)$ is single-valued and bounded.
(b) $f(x)$ is piecewise continuous; that is, it has only a finite number of discontinuities in the (closed) period.
(c) $f(x)$ is piecewise monotone; that is, it has only a finite number of maxima and minima in one period.

The $f(x)$ can have different analytic representations in different parts of the fundamental period.

Dirichlet's method of proof was to make a direct summation of n terms and to investigate what happens as n becomes infinite. He proved that for any given value of x the sum of the series is $f(x)$ provided $f(x)$ is continuous at that value of x and is $(1/2)[f(x-0) + f(x+0)]$ if $f(x)$ is discontinuous at that value of x.

In his proof Dirichlet had to give a careful discussion of the limiting values of the integrals

$$\int_0^a f(x) \frac{\sin \mu x}{\sin x}\, dx, \qquad a > 0$$

$$\int_a^b f(x) \frac{\sin \mu x}{\sin x}\, dx, \qquad b > a > 0$$

as μ increases indefinitely. These are still called the Dirichlet integrals.

It was in connection with this work that he gave the function which is c for rational values of x and d for irrational values of x (sec. 2). He had hoped to generalize the notion of integral so that a larger class of functions could still be representable by Fourier series converging to these functions, but the particular function just noted was intended as an example of one which could not be included in a broader notion of integral.

Riemann studied for a while under Dirichlet in Berlin and acquired an interest in Fourier series. In 1854 he took up the subject in his *Habilitations-schrift* (probationary essay) at Göttingen,[56] "Über die Darstellbarkeit einer Function durch eine trigonometrische Reihe," which aimed to find necessary and sufficient conditions that a function must satisfy so that at a point x in the interval $[-\pi, \pi]$ the Fourier series for $f(x)$ should converge to $f(x)$.

Riemann did prove the fundamental theorem that if $f(x)$ is bounded and integrable in $[-\pi, \pi]$ then the Fourier coefficients

(3) $\qquad a_n = \dfrac{1}{\pi} \displaystyle\int_{-\pi}^{\pi} f(x) \cos nx\, dx, \qquad b_n = \dfrac{1}{\pi} \displaystyle\int_{-\pi}^{\pi} f(x) \sin nx\, dx$

56. *Abh. der Ges. der Wiss. zu Gött.*, 13, 1868, 87–132 = *Werke*, 227–64.

THE INSTILLATION OF RIGOR IN ANALYSIS

approach zero as n tends to infinity. The theorem showed too that for bounded and integrable $f(x)$ the convergence of its Fourier series at a point in $[-\pi, \pi]$ depends only on the behavior of $f(x)$ in the neighborhood of that point. However, the problem of finding necessary *and* sufficient conditions on $f(x)$ so that its Fourier series converges to $f(x)$ was not and has not been solved.

Riemann opened up another line of investigation. He considered *trigonometric* series but did not require that the coefficients be determined by the formula (3) for the Fourier coefficients. He starts with the series

$$(4) \qquad \sum_{1}^{\infty} a_n \sin nx + \frac{b_0}{2} + \sum_{1}^{\infty} b_n \cos nx$$

and defines

$$A_0 = \frac{1}{2} b_0, \qquad A_n(x) = a_n \sin nx + b_n \cos nx.$$

Then the series (4) is equal to

$$f(x) = \sum_{n=0}^{\infty} A_n(x).$$

Of course $f(x)$ has a value only for those values of x for which the series converges. Let us refer to the series itself by Ω. Now the terms of Ω may approach zero for all x or for some x. These two cases are treated separately by Riemann.

If a_n and b_n approach zero, the terms of Ω approach zero for all x. Let $F(x)$ be the function

$$F(x) = C + C'x + A_0 \frac{x^2}{2} - A_1 - \frac{A_2}{4} - \cdots - \frac{A_n}{n^2} \cdots$$

which is obtained by two successive integrations of Ω. Riemann shows that $F(x)$ converges for all x and is continuous in x. Then $F(x)$ can itself be integrated. Riemann now proves a number of theorems about $F(x)$ which lead in turn to necessary and sufficient conditions for a series of the form (4) to converge to a given function $f(x)$ of period 2π. He then gives a necessary and sufficient condition that the trigonometric series (4) converge at a particular value of x, with a_n and b_n still approaching 0 as n approaches ∞.

Next he considers the alternate case where $\lim_{n \to \infty} A_n$ depends on the value of x and gives conditions which hold when the series Ω is convergent for particular values of x and a criterion for convergence at particular values of x.

He also shows that a given $f(x)$ may be integrable and yet not have a Fourier series representation. Further there are nonintegrable functions to which the series Ω converges for an infinite number of values of x taken between arbitrarily close limits. Finally a trigonometric series can converge for an infinite number of values of x in an arbitrarily small interval even though a_n and b_n do not approach zero for all x.

The nature of the convergence of Fourier series received further attention after the introduction of the concept of uniform convergence by Stokes and Seidel. It had been known since Dirichlet's time that the series were, in general, only conditionally convergent, if at all, and that their convergence depended upon the presence of positive and negative terms. Heine noted in a paper of 1870[57] that the usual proof that a bounded $f(x)$ is uniquely represented between $-\pi$ and π by a Fourier series is incomplete because the series may not be uniformly convergent and so cannot be integrated term by term. This suggested that there may nevertheless exist nonuniformly converging trigonometric series which do represent a function. Moreover, a continuous function might be representable by a Fourier series and yet the series might not be uniformly convergent. These problems gave rise to a new series of investigations seeking to establish the uniqueness of the representation of a function by a trigonometric series and whether the coefficients are necessarily the Fourier coefficients. Heine in the above-mentioned paper proved that a Fourier series which represents a bounded function satisfying the Dirichlet conditions is uniformly convergent in the portions of the interval $[-\pi, \pi]$ which remain when arbitrarily small neighborhoods of the points of discontinuity of the function are removed from the interval. In these neighborhoods the convergence is necessarily nonuniform. Heine then proved that if the uniform convergence just specified holds for a trigonometric series which represents a function, then the series is unique.

The second result, on uniqueness, is equivalent to the statement that if a trigonometric series of the form

(5)
$$\frac{a_0}{2} + \sum_{n=1}^{\infty} (a_n \cos nx + b_n \sin nx)$$

is uniformly convergent and represents zero where convergent, that is, except on a finite set P of points, then the coefficients are all zero and of course then the series represents zero throughout $[-\pi, \pi]$.

The problems associated with the uniqueness of trigonometric and Fourier series attracted Georg Cantor, who studied Heine's work. Cantor began his investigations by seeking uniqueness criteria for trigonometric series representations of functions. He proved[58] that when $f(x)$ is represented by a trigonometric series convergent for all x, there does not exist a different series of the same form which likewise converges for every x and

57. *Jour. für Math.*, 71, 1870, 353–65.
58. *Jour. für Math.*, 72, 1870, 139–42 = *Ges. Abh.*, 80–83.

represents the same $f(x)$. Another paper[59] gave a better proof for this last result.

The uniqueness theorem he proved can be restated thus: If, for all x, there is a convergent representation of zero by a trigonometric series, then the coefficients a_n and b_n are zero. Then Cantor demonstrates in the 1871 paper that the conclusion still holds even if the convergence is renounced for a finite number of x values. This paper was the first of a sequence of papers in which Cantor treats the sets of exceptional values of x. He extended[60] the uniqueness result to the case where an infinite set of exceptional values is permitted. To describe this set he first defined a point p to be a limit of a set of points S if every interval containing p contains infinitely many points of S. Then he introduced the notion of the derived set of a set of points. This derived set consists of the limit points of the original set. There is, then, a second derived set, that is, the derived set of the derived set, and so forth. If the nth derived set of a given set is a finite set of points then the given set is said to be of the nth kind or nth order (or of the first species). Cantor's final answer to the question of whether a function can have two different trigonometric series representations in the interval $[-\pi, \pi]$ or whether zero can have a non-zero Fourier representation is that if in the interval a trigonometric series adds up to zero for all x except those of a point set of the nth kind (at which one knows nothing more about the series) then all the coefficients of the series must be zero. In this 1872 paper Cantor laid the foundation of the theory of point sets which we shall consider in a later chapter. The problem of uniqueness was pursued by many other men in the last part of the nineteenth century and the early part of the twentieth.[61]

For about fifty years after Dirichlet's work it was believed that the Fourier series of any function continuous in $[-\pi, \pi]$ converged to the function. But Du Bois-Reymond[62] gave an example of a function continuous in $(-\pi, \pi)$ whose Fourier series did not converge at a particular point. He also constructed another continuous function whose Fourier series fails to converge at the points of an everywhere dense set. Then in 1875[63] he proved that if a trigonometric series of the form

$$a_0 + \sum_1^\infty (a_n \cos nx + b_n \sin nx)$$

converged to $f(x)$ in $[-\pi, \pi]$ and if $f(x)$ is integrable (in a sense even more general than Riemann's in that $f(x)$ can be unbounded on a set of the first

59. *Jour. für Math.*, 73, 1871, 294–6 = *Ges. Abh.*, 84–86.
60. *Math. Ann.*, 5, 1872, 123–32 = *Ges. Abh.*, 92–102.
61. Details can be found in E. W. Hobson, *The Theory of Functions of a Real Variable*, Vol. 2, 656–98.
62. *Nachrichten König. Ges. der Wiss. zu Gött.*, 1873, 571–82.
63. *Abh. der Bayer. Akad. der Wiss.*, 12, 1876, 117–66.

species) then the series must be the Fourier series for $f(x)$. He also showed[64] that any Fourier series of a function that is Riemann integrable can be integrated term by term even though the series is not uniformly convergent.

Many men then took up the problem already answered in one way by Dirichlet, namely, to give sufficient conditions that a function $f(x)$ have a Fourier series which converges to $f(x)$. Several results are classical. Jordan gave a sufficient condition in terms of the concept of a function of bounded variation, which he introduced.[65] Let $f(x)$ be bounded in $[a, b]$ and let $a = x_0, x_1, \ldots, x_{n-1}, x_n = b$ be a mode of division (partition) of this interval. Let $y_0, y_1, \ldots, y_{n-1}, y_n$ be the values of $f(x)$ at these points. Then, for every partition

$$\sum_0^{n-1} (y_{r+1} - y_r) = f(b) - f(a)$$

let t denote

$$\sum_0^{n-1} |y_{r+1} - y_r|.$$

To every mode of subdividing $[a, b]$ there is a t. When corresponding to all possible modes of division of $[a, b]$, the sums t have a least upper bound then f is defined to be of bounded variation in $[a, b]$.

Jordan's sufficient condition states that the Fourier series for the integrable function $f(x)$ converges to

$$\frac{1}{2} [f(x + 0) + f(x - 0)]$$

at every point for which there is a neighborhood in which $f(x)$ is of bounded variation.[66]

During the 1860s and 1870s the properties of the Fourier coefficients were also examined and among the important results obtained were what is called Parseval's theorem (who stated it under more restricted conditions, Chap. 29, sec. 3) according to which if $f(x)$ and $[f(x)]^2$ are Riemann integrable in $[-\pi, \pi]$ then

$$\frac{1}{\pi} \int_{-\pi}^{\pi} [f(x)]^2 \, dx = 2a_0^2 + \sum_1^{\infty} (a_n^2 + b_n^2),$$

64. *Math. Ann.*, 22, 1883, 260–68.
65. *Comp. Rend.*, 92, 1881, 228–30 = *Œuvres*, 4, 393–95 and *Cours d'analyse*, 2, 1st ed., 1882, Ch. V.
66. *Cours d'analyse*, 2nd ed., 1893, 1, 67–72.

and if $f(x)$ and $g(x)$ and their squares are Riemann integrable then

$$\frac{1}{\pi} \int_{-\pi}^{\pi} f(x) g(x) \, dx = 2a_0\alpha_0 + \sum_{1}^{\infty} (a_n\alpha_n + b_n\beta_n)$$

where a_n, b_n, α_n, and β_n are the Fourier coefficients of $f(x)$ and $g(x)$ respectively.

7. The Status of Analysis

The work of Bolzano, Cauchy, Weierstrass, and others supplied the rigor in analysis. This work freed the calculus and its extensions from all dependence upon geometrical notions, motion, and intuitive understandings. From the outset these researches caused a considerable stir. After a scientific meeting at which Cauchy presented his theory on the convergence of series Laplace hastened home and remained there in seclusion until he had examined the series in his *Mécanique céleste*. Luckily every one was found to be convergent. When Weierstrass's work became known through his lectures, the effect was even more noticeable. The improvements in rigor can be seen by comparing the first edition of Jordan's *Cours d'analyse* (1882–87) with the second (1893–96) and the third edition (3 vols., 1909–15). Many other treatises incorporated the new rigor.

The rigorization of analysis did not prove to be the end of the investigation into the foundations. For one thing, practically all of the work presupposed the real number system but this subject remained unorganized. Except for Weierstrass who, as we shall see, considered the problem of the irrational number during the 1840s, all the others did not believe it necessary to investigate the logical foundations of the number system. It would appear that even the greatest mathematicians must develop their capacities to appreciate the need for rigor in stages. The work on the logical foundations of the real number system was to follow shortly (Chap. 41).

The discovery that continuous functions need not have derivatives, that discontinuous functions can be integrated, the new light on discontinuous functions shed by Dirichlet's and Riemann's work on Fourier series, and the study of the variety and the extent of the discontinuities of functions made the mathematicians realize that the rigorous study of functions extends beyond those used in the calculus and the usual branches of analysis where the requirement of differentiability usually restricts the class of functions. The study of functions was continued in the twentieth century and resulted in the development of a new branch of mathematics known as the theory of functions of a real variable (Chap. 44).

Like all new movements in mathematics, the rigorization of analysis did not go unopposed. There was much controversy as to whether the refinements in analysis should be pursued. The peculiar functions that were introduced were attacked as curiosities, nonsensical functions, funny functions, and as mathematical toys perhaps more intricate but of no more consequence than magic squares. They were also regarded as diseases or part of the morbid pathology of functions and as having no bearing on the important problems of pure and applied mathematics. These new functions, violating laws deemed perfect, were looked upon as signs of anarchy and chaos which mocked the order and harmony previous generations had sought. The many hypotheses which now had to be made in order to state a precise theorem were regarded as pedantic and destructive of the elegance of the eighteenth-century classical analysis, "as it was in paradise," to use Du Bois-Reymond's phrasing. The new details were resented as obscuring the main ideas.

Poincaré, in particular, distrusted this new research. He said:[67]

> Logic sometimes makes monsters. For half a century we have seen a mass of bizarre functions which appear to be forced to resemble as little as possible honest functions which serve some purpose. More of continuity, or less of continuity, more derivatives, and so forth. Indeed from the point of view of logic, these strange functions are the most general; on the other hand those which one meets without searching for them, and which follow simple laws appear as a particular case which does not amount to more than a small corner.
>
> In former times when one invented a new function it was for a practical purpose; today one invents them purposely to show up defects in the reasoning of our fathers and one will deduce from them only that.

Charles Hermite said in a letter to Stieltjes, "I turn away with fright and horror from this lamentable evil of functions which do not have derivatives."

Another kind of objection was voiced by Du Bois-Reymond.[68] His concern was that the arithmetization of analysis separated analysis from geometry and consequently from intuition and physical thinking. It reduced analysis "to a simple game of symbols where the written signs take on the arbitrary significance of the pieces in a chess or card game."

The issue that provoked the most controversy was the banishment of divergent series notably by Abel and Cauchy. In a letter to Holmboë, written in 1826, Abel says,[69]

> The divergent series are the invention of the devil, and it is a shame to base on them any demonstration whatsoever. By using them, one may draw any conclusion he pleases and that is why these series have produced so many fallacies and so many paradoxes. . . . I have become prodigiously attentive to all this, for with the exception of the geometrical series, there does not exist in all of mathematics a single infinite series the

67. *L'Enseignement mathématique*, 11, 1899, 157–62 = *Œuvres*, 11, 129–34.
68. *Théorie générale des fonctions*, 1887, 61.
69. *Œuvres*, 2, 256.

sum of which has been determined rigorously. In other words, the things which are most important in mathematics are also those which have the least foundation.

However, Abel showed some concern about whether a good idea had been overlooked because he continues his letter thus: "That most of these things are correct in spite of that is extraordinarily surprising. I am trying to find a reason for this; it is an exceedingly interesting question." Abel died young and so never did pursue the matter.

Cauchy, too, had some qualms about ostracizing divergent series. He says in the introduction of his *Cours* (1821), "I have been forced to admit diverse propositions which appear somewhat deplorable, for example, that a divergent series cannot be summed." Despite this conclusion Cauchy continued to use divergent series as in notes appended to the publication in 1827[70] of a prize paper written in 1815 on water waves. He decided to look into the question of why divergent series proved so useful and as a matter of fact he ultimately did come close to recognizing the reason (Chap. 47).

The French mathematicians accepted Cauchy's banishment of divergent series. But the English and the Germans did not. In England the Cambridge school defended the use of divergent series by appealing to the principle of permanence of form (Chap. 32, sec. 1). In connection with divergent series the principle was first used by Robert Woodhouse (1773–1827). In *The Principles of Analytic Calculation* (1803, p. 3) he points out that in the equation

(6) $$\frac{1}{1-r} = 1 + r + r^2 + \cdots$$

the equality sign has "a more extended signification" than just numerical equality. Hence the equation holds whether the series diverges or not.

Peacock, too, applied the principle of permanence of forms to operations with divergent series.[71] On page 267 he says, "Thus since for $r < 1$, (6) above holds, then for $r = 1$ we do get $\infty = 1 + 1 + 1 + \cdots$. For $r > 1$ we get a negative number on the left and, because the terms on the right continually increase, a quantity more than ∞ on the right." This Peacock accepts. The point he tries to make is that the series can represent $1/(1-r)$ for all r. He says,

> If the operations of algebra be considered as general, and the symbols which are subject to them as unlimited in value, it will be impossible to avoid the formation of divergent as well as convergent series; and if such series be considered as the results of operations which are definable, apart from the series themselves, then it will not be very important to

70. *Mém. des sav. étrangers*, 1, 1827, 3–312; see *Œuvres*, (1), 1, 238, 277, 286.

71. *Report on the Recent Progress and Present State of Certain Branches of Analysis*, Brit. Assn. for Adv. of Science, 3, 1833, 185–352.

enter into such an examination of the relation of the arithmetical values of the successive terms as may be necessary to ascertain their convergence or divergence; for under such circumstances, they must be considered as equivalent forms representing their generating function, and as possessing for the purposes of such operations, equivalent properties. . . . The attempt to exclude the use of divergent series in symbolical operations would necessarily impose a limit upon the universality of algebraic formulas and operations which is altogether contrary to the spirit of science. . . . It would necessarily lead to a great and embarrassing multiplication of cases: it would deprive almost all algebraical operations of much of their certainty and simplicity.

Augustus De Morgan, though much sharper and more aware than Peacock of the difficulties in divergent series, was nevertheless under the influence of the English school and also impressed by the results obtained by the use of divergent series despite the difficulties in them. In 1844 he began an acute and yet confused paper on "Divergent Series"[72] with these words, "I believe it will be generally admitted that the heading of this paper describes the only subject yet remaining, of an elementary character, on which a serious schism exists among mathematicians as to absolute correctness or incorrectness of results." The position De Morgan took he had already declared in his *Differential and Integral Calculus*:[73] "The history of algebra shows us that nothing is more unsound than the rejection of any method which naturally arises, on account of one or more apparently valid cases in which such a method leads to erroneous results. Such cases should indeed teach caution, but not rejection; if the latter had been preferred to the former, negative quantities, and still more their square roots, would have been an effectual bar to the progress of algebra . . . and those immense fields of analysis over which even the rejectors of divergent series now range without fear, would have been not so much as discovered, much less cultivated and settled. . . . The motto which I should adopt against a course which seems to me calculated to stop the progress of discovery would be contained in a word and a symbol—remember $\sqrt{-1}$." He distinguishes between the arithmetic and algebraic significance of a series. The algebraic significance holds in all cases. To account for some of the false conclusions obtained with divergent series he says in the 1844 paper (p. 187) that integration is an arithmetic and not an algebraic operation and so could not be applied without further thought to divergent series. But the derivation of

$$\frac{1}{1-r} = 1 + r + r^2 + \cdots$$

72. *Trans. Camb. Philo. Soc.*, 8, Part II, 1844, 182–203, pub. 1849.
73. London, 1842, p. 566.

by starting with $y = 1 + ry$, replacing y on the right by $1 + ry$, and continuing thus, he accepts because it is algebraic. Likewise from $z = 1 + 2z$ one gets $z = 1 + 2 + 4 + \cdots$. Hence $-1 = 1 + 2 + 4 + \cdots$ and this is right. He accepts the entire theory (as of that date) of trigonometric series but would be willing to reject it if one could give one instance where $1 - 1 + 1 - 1 + \cdots$ does not equal $1/2$ (see Chap. 20).

Many other prominent English mathematicians gave other kinds of justification for the acceptance of divergent series, some going back to an argument of Nicholas Bernoulli (Chap. 20, sec. 7) that the series (6) contains a remainder r^∞ or $r^\infty/(1 - r^\infty)$. This must be taken into account (though they did not indicate how). Others said the sum of a divergent series is algebraically true but arithmetically false.

Some German mathematicians used the same arguments as Peacock though they used different words, such as syntactical operations as opposed to arithmetical operations or literal as opposed to numerical. Martin Ohm[74] said, "An infinite series (leaving aside any question of convergence or divergence) is completely suited to represent a given expression if one can be sure of having the correct law of development of the series. Of the *value* of an infinite series one can speak only if it converges." Arguments in Germany in favor of the legitimacy of divergent series were advanced for several more decades.

The defense of the use of divergent series was not nearly so foolish as it might seem, though many of the arguments given in behalf of the series were, perhaps, far-fetched. For one thing in the whole of eighteenth-century analysis the attention to rigor or proof was minimal and this was acceptable because the results obtained were almost always correct. Hence mathematicians became accustomed to loose procedures and arguments. But, even more to the point, many of the concepts and operations which had caused perplexities, such as complex numbers, were shown to be correct after they were fully understood. Hence mathematicians thought that the difficulties with divergent series would also be cleared up when a better understanding was obtained and that divergent series would prove to be legitimate. Further the operations with divergent series were often bound up with other little understood operations of analysis such as the interchange of the order of limits, integration over discontinuities of an integrand and integration over an infinite interval, so that the defenders of divergent series could maintain that the false conclusions attributed to the use of divergent series came from other sources of trouble.

One argument which might have been brought up is that when an analytic function is expressed in some domain by a power series, what Weierstrass called an element, this series does indeed carry with it the

74. *Aufsätze aus dem Gebiet der höheren Mathematik* (Essays in the Domain of Higher Mathematics, 1823).

"algebraic" or "syntactical" properties of the function and these properties are carried beyond the domain of convergence of the element. The process of analytic continuation uses this fact. Actually there was sound mathematical substance in the concept of divergent series which accounted for their usefulness. But the recognition of this substance and the final acceptance of divergent series had to await a new theory of infinite series (Chap. 47).

Bibliography

Abel, N. H.: *Œuvres complètes*, 2 vols., 1881, Johnson Reprint Corp., 1964.

————: *Mémorial publié à l'occasion du centenaire de sa naissance*, Jacob Dybwad, 1902. Letters to and from Abel.

Bolzano, B.: *Rein analytischer Beweis des Lehrsatzes, dass zwischen je zwei Werthen, die ein entgegengesetztes Resultat gewähren, wenigstens eine reele Wurzel der Gleichung liege*, Gottlieb Hass, Prague, 1817 = *Abh. Königl. Böhm. Ges. der Wiss.*, (3), 5, 1814–17, pub. 1818 = *Ostwald's Klassiker der exakten Wissenschaften* #153, 1905, 3–43. Not contained in Bolzano's *Schriften*.

————: *Paradoxes of the Infinite*, Routledge and Kegan Paul, 1950. Contains a historical survey of Bolzano's work.

————: *Schriften*, 5 vols., Königlichen Böhmischen Gesellschaft der Wissenschaften, 1930–48.

Boyer, Carl B.: *The Concepts of the Calculus*, Dover (reprint), 1949, Chap. 7.

Burkhardt, H.: "Über den Gebrauch divergenter Reihen in der Zeit von 1750–1860," *Math. Ann.*, 70, 1911, 169–206.

————: "Trigonometrische Reihe und Integrale," *Encyk. der Math. Wiss.*, II A12, 819–1354, B. G. Teubner, 1904–16.

Cantor, Georg: *Gesammelte Abhandlungen* (1932), Georg Olms (reprint), 1962.

Cauchy, A. L.: *Œuvres*, (2), Gauthier-Villars, 1897–99, Vols. 3 and 4.

Dauben, J. W.: "The Trigonometric Background to Georg Cantor's Theory of Sets," *Archive for History of Exact Sciences*, 7, 1971, 181–216.

Dirichlet, P. G. L.: *Werke*, 2 vols. Georg Reimer, 1889–97, Chelsea (reprint), 1969.

Du Bois-Reymond, Paul: *Zwei Abhandlungen über unendliche und trigonometrische Reihen* (1871 and 1874), Ostwald's Klassiker #185; Wilhelm Engelmann, 1913.

Freudenthal, H.: "Did Cauchy Plagiarize Bolzano?," *Archive for History of Exact Sciences*, 7, 1971, 375–92.

Gibson, G. A.: "On the History of Fourier Series," *Proc. Edinburgh Math. Soc.*, 11, 1892/93, 137–66.

Grattan-Guinness, I.: "Bolzano, Cauchy and the 'New Analysis' of the Nineteenth Century," *Archive for History of Exact Sciences*, 6, 1970, 372–400.

————: *The Development of the Foundations of Mathematical Analysis from Euler to Riemann*, Massachusetts Institute of Technology Press, 1970.

Hawkins, Thomas W., Jr.: *Lebesgue's Theory of Integration: Its Origins and Development*, University of Wisconsin Press, 1970, Chaps. 1–3.

Manheim, Jerome H.: *The Genesis of Point Set Topology*, Macmillan, 1964, Chaps. 1–4.

Pesin, Ivan N.: *Classical and Modern Integration Theories*, Academic Press, 1970, Chap. 1.

Pringsheim, A.: "Irrationalzahlen und Konvergenz unendlichen Prozesse," *Encyk. der Math. Wiss.*, IA3, 47–147, B. G. Teubner, 1898–1904.

Reiff, R.: *Geschichte der unendlichen Reihen*, H. Lauppsche Buchhandlung, 1889; Martin Sändig (reprint), 1969.

Riemann, Bernhard: *Gesammelte mathematische Werke*, 2nd. ed. (1902), Dover (reprint), 1953.

Schlesinger, L.: "Über Gauss' Arbeiten zur Funktionenlehre," *Nachrichten König. Ges. der Wiss. zu Gött.*, 1912, Beiheft, 1–43. Also in Gauss's *Werke*, 10_2, 77 ff.

Schoenflies, Arthur M.: "Die Entwicklung der Lehre von den Punktmannig-faltigkeiten," *Jahres. der Deut. Math.-Verein.*, 8_2, 1899, 1–250.

Singh, A. N.: "The Theory and Construction of Non-Differentiable Functions," in E. W. Hobson: *Squaring the Circle and Other Monographs*, Chelsea (reprint), 1953.

Smith, David E.: *A Source Book in Mathematics*, Dover (reprint), 1959, Vol. 1, 286–91, Vol. 2, 635–37.

Stolz, O.: "B. Bolzanos Bedeutung in der Geschichte der Infinitesimalrechnung," *Math. Ann.*, 18, 1881, 255–79.

Weierstrass, Karl: *Mathematische Werke*, 7 vols., Mayer und Müller, 1894–1927.

Young, Grace C.: "On Infinite Derivatives," *Quart. Jour. of Math.*, 47, 1916, 127–75.

41

The Foundations of the Real and Transfinite Numbers

God made the integers; all else is the work of man.
LEOPOLD KRONECKER

1. Introduction

One of the most surprising facts in the history of mathematics is that the logical foundation of the real number system was not erected until the late nineteenth century. Up to that time not even the simplest properties of positive and negative rational numbers and irrational numbers were logically established, nor were these numbers defined. Even the logical foundation of complex numbers had not been long in existence (Chap. 32, sec. 1), and that foundation presupposed the real number system. In view of the extensive development of algebra and analysis, all of which utilized the real numbers, the failure to consider the precise structure and properties of the real numbers shows how illogically mathematics progresses. The intuitive understanding of these numbers seemed adequate and mathematicians were content to operate on this basis.

The rigorization of analysis forced the realization that the lack of clarity in the number system itself had to be remedied. For example, Bolzano's proof (Chap. 40, sec. 2) that a continuous function that is negative for $x = a$ and positive for $x = b$ is zero for some value of x between a and b floundered at a critical point because he lacked an adequate understanding of the structure of the real number system. The closer study of limits also showed the need to understand the real number system, for rational numbers can have an irrational limit and conversely. Cauchy's inability to prove the sufficiency of his criterion for the convergence of a sequence likewise resulted from his lack of understanding of the structure of the number system. The study of discontinuities of functions representable by Fourier series revealed the same deficiency. It was Weierstrass who first pointed out that to establish carefully the properties of continuous functions he needed the theory of the arithmetic continuum.

979

Another motivation to erect the foundations of the number system was the desire to secure the truth of mathematics. One consequence of the creation of non-Euclidean geometry was that geometry had lost its status as truth (Chap. 36, sec. 8), but it still seemed that the mathematics built on the ordinary arithmetic must be unquestionable reality in some philosophical sense. As far back as 1817, in his letter to Olbers, Gauss[1] had distinguished arithmetic from geometry in that only the former was purely a priori. In his letter to Bessel of April 9, 1830,[2] he repeats the assertion that only the laws of arithmetic are necessary and true. However, the foundation of the number system that would dispel any doubts about the truth of arithmetic and of the algebra and analysis built on that base was lacking.

It is very much worth noting that before the mathematicians appreciated that the number system itself must be analyzed, the problem that had seemed most pertinent was to build the foundations of *algebra*, and in particular to account for the fact that one could use letters to represent real and complex numbers and yet operate with letters by means of properties accepted as true for the positive integers. To Peacock, De Morgan, and Duncan Gregory, algebra in the early nineteenth century was an ingenious but also an ingenuous complex of manipulatory schemes with some rhyme but very little reason; it seemed to them that the crux of the current confusion lay in the inadequate foundation for algebra. We have already seen how these men resolved this problem (Chap. 32, sec. 1). The late nineteenth-century men realized that they must probe deeper on behalf of analysis and clarify the structure of the entire real number system. As a by-product they would also secure the logical structure of algebra, for it was already intuitively clear that the various types of numbers possessed the same formal properties. Hence if they could establish these properties on a sound foundation, they could apply them to letters that stood for any of these numbers.

2. Algebraic and Transcendental Numbers

A step in the direction of an improved understanding of irrational numbers was the mid-nineteenth-century work on algebraic and transcendental irrationals. The distinction between algebraic and transcendental irrationals had been made in the eighteenth century (Chap. 25, sec. 1). The interest in this distinction was heightened by the nineteenth-century work on the solution of equations, because this work revealed that not all algebraic irrationals could be obtained by algebraic operations on rational numbers. Moreover, the problem of determining whether e and π were algebraic or transcendental continued to attract mathematicians.

1. *Werke*, 8, 177.
2. *Werke*, 8, 201.

Up to 1844, the question of whether there were any transcendental irrationals was open. In that year Liouville[3] showed that any number of the form

$$\frac{a_1}{10} + \frac{a_2}{10^{2!}} + \frac{a_3}{10^{3!}} + \cdots,$$

where the a_i are arbitrary integers from 0 to 9, is transcendental.

To prove this, Liouville first proved some theorems on the approximation of algebraic irrationals by rational numbers. By definition (Chap. 25, sec. 1) an algebraic number is any number, real or complex, that satisfies an algebraic equation

$$a_0 x^n + a_1 x^{n-1} + \cdots + a_n = 0$$

where the a_i are integers. A root is said to be an algebraic number of degree n if it satisfies an equation of the nth degree but of no lower degree. Some algebraic numbers are rational; these are of degree one. Liouville proved that if p/q is any approximation to an algebraic irrational x of degree n, with p and q integral, then there exists a positive number M such that

$$\left| x - \frac{p}{q} \right| > \frac{M}{q^n}.$$

This means that any rational approximation to an algebraic irrational of degree n by any p/q must be less accurate than M/q^n. Alternatively we may say that if x is an algebraic irrational of degree n, there is a positive number M such that the inequality

$$\left| x - \frac{p}{q} \right| < \frac{M}{q^\mu}$$

has no solutions in integers p and q for $\mu = n$ and hence for $\mu \leq n$. Then x is transcendental if for a fixed M and for every positive integer μ the inequality has a solution p/q. By showing that his irrationals satisfy this last criterion, Liouville proved they were transcendental.

The next big step in the recognition of specific transcendental numbers was Hermite's proof in 1873[4] that e is transcendental. After obtaining this result, Hermite wrote to Carl Wilhelm Borchardt (1817–80), "I do not dare to attempt to show the transcendence of π. If others undertake it, no one will be happier than I about their success, but believe me, my dear friend, this cannot fail to cost them some effort."

That π is transcendental had already been suspected by Legendre (Chap. 25, sec. 1). Ferdinand Lindemann (1852–1939) proved this in

3. *Comp. Rend.*, 18, 1844, 910–11, and *Jour. de Math.*, (1), 16, 1851, 133–42.
4. *Comp. Rend.*, 77, 1873, 18–24, 74–79, 226–33, 285–93 = *Œuvres*, 2, 150–81.

1882[5] by a method that does not differ essentially from Hermite's. Lindemann established that if $x_1, x_2, \ldots, x_n$ are distinct algebraic numbers, real or complex, and $p_1, p_2, \ldots, p_n$ are algebraic numbers and not all zero, then the sum

$$p_1 e^{x_1} + p_2 e^{x_2} + \cdots + p_n e^{x_n}$$

cannot be 0. If we take $n = 2$, $p_1 = 1$, and $x_2 = 0$, we see that e^{x_1} cannot be algebraic for an x_1 that is algebraic and nonzero. Since x_1 can be 1, e is transcendental. Now it was known that $e^{i\pi} + 1 = 0$; hence the number $i\pi$ cannot be algebraic. Then π is not, because i is, and the product of two algebraic numbers is algebraic. The proof that π is transcendental disposed of the last point in the famous construction problems of geometry, for all constructible numbers are algebraic.

One mystery about a fundamental constant remains. Euler's constant (Chap. 20, sec. 4)

$$C = \lim_{n \to \infty} \left(1 + \frac{1}{2} + \cdots + \frac{1}{n} - \log n \right),$$

which is approximately 0.577216 and which plays an important role in analysis, notably in the study of the gamma and zeta functions, is not known to be rational or irrational.

3. The Theory of Irrational Numbers

By the latter part of the nineteenth century the question of the logical structure of the real number system was faced squarely. The irrational numbers were considered to be the main difficulty. However, the development of the meaning and properties of irrational numbers presupposes the establishment of the rational number system. The various contributors to the theory of irrational numbers either assumed that the rational numbers were so assuredly known that no foundation for them was needed or gave some hastily improvised scheme.

Curiously enough, the erection of a theory of irrationals required not much more than a new point of view. Euclid in Book V of the *Elements* had treated incommensurable ratios of magnitudes and had defined the equality and inequality of such ratios. His definition of equality (Chap. 4, sec. 5) amounted to dividing the rational numbers m/n into two classes, those for which m/n is less than the incommensurable ratio a/b of the magnitudes a and b and those for which m/n is greater. It is true that Euclid's logic was deficient because he never defined an incommensurable ratio. Moreover, Euclid's development of the theory of proportion, the equality of two incom-

5. *Math. Ann.*, 20, 1882, 213–25.

mensurable ratios, was applicable only to geometry. Nevertheless, he did have the essential idea that could have been used sooner to define irrational numbers. Actually Dedekind did make use of Euclid's work and acknowledged this debt;[6] Weierstrass too may have been guided by Euclid's theory. However, hindsight is easier than foresight. The long delay in taking advantage of some reformulation of Euclid's ideas is readily accounted for. Negative numbers had to be fully accepted so that the complete rational number system would be available. Moreover, the need for a theory of irrationals had to be felt, and this happened only when the arithmetization of analysis had gotten well under way.

In two papers, read before the Royal Irish Academy in 1833 and 1835 and published as "Algebra as the Science of Pure Time," William R. Hamilton offered the first treatment of irrational numbers.[7] He based his notion of all the numbers, rationals and irrationals, on time, an unsatisfactory basis for mathematics (though regarded by many, following Kant, as a basic intuition). After presenting a theory of rational numbers he introduced the idea of partitioning the rationals into two classes (the idea will be described more fully in connection with Dedekind's work) and defined an irrational number as such a partition. He did not complete the work.

Apart from this unfinished work all pre-Weierstrassian introductions of irrationals used the notion that an irrational is the limit of an infinite sequence of rationals. But the limit, if irrational, does not exist logically until irrationals are defined. Cantor[8] points out that this logical error escaped notice for some time because the error did not lead to subsequent difficulties. Weierstrass, in lectures at Berlin beginning in 1859, recognized the need for and gave a theory of irrational numbers. A publication by H. Kossak, *Die Elemente der Arithmetik* (1872), claimed to present this theory but Weierstrass disowned it.

In 1869 Charles Méray (1835–1911), an apostle of the arithmetization of mathematics and the French counterpart of Weierstrass, gave a definition of the irrationals based on the rationals.[9] Georg Cantor also gave a theory, which he needed to clarify the ideas on point sets that he used in his 1871 work on Fourier series. This was followed one year later by the theory of (Heinrich) Eduard Heine, which appeared in the *Journal für Mathematik*,[10] and one by Dedekind, which he published in *Stetigkeit und irrationale Zahlen*.[11]

The various theories of irrational numbers are in essence very much alike; we shall therefore confine ourselves to giving some indication of the

6. *Essays*, p. 40.
7. *Trans. Royal Irish Academy*, 17, 1837, 293–422 = *Math. Papers*, 3, 3–96.
8. *Math. Ann.*, 21, 1883, p. 566.
9. *Revue des Sociétés Savants*, 4, 1869, 280–89.
10. *Jour. für Math.*, 74, 1872, 172–88.
11. Continuity and Irrational Numbers, 1872 = *Werke*, 3, 314–34.

theories of Cantor and Dedekind. Cantor[12] starts with the rational numbers. In his 1883 paper,[13] wherein he gives more details about his theory of irrational numbers, he says (p. 565) that it is not necessary to enter into the rational numbers because this had been done by Hermann Grassmann in his *Lehrbuch der Arithmetik* (1861) and J. H. T. Muller (1797–1862) in his *Lehrbuch der allgemeinen Arithmetik* (1855). Actually these presentations did not prove definitive. Cantor introduced a new class of numbers, the real numbers, which contain rational real and irrational real numbers. He builds the real numbers on the rationals by starting with any sequence of rationals that obeys the condition that for any given ε, all the members except a finite number differ from each other by less than ε, or that

$$\lim_{n \to \infty} (a_{n+m} - a_n) = 0$$

for arbitrary m. Such a sequence he calls a fundamental sequence. Each such sequence is, by definition, a real number that we can denote by b. Two such sequences (a_ν) and (b_ν) are the same real number if and only if $|a_\nu - b_\nu|$ approaches zero as ν becomes infinite.

For such sequences three possibilities present themselves. Given any arbitrary rational number, the members of the sequence for sufficiently large ν are all smaller in absolute value than the given number; or, from a given ν on, the members are all larger than some definite positive rational member ρ; or, from a given ν on, the members are all smaller than some definite negative rational number $-\rho$. In the first case $b = 0$; in the second $b > 0$; in the third $b < 0$.

If (a_ν) and (a'_ν) are two fundamental sequences, denoted by b and b', then one can show that $(a_\nu \pm a'_\nu)$ and $(a_\nu \cdot a'_\nu)$ are fundamental sequences. These define $b \pm b'$ and $b \cdot b'$. Moreover, if $b \neq 0$, then (a'_ν/a_ν) is also a fundamental sequence that defines b'/b.

The rational real numbers are included in the above definition of real numbers, because, for example, any sequence (a_ν) with a_ν equal to the rational number a for each ν defines the rational real number a.

Now one can define the equality and inequality of any two real numbers. Indeed $b = b'$, $b > b'$, or $b < b'$, according as $b - b'$ equals 0, is greater than 0 or is less than 0.

The next theorem is crucial. Cantor proves that if (b_ν) is any sequence of real numbers (rational or irrational), and if $\lim_{\nu \to \infty} (b_{\nu+\mu} - b_\nu) = 0$ for arbitrary μ, then there is a unique real number b, determined by a fundamental sequence (a_ν) of rational a_ν such that

$$\lim_{\nu \to \infty} b_\nu = b.$$

12. *Math. Ann.*, 5, 1872, 123–32 = *Ges. Abh.*, 92–102.
13. *Math. Ann.*, 21, 1883, 545–91 = *Ges. Abh.*, 165–204.

That is, the formation of fundamental sequences of real numbers does not create the need for still newer types of numbers that can serve as limits of these fundamental sequences, because the already existing real numbers suffice to provide the limits. In other words, from the standpoint of providing limits for fundamental sequences (or what amounts to the same thing, sequences which satisfy Cauchy's criterion of convergence), the real numbers are a complete system.

Dedekind's theory of irrational numbers, presented in his book of 1872 mentioned above, stems from ideas he had in 1858. At that time he had to give lectures on the calculus and realized that the real number system had no logical foundation. To prove that a monotonically increasing quantity that is bounded approaches a limit he, like the other authors, had to resort to geometrical evidence. (He says that this is still the way to do it in the first treatment of the calculus, especially if one does not wish to lose much time.) Moreover, many basic arithmetic theorems were not proven. He gives as an example the fact that $\sqrt{2} \cdot \sqrt{3} = \sqrt{6}$ had not yet been strictly demonstrated.

He then states that he presupposes the development of the rational numbers, which he discusses briefly. To approach the irrational numbers he asks first what is meant by geometrical continuity. Contemporary and earlier thinkers—for example, Bolzano—believed that continuity meant the existence of at least one other number between any two, the property now known as denseness. But the rational numbers in themselves form a dense set. Hence denseness is not continuity.

Dedekind obtained the suggestion for the definition of an irrational number by noting that in every division of the line into two classes of points such that every point in one class is to the left of each point in the second, there is *one and only one point* that produces the division. This fact makes the line continuous. For the line it is an axiom. He carried this idea over to the number system. Let us consider, Dedekind says, any division of the rational numbers into two classes such that any number in the first class is less than any number in the second. Such a division of the rational numbers he calls a cut. If the classes are denoted by A_1 and A_2, then the cut is denoted by (A_1, A_2). From some cuts, namely, those determined by a rational number, there is either a largest number in A_1 or a smallest number in A_2. Conversely every cut in the rationals in which there is a largest number in the first class or a smallest in the second is determined by a rational number.

But there are cuts that are not determined by rational numbers. If we put into the first class all negative rational numbers and all positive ones whose squares are less than 2, and put into the second class all the other rationals, then this cut is not determined by a rational number. To each such cut "we create a new irrational member α which is fully defined by this cut; we will say that the number α corresponds to this cut or that it brings about this cut." Hence there corresponds to each cut one and only one either rational or irrational number.

Dedekind's language in introducing irrational numbers leaves a little to be desired. He introduces the irrational α as corresponding to the cut and defined by the cut. But he is not too clear about where α comes from. He should say that the irrational number α is no more than the cut. In fact Heinrich Weber told Dedekind this, and in a letter of 1888 Dedekind replied that the irrational number α is not the cut itself but is something distinct, which corresponds to the cut and which brings about the cut. Likewise, while the rational numbers generate cuts, they are *not* the same as the cuts. He says we have the mental power to create such concepts.

He then defines when one cut (A_1, A_2) is less than or greater than another cut (B_1, B_2). After having defined inequality, he points out that the real numbers possess three provable properties: (1) If $\alpha > \beta$ and $\beta > \gamma$, then $\alpha > \gamma$. (2) If α and γ are two different real numbers, then there is an infinite number of different numbers which lie between α and γ. (3) If α is any real number, then the real numbers are divided into two classes A_1 and A_2, each of which contains an infinite number of members, and each member of A_1 is less than α and each member of A_2 is greater than α. The number α itself can be assigned to either class. The class of real numbers now possesses *continuity*, which he expresses thus: If the set of all real numbers is divided into two classes A_1 and A_2 such that each member of A_1 is less than all members of A_2, then there exists one and only one number α which brings about this division.

He defines next the operations with real numbers. Addition of the cuts (A_1, A_2) and (B_1, B_2) is defined thus: If c is any rational number, then we put it in the class C_1 if there is a number a_1 in A_1 and a number b_1 in B_1 such that $a_1 + b_1 \geq c$. All other rational numbers are put in the class C_2. This pair of classes C_1 and C_2 forms a cut (C_1, C_2) because every member of C_1 is less than every member of C_2. The cut (C_1, C_2) is the sum of (A_1, A_2) and (B_1, B_2). The other operations, he says, are defined analogously. He can now establish properties such as the associative and commutative properties of addition and multiplication. Though Dedekind's theory of irrational numbers, with minor modifications such as the one indicated above, is logically satisfactory, Cantor criticized it because cuts do not appear naturally in analysis.

There are other approaches to the theory of irrational numbers beyond those already mentioned or described. For example, Wallis in 1696 had identified rational numbers and periodic decimal numbers. Otto Stolz (1842–1905), in his *Vorlesungen über allgemeine Arithmetik*,[14] showed that every irrational number can be represented as a nonperiodic decimal and this fact can be used as a defining property.

14. 1886, 1, 109–19.

It is apparent from these various approaches that the logical definition of the irrational number is rather sophisticated. Logically an irrational number is not just a single symbol or a pair of symbols, such as a ratio of two integers, but an infinite collection, such as Cantor's fundamental sequence or Dedekind's cut. The irrational number, logically defined, is an intellectual monster, and we can see why the Greeks and so many later generations of mathematicians found such numbers difficult to grasp.

Advances in mathematics are not greeted with universal approbation. Hermann Hankel, himself the creator of a logical theory of rational numbers, objected to the theories of irrational numbers.[15] "Every attempt to treat the irrational numbers formally and without the concept of [geometric] magnitude must lead to the most abstruse and troublesome artificialities, which, even if they can be carried through with complete rigor, as we have every right to doubt, do not have a higher scientific value."

4. The Theory of Rational Numbers

The next step in the erection of foundations for the number system was the definition and deduction of the properties of the rational numbers. As already noted, one or two efforts in this direction preceded the work on irrational numbers. Most of the workers on the rational numbers assumed that the nature and properties of the ordinary integers were known and that the problem was to establish logically the negative numbers and fractions.

The first such effort was made by Martin Ohm (1792–1872), a professor in Berlin and brother of the physicist, in his *Versuch eines vollkommen consequenten Systems der Mathematik* (Study of a Complete Consistent System of Mathematics, 1822). Then Weierstrass, in lectures given during the 1860s, derived the rational numbers from the natural numbers by introducing the positive rationals as couples of natural numbers, the negative integers as another type of couple of natural numbers, and the negative rationals as couples of negative integers. This idea was utilized independently by Peano, and so we shall present it in more detail later in connection with his work. Weierstrass did not feel the need to clarify the logic of the integers. Actually his theory of rational numbers was not free of difficulties. However, in his lectures from 1859 on, he did affirm correctly that once the whole numbers were admitted there was no need for further axioms to build up the real numbers.

The key problem in building up the rational number system was the founding of the ordinary integers by some process and the establishment of the properties of the integers. Among those who worked on the theory of the

15. *Theorie der complexen Zahlensystem*, 1867, p. 46–47.

integers, a few believed that the whole numbers were so fundamental that no logical analysis of them could be made. This position was taken by Kronecker, who was motivated by philosophical considerations into which we shall enter more deeply later. Kronecker too wished to arithmetize analysis, that is, to found analysis on the integers; but he thought one could not go beyond recognition of knowledge of the integers. Of these man has a fundamental intuition. "God made the integers," he said, "all else is the work of man."

Dedekind gave a theory of the integers in his *Was sind und was sollen die Zahlen*.[16] Though published in 1888, the work dates from 1872 to 1878. He used set-theoretic ideas, which by this time Cantor had already advanced, and which were to assume great importance. Nevertheless, Dedekind's approach was so complicated that it was not accorded much attention.

The approach to the integers that best suited the axiomatizing proclivities of the late nineteenth century was to introduce them entirely by a set of axioms. Giuseppe Peano (1858–1932) also accomplished this in his *Arithmetices Principia Nova Methodo Exposita* (1889).[17] Presumably he was influenced by Dedekind's 1888 work but he gave his own slant to the ideas. Since Peano's approach is very widely used we shall review it.

Peano used a great deal of symbolism because he wished to sharpen the reasoning. Thus ε means to belong to; $\supset$ means implies; N_0 means the class of natural numbers; and $a+$ denotes the next natural number after a. Peano used this symbolism in his presentation of all of mathematics, notably in his *Formulario mathematico* (5 vols., 1895–1908). He used it also in his lectures, and the students rebelled. He tried to satisfy them by passing all of them, but that did not work, and he was obliged to resign his professorship at a military academy and he remained at the University of Turin.

Though Peano's work influenced the further development of symbolic logic and the later movement of Frege and Russell to build mathematics on logic, his work must be distinguished from that of Frege and Russell. Peano did *not* wish to build mathematics on logic. To him, logic was the servant of mathematics.

Peano started with the undefined concepts (cf. Chap. 42, sec. 2) of "set," "natural numbers," "successor," and "belong to." His five axioms for the natural numbers are:

(1) 1 is a natural number.
(2) 1 is not the successor of any other natural number.
(3) Each natural number a has a successor.
(4) If the successors of a and b are equal then so are a and b.

16. The Nature and Meaning of Numbers = *Werke*, 3, 335–91.
17. *Opere scelte*, 2, 20–55, and *Rivista di Matematica*, 1, 1891, 87–102, 256–57 = *Opere scelte*, 3, 80–109.

(5) If a set S of natural numbers contains 1 and if when S contains any number a it also contains the successor of a, then S contains all the natural numbers.

This last axiom is the axiom of mathematical induction.

Peano also adopted the reflexive, symmetric, and transitive axioms for equality. That is, $a = a$; if $a = b$, then $b = a$; and if $a = b$ and $b = c$, then $a = c$. He defined addition by the statements that for each pair of natural numbers a and b there is a unique sum such that

$$a + 1 = a+$$
$$a + (b+) = (a+b)+.$$

Likewise, multiplication was defined by the statement that to each pair of natural numbers a and b there is a unique product such that

$$a \cdot 1 = a$$
$$a \cdot b+ = (a \cdot b) + a.$$

He then established all the familiar properties of natural numbers.

From the natural numbers and their properties others found it simple to define and establish the properties of the negative whole numbers and the rational numbers. One can first define the positive and negative integers as a new class of numbers, each an ordered pair of natural numbers. Thus (a, b) where a and b are natural numbers is an integer. The intuitive meaning of (a, b) is $a - b$. Hence when $a > b$, the couple represents the usual positive integer, and when $a < b$, the couple represents the usual negative integer. Suitable definitions of the operations of addition and multiplication lead to the usual properties of the positive and negative integers.

Given the integers, one introduces the rational numbers as ordered couples of integers. Thus if A and B are integers, the ordered couple (A, B) is a rational number. Intuitively (A, B) is A/B. Again suitable definitions of the operations of addition and multiplication of the couples lead to the usual properties of the rational numbers.

Thus, once the logical approach to the natural numbers was attained, the problem of building up the foundations of the real number system was completed. As we have already noted, generally the men who worked on the theory of irrationals assumed that the rational numbers were so thoroughly understood that they could be taken for granted, or made only a minor gesture toward clarifying them. After Hamilton had grounded the complex numbers on the real numbers, and after the irrationals were defined in terms of the rational numbers, the logic of this last class was finally created. The historical order was essentially the reverse of the logical order required to build up the complex number system.

5. *Other Approaches to the Real Number System*

The essence of the approaches to the logical foundations of the real number system thus far described is to obtain the integers and their properties in some manner and, with these in hand, then to derive the negative numbers, the fractions, and finally the irrational numbers. The logical base of this approach is some series of assertions concerning the natural numbers only, for example, Peano's axioms. All the other numbers are constructed. Hilbert called the above approach the genetic method (he may not have known Peano's axioms at this time but he knew other approaches to the natural numbers). He grants that the genetic method may have pedagogical or heuristic value but, he says, it is logically more secure to apply the axiomatic method to the entire real number system. Before we state his reasons, let us look at his axioms.[18]

He introduces the undefined term, number, denoted by $a, b, c, \ldots$, and then gives the following axioms:

I. *Axioms of Connection*

I_1. From the number a and the number b there arises through addition a definite number c; in symbols

$$a + b = c \quad \text{or} \quad c = a + b.$$

I_2. If a and b are given numbers, then there exists one and only one number x and also one and only one number y so that

$$a + x = b \quad \text{and} \quad y + a = b.$$

I_3. There is a definite number, denoted by 0, so that for each a

$$a + 0 = a \quad \text{and} \quad 0 + a = a.$$

I_4. From the number a and the number b there arises by another method, by multiplication, a definite number c; in symbols

$$ab = c \quad \text{or} \quad c = ab.$$

I_5. If a and b are arbitrary given numbers and a not 0, then there exists one and only one number x, and also one and only one number y such that

$$ax = b \quad \text{and} \quad ya = b.$$

I_6. There exists a definite number, denoted by 1, such that for each a we have

$$a \cdot 1 = a \quad \text{and} \quad 1 \cdot a = a.$$

18. *Jahres. der Deut. Math.-Verein.*, 8, 1899, 180–84; this article is not in Hilbert's *Gesammelte Abhandlungen*. It is in his *Grundlagen der Geometrie*, 7th ed., Appendix 6.

II. *Axioms of Calculation*

II_1. $a + (b + c) = (a + b) + c.$

II_2. $a + b = b + a.$

II_3. $a(bc) = (ab)c.$

II_4. $a(b + c) = ab + ac.$

II_5. $(a + b)c = ac + bc.$

II_6. $ab = ba.$

III. *Axioms of Order*

III_1. If a and b are any two different numbers, then one of these is always greater than the other; the latter is said to be smaller; in symbols

$$a > b \quad \text{and} \quad b < a.$$

III_2. If $a > b$ and $b > c$ then $a > c.$

III_3. If $a > b$ then it is always true that

$$a + c > b + c \quad \text{and} \quad c + a > c + b.$$

III_4. If $a > b$ and $c > 0$ then $ac > bc$ and $ca > cb.$

IV. *Axioms of Continuity*

IV_1. (Axiom of Archimedes) If $a > 0$ and $b > 0$ are two arbitrary numbers then it is always possible to add a to itself a sufficient number of times so that

$$a + a + \cdots + a > b.$$

IV_2. (Axiom of Completeness) It is not possible to adjoin to the system of numbers any collection of things so that in the combined collection the preceding axioms are satisfied; that is, briefly put, the numbers form a system of objects which cannot be enlarged with the preceding axioms continuing to hold.

Hilbert points out that these axioms are not independent; some can be deduced from the others. He then affirms that the objections against the existence of infinite sets (sec. 6) are not valid for the above conception of the real numbers. For, he says, we do not have to think about the collection of all possible laws in accordance with which the elements of a fundamental sequence (Cantor's sequences of rational numbers) can be formed. We have but to consider a closed system of axioms and conclusions that can be deduced from them by a finite number of logical steps. He does point out that it is necessary to prove the consistency of this set of axioms, but when this is done the objects defined by it, the real numbers, exist in the mathematical sense. Hilbert was not aware at this time of the difficulty of proving the consistency of axioms for real numbers.

To Hilbert's claim that his axiomatic method is superior to the genetic method, Bertrand Russell replied that the former has the advantage of theft over honest toil. It assumes at once what can be built up from a much smaller set of axioms by deductive arguments.

As in the case of almost every significant advance in mathematics, the creation of the theory of real numbers met with opposition. Du Bois-Reymond, whom we have already cited as opposing the arithmetization of analysis, said in his *Théorie générale des fonctions* of 1887, [19]

> No doubt with help from so-called axioms, from conventions, from philosophic propositions contributed *ad hoc*, from unintelligible extensions of originally clear concepts, a system of arithmetic can be constructed which resembles in every way the one obtained from the concept of magnitude, in order thus to isolate the computational mathematics, as it were, by a cordon of dogmas and defensive definitions. . . . But in this way one can also invent other arithmetic systems. Ordinary arithmetic is just the one which corresponds to the concept of linear magnitude.

Despite attacks such as this, the completion of the work on the real numbers seemed to mathematicians to resolve all the logical problems the subject had faced. Arithmetic, algebra, and analysis were by far the most extensive part of mathematics and this part was now securely grounded.

6. *The Concept of an Infinite Set*

The rigorization of analysis had revealed the need to understand the structure of sets of real numbers. To treat this problem Cantor had already introduced (Chap. 40, sec. 6) some notions about infinite sets of points, especially the sets of the first species. Cantor decided that the study of infinite sets was so important that he undertook to study infinite sets as such. This study, he expected, would enable him to distinguish clearly the different infinite sets of discontinuities.

The central difficulty in the theory of sets is the very concept of an infinite set. Such sets had naturally come to the attention of mathematicians and philosophers from Greek times onward, and their very nature and seemingly contradictory properties had thwarted any progress in understanding them. Zeno's paradoxes are perhaps the first indication of the difficulties. Neither the infinite divisibility of the straight line nor the line as an infinite set of discrete points seemed to permit rational conclusions about motion. Aristotle considered infinite sets, such as the set of whole numbers, and denied the existence of an infinite set of objects as a fixed entity. For him, sets could be only potentially infinite (Chap. 3, sec. 10).

19. Page 62, French ed. of *Die allgemeine Funktionentheorie*, 1882.

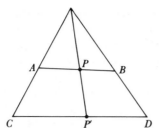

Figure 41.1

Proclus, the commentator on Euclid, noted that since a diameter of a circle divides it into halves and since there is an infinite number of diameters, there must be twice that number of halves. This seems to be a contradiction to many, Proclus says, but he resolves it by saying that one cannot speak of an actual infinity of diameters or parts of a circle. One can speak only of a larger and larger number of diameters or parts of a circle. In other words, Proclus accepted Aristotle's concept of a potential infinity but not an actual infinity. This avoids the problem of a double infinity equaling an infinity.

Throughout the Middle Ages philosophers took one side or the other on the question of whether there can be an actual infinite collection of objects. It was noted that the points of two concentric circles could be put into one-to-one correspondence with each other by associating points on a common radius. Yet one circumference was longer than the other.

Galileo struggled with infinite sets and rejected them because they were not amenable to reason. In his *Two New Sciences* (pp. 18–40 of the English translation), he notes that the points of two unequal lengths AB and CD (Fig. 41.1) can be put into one-to-one correspondence with each other and so presumably contain the same number of points. He also notes that the whole numbers can be put into one-to-one correspondence with their squares merely by assigning each number to its square. But this leads to different "amounts" of infinities, which Galileo says cannot be. All infinite quantities are the same and cannot be compared.

Gauss, in his letter to Schumacher of July 12, 1831,[20] says, "I protest against the use of an infinite quantity as an actual entity; this is never allowed in mathematics. The infinite is only a manner of speaking, in which one properly speaks of limits to which certain ratios can come as near as desired, while others are permitted to increase without bound." Cauchy, like others before him, denied the existence of infinite sets because the fact that a part can be put into one-to-one correspondence with the whole seemed contradictory to him.

The polemics on the various problems involving sets were endless and

20. *Werke*, 8, 216.

involved metaphysical and even theological arguments. The attitude of most mathematicians toward this problem was to ignore what they could not solve. On the whole they also avoided the explicit recognition of actually infinite sets, though they used infinite series, for example, and the real number system. They would speak of points of a line and yet avoid saying that the line is composed of an infinite number of points. This avoidance of troublesome problems was hypocritical, but it did suffice to build classical analysis. However, when the nineteenth century faced the problem of instituting rigor in analysis, it could no longer side-step many questions about infinite sets.

7. The Foundation of the Theory of Sets

Bolzano, in his *Paradoxes of the Infinite* (1851), which was published three years after his death, was the first to take positive steps toward a definitive theory of sets. He defended the existence of actually infinite sets and stressed the notion of equivalence of two sets, by which he meant what was later called the one-to-one correspondence between the elements of the two sets. This notion of equivalence applied to infinite sets as well as finite sets. He noted that in the case of infinite sets a part or subset could be equivalent to the whole and insisted that this must be accepted. Thus the real numbers between 0 and 5 can be put into one-to-one correspondence with the real numbers between 0 and 12 through the formula $y = 12x/5$, despite the fact that the second set of numbers contains the first set. Numbers could be assigned to infinite sets and there would be different transfinite numbers for different infinite sets, though Bolzano's assignment of transfinite numbers was incorrect according to the later theory of Cantor.

Bolzano's work on the infinite was more philosophical than mathematical and did not make sufficiently clear the notion of what was called later the power of a set or the cardinal number of a set. He, too, encountered properties that appeared paradoxical to him, and these he cites in his book. He decided that transfinite numbers were not needed to found the calculus and so did not pursue them farther.

The creator of the theory of sets is Georg Cantor (1845–1918) who was born in Russia of Danish-Jewish parentage but moved to Germany with his parents. His father urged him to study engineering and Cantor entered the University of Berlin in 1863 with that intention. There he came under the influence of Weierstrass and turned to pure mathematics. He became *Privatdozent* at Halle in 1869 and professor in 1879. When he was twenty-nine he published his first revolutionary paper on the theory of infinite sets in the *Journal für Mathematik*. Although some of its propositions were deemed faulty by the older mathematicians, its overall originality and brilliance attracted attention. He continued to publish papers on the theory of sets and on transfinite numbers until 1897.

Cantor's work which resolved age-old problems and reversed much previous thought, could hardly be expected to receive immediate acceptance. His ideas on transfinite ordinal and cardinal numbers aroused the hostility of the powerful Leopold Kronecker, who attacked Cantor's ideas savagely over more than a decade. At one time Cantor suffered a nervous breakdown, but resumed work in 1887. Even though Kronecker died in 1891, his attacks left mathematicians suspicious of Cantor's work.

Cantor's theory of sets is spread over many papers and so we shall not attempt to indicate the specific papers in which each of his notions and theorems appear. These papers are in the *Mathematische Annalen* and the *Journal für Mathematik* from 1874 on.[21] By a set Cantor meant a collection of definite and separate objects which can be entertained by the mind and to which we can decide whether or not a given object belongs. He says that those who argue for only potentially infinite sets are wrong, and he refutes the earlier arguments of mathematicians and philosophers against actually infinite sets. For Cantor a set is infinite if it can be put into one-to-one correspondence with part of itself. Some of his set-theoretic notions, such as limit point of a set, derived set, and set of the first species, were defined and used in a paper on trigonometric series;[22] these we have already described in the preceding chapter (sec. 6). A set is closed if it contains all its limit points. It is open if every point is an interior point, that is, if each point may be enclosed in an interval that contains only points of the set. A set is perfect if each point is a limit point and the set is closed. He also defined the union and intersection of sets. Though Cantor was primarily concerned with sets of points on a line or sets of real numbers, he did extend these notions of set theory to sets of points in n-dimensional Euclidean space.

He sought next to distinguish infinite sets as to "size" and, like Bolzano, decided that one-to-one correspondence should be the basic principle. Two sets that can be put into one-to-one correspondence are equivalent or have the same power. (Later the term "power" became "cardinal number.") Two sets may be unequal in power. If of two sets of objects M and N, N can be put into one-to-one correspondence with a subset of M but M cannot be put into one-to-one correspondence with a subset of N, the power of M is larger than that of N.

Sets of numbers were of course the most important, and so Cantor illustrates his notion of equivalence or power with such sets. He introduces the term "enumerable" for any set that can be put into one-to-one correspondence with the positive integers. This is the smallest infinite set. Then

21. *Ges. Abh.*, 115–356.
22. *Math. Ann.*, 5, 1872, 122–32 = *Ges. Abh.*, 92–102.

Cantor proved that the set of rational numbers is enumerable. He gave one proof in 1874.[23] However, his second proof[24] is the one now most widely used and we shall describe it.

The rational numbers are arranged thus:

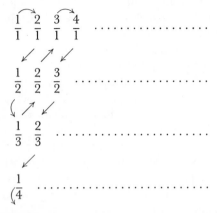

It will be noted that all those in any one diagonal have the same sum of numerator and denominator. Now one starts with 1/1 and follows the arrows assigning the number 1 to 1/1, 2 to 2/1, 3 to 1/2, 4 to 1/3, and so on. Every rational number will be reached at some stage and to each one a finite integer will be assigned. Hence the above set of rational numbers (in which some appear many times) is in one-to-one correspondence with the integers. Then if duplicates are eliminated the set of rational numbers will still be infinite and necessarily enumerable since this is the smallest infinite set.

Still more surprising is Cantor's proof in the 1874 paper just referred to that the set of all algebraic numbers, that is, the set of all numbers that are solutions of all algebraic equations

$$a_0 x^n + a_1 x^{n-1} + \cdots + a_n = 0,$$

where the a_i are integers, is also enumerable.

To prove this, he assigns to any algebraic equation of degree n the height N defined by

$$N = n - 1 + |a_0| + |a_1| + \cdots + |a_n|,$$

where the a_i are the coefficients of the equation. The height N is an integer. To each N there corresponds only a finite number of algebraic equations and hence only a finite number of algebraic numbers, say $\phi(N)$. Thus $\phi(1) = 1$; $\phi(2) = 2$; $\phi(3) = 4$. He starts with $N = 1$ and labels the corresponding algebraic numbers from 1 to n_1; the algebraic numbers that have

23. *Jour. für Math.*, 77, 1874, 258–62 = *Ges. Abh.*, 115–18.
24. *Math. Ann.*, 46, 1895, 481–512 = *Ges. Abh.*, 283–356, pp. 294–95 in particular.

height 2 are labeled from $n_1 + 1$ to n_2; and so forth. Because each algebraic number will be reached and be assigned to one and only one integer, the set of algebraic numbers is enumerable.

In his correspondence with Dedekind in 1873, Cantor posed the question of whether the set of real numbers can be put into one-to-one correspondence with the integers, and some weeks later he answered in the negative. He gave two proofs. The first (in the 1874 article just referred to) is more complicated than the second,[25] which is the one most often used today. It also has the advantage, as Cantor pointed out, of being independent of technical considerations about irrational numbers.

Cantor's second proof that the real numbers are uncountable (non-enumerable) begins by assuming that the real numbers between 0 and 1 are countable (enumerable). Let us write each as a decimal and let us agree that a number such as 1/2 will be written as .4999.... If these real numbers are countable, then we can assign each one to an integer n, thus:

$$1 \leftrightarrow 0. \ a_{11} \ a_{12} \ a_{13} \ \cdots$$
$$2 \leftrightarrow 0. \ a_{21} \ a_{22} \ a_{23} \ \cdots$$
$$3 \leftrightarrow 0. \ a_{31} \ a_{32} \ a_{33} \ \cdots$$

Now let us define a real number between 0 and 1 thus: Let $b = 0.b_1 b_2 b_3 \ldots$, where $b_k = 9$ if $a_{kk} = 1$ and $b_k = 1$ if $a_{kk} \neq 1$. This real number differs from any of those in the above correspondence. However, this was supposed to contain all the real numbers between 0 and 1. Hence there is a contradiction.

Since the real numbers are uncountable and the algebraic numbers are countable, there must be transcendental irrationals. This is Cantor's nonconstructive existence proof, which should be compared with Liouville's actual construction of transcendental irrationals (sec. 2).

In 1874 Cantor occupied himself with the equivalence of the points of a line and the points of R^n (n-dimensional space) and sought to prove that a one-to-one correspondence between these two sets was impossible. Three years later he proved that there is such a correspondence. He wrote to Dedekind,[26] "I see it but I do not believe it."

The idea[27] used to set up this one-to-one correspondence can be exhibited readily if we set up such a correspondence between the points of the unit square and the points of the segment $(0, 1)$. Let (x, y) be a point of the unit square and z a point of the unit interval. Let x and y be represented by infinite decimals so that in a finite decimal terminating in zero, we replace the 0 by an infinite sequence of 9's. We now break up x and y

25. *Jahres. der Deut. Math.-Verein.*, 1, 1890/91, 75–78 = *Ges. Abh.*, 278–81.
26. *Briefwechsel Cantor-Dedekind*, p. 34.
27. *Jour. für Math.*, 84, 1878, 242–58 = *Ges. Abh.*, 119–33.

into groups of decimals, each group ending with the first nonzero digit encountered. Thus

$$x = .3\ 002\ 03\ 04\ 6\ \cdots$$
$$y = .01\ 6\ 07\ 8\ 09\ \cdots.$$

Form

$$z = .3\ 01\ 002\ 6\ 03\ 07\ 04\ 8\ 6\ 09\ \cdots$$

by choosing as the groups in z the first group from x, then the first from y, and so forth. If two x's or y's differ in some digit then the corresponding z's will differ. Hence to each (x, y) there is a unique z. Given a z, one breaks up its decimal representation into the groups just described and forms the x and y by reversing the above process. Again two different z's will yield two different pairs (x, y) so that to each z there is a unique (x, y). The one-to-one correspondence just described is not continuous; roughly stated, this means that neighboring z-points do not necessarily go into neighboring (x, y)-points, nor conversely.

Du Bois-Reymond objected to this proof.[28] "It appears repugnant to common sense. The fact is that this is simply the conclusion of a type of reasoning which allows the intervention of idealistic fictions, wherein one lets them play the role of genuine quantities even though they are not even limits of representations of quantities. This is where the paradox resides."

8. *Transfinite Cardinals and Ordinals*

Having demonstrated the existence of sets with the same power and different powers, Cantor pursued this concept of the power of a set and introduced a theory of cardinal and ordinal numbers in which the transfinite cardinals and ordinals are the striking elements. Cantor developed this work in a series of papers in the *Mathematische Annalen* from 1879 to 1884, all under the title "Über unendliche lineare Punktmannichfaltigkeiten" (On Infinite Linear Aggregates of Points). Then he wrote two definitive papers in 1895 and 1897 in the same journal.[29]

In the fifth paper on linear aggregates[30] Cantor opens with the observation,

> The description of my investigations in the theory of aggregates has reached a stage where their continuation has become dependent on a generalization of real positive integers beyond the present limits; a

28. Page 167, French ed. (1887) of his *Die allgemeine Funktionentheorie* (1882).
29. *Math. Ann.*, 46, 1895, 481–512, and 49, 1897, 207–46 = *Ges. Abh.*, 282–351; an English translation of these two papers may be found in Georg Cantor, *Contributions to the Founding of a Theory of Transfinite Numbers*, Dover (reprint), no date.
30. 1883, *Ges. Abh.*, 165.

generalization which takes a direction in which, as far as I know, nobody has yet looked.

I depend on this generalization of the number concept to such an extent that without it I could not freely take even small steps forward in the theory of sets. I hope that this situation justifies or, if necessary, excuses the introduction of seemingly strange ideas into my arguments. In fact the purpose is to generalize or extend the series of real integers beyond infinity. Daring as this might appear, I express not only the hope but also the firm conviction that in due course this generalization will be acknowledged as a quite simple, appropriate, and natural step. Still I am well aware that by adopting such a procedure I am putting myself in opposition to widespread views regarding infinity in mathematics and to current opinions on the nature of number.

He points out that his theory of infinite or transfinite numbers is distinct from the concept of infinity wherein one speaks of a variable becoming infinitely small or infinitely large. Two sets which are in one-to-one correspondence have the same power or cardinal number. For finite sets the cardinal number is the usual number of objects in the set. For infinite sets new cardinal numbers are introduced. The cardinal number of the set of whole numbers he denoted by $\aleph_0$. Since the real numbers cannot be put into one-to-one correspondence with the whole numbers, the set of real numbers must have another cardinal number which is denoted by c, the first letter of continuum. Just as in the case of the concept of power, if two sets M and N are such that N can be put into one-to-one correspondence with a subset of M but M cannot be put into one-to-one correspondence with a subset of N, the cardinal number of M is greater than that of N. Then $c > \aleph_0$.

To obtain a cardinal number larger than a given one,[31] one considers any set M which the second cardinal number represents and considers the set N of all the subsets of M. Among the subsets of M are the individual elements of M, all pairs of elements of M, and so on. Now it is certainly possible to set up a one-to-one correspondence between M and a subset of N because one subset of N consists of all the individual members of N (each regarded as a set and as a member of N), and these members are the very members of M. It is not possible to set up a one-to-one correspondence between M and all the members of N. For suppose such a one-to-one correspondence were set up. Let m be any member of M and let us consider all the m's such that the subsets (members) of N that are associated with m's of M do *not* contain the m to which they correspond under this supposed one-to-one correspondence. Let η be the set of all these m's. η is of course a member of N. Cantor affirms that η is not included in the supposed one-to-one correspondence. For if η corresponded to some m of M, and η contained m,

31. Cantor, *Ges. Abh.*, 278–80.

that would be a contradiction of the very definition of η. On the other hand if η did not contain m, it should because η was by definition the set of all m's which were not contained in the corresponding subsets of N. Hence the assumption that there is a one-to-one correspondence between the elements m of M and of N, which consists of all the subsets of M, leads to a contradiction. Then the cardinal number of the set consisting of all the subsets of a given set is larger than that of the given set.

Cantor defined the sum of two cardinal numbers as the cardinal number of the set that is the union of the (disjoined) sets represented by the summands. Cantor also defined the product of any two cardinal numbers. Given two cardinal numbers α and β, one takes a set M represented by α and a set N represented by β and forms the pairs of elements (m, n) where m is any element of M and n of N. Then the cardinal number of the set of all possible pairs is the product of α and β.

Powers of cardinal numbers are also defined. If we have a set M of m objects and a set N of n objects, Cantor defines the set m^n, and this amounts to the set of permutations of m objects n at a time with repetitions of the m objects permitted. Thus if $m = 3$ and $n = 2$ and we have m_1, m_2, m_3, the permutations are

$$m_1 m_1 \quad m_2 m_2 \quad m_3 m_1$$

$$m_1 m_2 \quad m_2 m_1 \quad m_3 m_2$$

$$m_1 m_3 \quad m_2 m_3 \quad m_3 m_3.$$

Cantor defines the cardinal number of this number of permutations to be α^β where α is the cardinal number of M and β that of N. He then proves that $2^{\aleph_0} = c$.

Cantor calls attention to the fact that his theory of cardinal numbers applies in particular to finite cardinals, and so he has given "the most natural, shortest, and most rigorous foundation for the theory of finite numbers."

The next concept is that of ordinal number. He has already found the need for this concept in his introduction of successive derived sets of a given point set. He now introduces it abstractly. A set is simply ordered if any two elements have a definite order so that given m_1 and m_2 either m_1 precedes m_2 or m_2 precedes m_1; the notation is $m_1 < m_2$ or $m_2 < m_1$. Further if $m_1 < m_2$ and $m_2 < m_3$, then simple order also implies $m_1 < m_3$; that is, the order relationship is transitive. An ordinal number of an ordered set M is the order type of the order in the set. Two ordered sets are similar if there is a one-to-one correspondence between them and if, when m_1 corresponds to n_1 and m_2 to n_2 and $m_1 < m_2$, then $n_1 < n_2$. Two similar sets have the same type or ordinal number. As examples of ordered sets, we may use any finite set of numbers in any given order. For a finite set, no matter what the order is, the ordinal number is the same and the symbol for it can be taken to be the cardinal number of the set of numbers in the set. The ordinal number of the set of positive integers in their natural order is denoted by ω. On the other hand, the set of positive integers in decreasing order, that is

$$\ldots, 4, 3, 2, 1$$

is denoted by $*\omega$. The set of positive and negative integers and zero in the usual order has the ordinal number $*\omega + \omega$.

Then Cantor defines the addition and multiplication of ordinal numbers. The sum of two ordinal numbers is the ordinal number of the first ordered set plus the ordinal number of the second ordered set taken in that specific order. Thus the set of positive integers followed by the first five integers, that is,

$$1, 2, 3, \cdots, 1, 2, 3, 4, 5,$$

has the ordinal number $\omega + 5$. Also the equality and inequality of ordinal numbers is defined in a rather obvious way.

And now he introduces the full set of transfinite ordinals, partly for their own value and partly to define precisely higher transfinite cardinal numbers. To introduce these new ordinals he restricts the simply ordered sets to well-ordered sets.[32] A set is well-ordered if it has a first element in the ordering and if every subset has a first element. There is a hierarchy of ordinal numbers and cardinal numbers. In the first class, denoted by Z_1, are the finite ordinals

$$1, 2, 3, \cdots.$$

In the second class, denoted by Z_2, are the ordinals

$$\omega, \omega + 1, \omega + 2, \cdots, 2\omega, 2\omega + 1, \cdots, 3\omega, 3\omega + 1, \cdots, \omega^2, \omega^3, \cdots, \omega^\omega, \cdots$$

Each of these ordinals is the ordinal of a set whose cardinal number is $\aleph_0$.

The set of ordinals in Z_2 has a cardinal number. The set is not enumerable and so Cantor introduces a new cardinal number $\aleph_1$ as the cardinal number of the set Z_2. $\aleph_1$ is then shown to be the next cardinal after $\aleph_0$.

The ordinals of the third class, denoted by Z_3, are

$$\Omega, \Omega + 1, \Omega + 2, \cdots, \Omega + \Omega, \cdots.$$

These are the ordinal numbers of well-ordered sets, each of which has $\aleph_1$ elements. However, the set of ordinals Z_3 has more than $\aleph_1$ elements, and Cantor denotes the cardinal number of the set Z_3 by $\aleph_2$. This hierarchy of ordinals and cardinals can be continued indefinitely.

32. *Math. Ann.*, 21, 1883, 545–86 = *Ges. Abh.*, 165–204.

Now Cantor had shown that given any set, it is always possible to create a new set, the set of subsets of the given set, whose cardinal number is larger than that of the given set. If $\aleph_0$ is the given set, then the cardinal number of the set of subsets is $2^{\aleph_0}$. Cantor had proved that $2^{\aleph_0} = c$, where c is the cardinal number of the continuum. On the other hand he introduced $\aleph_1$ through the ordinal numbers and proved that $\aleph_1$ is the next cardinal after $\aleph_0$. Hence $\aleph_1 \leq c$, but the question as to whether $\aleph_1 = c$, known as the continuum hypothesis, Cantor, despite arduous efforts, could not answer. In a list of outstanding problems presented at the International Congress of Mathematicians in 1900, Hilbert included this (Chap. 43, sec. 5; see also Chap. 51, sec. 8).

For general sets M and N it is possible that M cannot be put into one-to-one correspondence with any subset of N and N cannot be put into one-to-one correspondence with a subset of M. In this case, though M and N have cardinal numbers α and β, say, it is not possible to say that $\beta = \alpha$, $\alpha < \beta$, or $\alpha > \beta$. That is, the two cardinal numbers are not comparable. For well-ordered sets Cantor was able to prove that this situation cannot arise. It seemed paradoxical that there should be non-well-ordered sets whose cardinal numbers cannot be compared. But this problem, too, Cantor could not solve.

Ernst Zermelo (1871–1953) took up the problem of what to do about the comparison of the cardinal numbers of sets that are not well-ordered. In 1904[33] he proved, and in 1908[34] gave a second proof, that every set can be well-ordered (in some rearrangement). To make the proof he had to use what is now known as the axiom of choice (Zermelo's axiom), which states that given any collection of nonempty, disjoined sets, it is possible to choose just one member from each set and so make up a new set. The axiom of choice, the well-ordering theorem, and the fact that any two sets may be compared as to size (that is, if their cardinal numbers are α and β, either $\alpha = \beta$, $\alpha < \beta$, or $\alpha > \beta$) are equivalent principles.

9. The Status of Set Theory by 1900

Cantor's theory of sets was a bold step in a domain that, as already noted, had been considered intermittently since Greek times. It demanded strict application of purely rational arguments, and it affirmed the existence of infinite sets of higher and higher power, which are entirely beyond the grasp of human intuition. It would be singular if these ideas, far more revolutionary than most others previously introduced, had not met with opposition. The doubts as to the soundness of this development were rein-

33. *Math. Ann.*, 59, 1904, 514–16.
34. *Math. Ann.*, 65, 1908, 107–28.

forced by questions raised by Cantor himself and by others. In letters to Dedekind of July 28 and August 28, 1899,[35] Cantor asked whether the set of all cardinal numbers is itself a set, because if it is it would have a cardinal number larger than any other cardinal. He thought he answered this in the negative by distinguishing between consistent and inconsistent sets. However, in 1897 Cesare Burali-Forti (1861–1931) pointed out that the sequence of *all* ordinal numbers, which is well-ordered, should have the greatest of all ordinal numbers as its ordinal number.[36] Then this ordinal number is greater than *all* the ordinal numbers. (Cantor had already noted this difficulty in 1895.) These and other unresolved problems, called paradoxes, were beginning to be noted by the end of the nineteenth century.

The opposition did make itself heard. Kronecker, as we have already observed, opposed Cantor's ideas almost from the start. Felix Klein was by no means in sympathy with them. Poincaré[37] remarked critically, "But it has happened that we have encountered certain paradoxes, certain apparent contradictions which would have pleased Zeno of Elea and the school of Megara. . . . I think for my part, and I am not the only one, that the important point is never to introduce objects that one cannot define completely in a finite number of words." He refers to set theory as an interesting "pathological case." He also predicted (in the same article) that "Later generations will regard [Cantor's] *Mengenlehre* as a disease from which one has recovered." Hermann Weyl spoke of Cantor's hierarchy of alephs as a fog on a fog.

However, many prominent mathematicians were impressed by the uses to which the new theory had already been put. At the first International Congress of Mathematicians in Zurich (1897), Adolf Hurwitz and Hadamard indicated important applications of the theory of transfinite numbers to analysis. Additional applications were soon made in the theory of measure (Chap. 44) and topology (Chap. 50). Hilbert spread Cantor's ideas in Germany, and in 1926[38] said, "No one shall expel us from the paradise which Cantor created for us." He praised Cantor's transfinite arithmetic as "the most astonishing product of mathematical thought, one of the most beautiful realizations of human activity in the domain of the purely intelligible."[39] Bertrand Russell described Cantor's work as "probably the greatest of which the age can boast."

35. *Ges. Abh.*, 445–48.
36. *Rendiconti del Circolo Matematico di Palermo*, 11, 1897, 154–64 and 260.
37. *Proceedings of the Fourth Internat. Cong. of Mathematicians*, Rome, 1908, 167–82; *Bull. des Sci. Math.*, (2), 32, 1908, 168–90 = extract in *Œuvres*, 5, 19–23.
38. *Math. Ann.*, 95, 1926, 170 = *Grundlagen der Geometrie*, 7th ed., 1930, 274.
39. *Math. Ann.*, 95, 1926, 167 = *Grundlagen der Geometrie*, 7th ed., 1930, 270. The article "Über das Unendliche," from which the above quotations were taken, appears also in French in *Acta Math.*, 48, 1926, 91–122. It is not included in Hilbert's *Gesammelte Abhandlungen*.

Bibliography

Becker, Oskar: *Grundlagen der Mathematik in geschichtlicher Entwicklung*, Verlag Karl Alber, 1954, pp. 217–316.

Boyer, Carl B.: *A History of Mathematics*, John Wiley and Sons, 1968, Chap. 25.

Cantor, Georg: *Gesammelte Abhandlungen*, 1932, Georg Olms (reprint), 1962.

————: *Contributions to the Founding of the Theory of Transfinite Numbers*, Dover (reprint), no date. This contains an English translation of Cantor's two key papers of 1895 and 1897 and a very helpful introduction by P. E. B. Jourdain.

Cavaillès, Jean: *Philosophie mathématique*, Hermann, 1962. Also contains the Cantor-Dedekind correspondence translated into French.

Dedekind, R.: *Essays on the Theory of Numbers*, Dover (reprint), 1963. Contains an English translation of Dedekind's "Stetigkeit und irrationale Zahlen" and "Was sind und was sollen die Zahlen." Both essays are in Dedekind's *Werke*, 3, 314–34 and 335–91.

Fraenkel, Abraham A.: "Georg Cantor," *Jahres. der Deut. Math.-Verein.*, 39, 1930, 189–266. A historical account of Cantor's work.

Helmholtz, Hermann von: *Counting and Measuring*, D. Van Nostrand, 1930. English translation of Helmholtz's *Zählen und Messen, Wissenschaftliche Abhandlungen*, 3, 356–91.

Manheim, Jerome H.: *The Genesis of Point Set Topology*, Macmillan, 1964, pp. 76–110.

Meschkowski, Herbert: *Ways of Thought of Great Mathematicians*, Holden-Day, 1964, pp. 91–104.

————: *Evolution of Mathematical Thought*, Holden-Day, 1965, Chaps. 4–5.

————: *Probleme des Unendlichen: Werk und Leben Georg Cantors*, F. Vieweg und Sohn, 1967.

Noether, E. and J. Cavaillès: *Briefwechsel Cantor-Dedekind*, Hermann, 1937.

Peano G.: *Opere scelte*, 3 vols., Edizioni Cremonese, 1957–59.

Schoenflies, Arthur M.: *Die Entwickelung der Mengenlehre und ihre Anwendungen*, two parts, B. G. Teubner, 1908, 1913.

Smith, David Eugene: *A Source Book in Mathematics*, Dover (reprint), 1959, Vol. 1, pp. 35–45, 99–106.

Stammler, Gerhard: *Der Zahlbegriff seit Gauss*, Georg Olms, 1965.

42
The Foundations of Geometry

> Geometry is nothing if it be not rigorous . . . . The methods
> of Euclid are, by almost universal consent, unexceptional in
> point of rigor.   H. J. S. SMITH (1873)

> It has been customary when Euclid, considered as a textbook,
> is attacked for his verbosity or his obscurity or his pedantry,
> to defend him on the ground that his logical excellence is
> transcendent, and affords an invaluable training to the
> youthful powers of reasoning. This claim, however, vanishes
> on a close inspection. His definitions do not always define,
> his axioms are not always indemonstrable, his demonstrations
> require many axioms of which he is quite unconscious. A
> valid proof retains its demonstrative force when no figure is
> drawn, but very many of Euclid's earlier proofs fail before
> this test. . . . The value of his work as a masterpiece of logic
> has been very grossly exaggerated.
>   BERTRAND RUSSELL (1902)

1. The Defects in Euclid

Criticism of Euclid's definitions and axioms (Chap. 4, sec. 10) dates back to
the earliest known commentators, Pappus and Proclus. When the Europeans
were first introduced to Euclid during the Renaissance, they too noted
flaws. Jacques Peletier (1517–82), in his *In Euclidis Elementa Geometrica
Demonstrationum* (1557), criticized Euclid's use of superposition to prove
theorems on congruence. Even the philosopher Arthur Schopenhauer said
in 1844 that he was surprised that mathematicians attacked Euclid's parallel
postulate rather than the axiom that figures which coincide are equal. He
argued that coincident figures are automatically identical or equal and hence
no axiom is needed, or coincidence is something entirely empirical, which
belongs not to pure intuition (*Anschauung*) but to external sensuous experience.
Moreover, the axiom presupposes the mobility of figures; but that which is
movable in space is matter, and hence outside geometry. In the nineteenth
century it was generally recognized that the method of superposition either

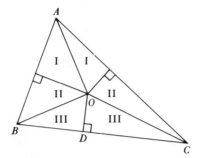

Figure 42.1

rested on unstated axioms or should be replaced by another approach to congruence.

Some critics did not like as an axiom the statement that all right angles are equal and sought to prove it, of course on the basis of the other axioms. Christophorus Clavius (1537–1612), an editor of Euclid's work, noted the absence of an axiom guaranteeing the existence of a fourth proportional to three given magnitudes (Chap. 4, sec. 5). Rightly Leibniz commented that Euclid relied upon intuition when he asserted (Book 1, Proposition 1) that two circles, each of which passes through the center of the other, have a point in common. Euclid assumed, in other words, that a circle is some kind of continuous structure and so must have a point where it is cut by another circle.

The shortcomings in Euclid's presentation of geometry were also noted by Gauss. In a letter to Wolfgang Bolyai of March 6, 1832,[1] Gauss pointed out that to speak of a part of a plane inside a triangle calls for the proper foundation. He also says, "In a complete development such words as 'between' must be founded on clear concepts, which can be done, but which I have not found anywhere." Gauss made additional criticisms of the definition of the straight line[2] and of the definition of the plane as a surface in which a line joining any two points of the plane must lie.[3]

It is well known that many "proofs" of false results can be made because Euclid's axioms do not dictate where certain points must lie in relation to others. There is, for example, the "proof" that every triangle is isoceles. One constructs the angle bisector at A of triangle ABC and the perpendicular bisector of side BC (Fig. 42.1). If these two lines are parallel, the angle bisector is perpendicular to BC and the triangle is isosceles. We suppose, then, that the lines meet at O, say, and we shall still "show" that the triangle is isosceles. We now draw the perpendiculars OF to AB and OE to AC.

1. *Werke*, 8, 222.
2. *Werke*, 8, 196.
3. *Werke*, 8, 193–95 and 200.

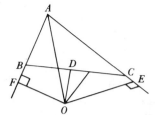

Figure 42.2

Then the triangles marked I are congruent, and $OF = OE$. The triangles marked III are also congruent, and $OB = OC$. Consequently the triangles marked II are congruent, and $FB = EC$. From the triangles marked I we have $AF = AE$. Then $AB = AC$ and the triangle is isosceles.

One might question the position of the point O, and indeed one can show that it must lie outside the triangle on the circumscribed circle. If, however, one draws Figure 42.2, it can again be "proved" that triangle ABC is isosceles. The flaw is that of the two points E and F one must lie inside and the other outside of the respective sides of the triangle. But this means that we must be able to determine the correct position of F with respect to A and B and E with respect to A and C before starting the proof. Of course one should not rely upon drawing a correct figure to determine the locations of E and F, but this was precisely what Euclid and the mathematicians up to 1800 did. Euclidean geometry was supposed to have offered accurate proofs of theorems suggested intuitively by figures, but actually it offered intuitive proofs of accurately drawn figures.

Though criticisms of the logical structure of Euclid's *Elements* were launched almost from the time it was written, they were not widely known or the defects were regarded as minor. The *Elements* was generally taken to be the model of rigor. However, the work on non-Euclidean geometry made mathematicians aware of the full extent of the deficiencies in Euclid's structure, for in carrying out proofs they had to be especially critical of what they were accepting. The recognition of the many deficiencies finally obliged the mathematicians to undertake the reconstruction of the foundations of Euclidean geometry and of other geometries that contained the same weaknesses. This activity became extensive in the last third of the nineteenth century.

2. Contributions to the Foundations of Projective Geometry

In the 1870s the work on projective geometry in relation to the metric geometries revealed that projective geometry is the fundamental one (Chap. 38). Perhaps for this reason, the foundational work began with projective geometry. However, nearly all of the writers were equally concerned to build up the metric geometries, either on the basis of projective geometry or independently. Hence the books and papers of the late nineteenth and early twentieth centuries dealing with the foundations of geometry cannot be segregated under distinct topics.

The work on non-Euclidean geometry had brought the realization that geometries are man-made constructions bearing upon physical space but not necessarily exact idealizations of it. This fact implied that several major changes had to be incorporated in any axiomatic approach to geometry. These were recognized and stressed by Moritz Pasch (1843–1930), who was the first to make major contributions to the foundations of geometry. His *Vorlesungen über neuere Geometrie* (1st ed., 1882; 2nd ed., revised by Max Dehn, 1926) is a groundbreaking work.

Pasch noted that Euclid's common notions, such as point and line, were really not defined. Defining a point as that which has no parts means little, for what is the meaning of parts? In fact, Pasch pointed out, as had Aristotle and a few later mathematicians such as Peacock and Boole, that some concepts must be undefined or else either the process of definition would be unending or mathematics would rest on physical concepts. Once some undefined concepts are selected, others can be defined in terms of these. Thus in geometry point, line, plane (Pasch also used, in his first edition, congruence of line segments) might be chosen as the undefined terms. The choice is not unique. Since there are undefined terms, the question arises as to what properties of these concepts can be used to make proofs about them. The answer given by Pasch is that the axioms make assertions about the undefined terms and these are the only assertions about them that we may use. As Gergonne put it as far back as 1818,[4] the undefined concepts are defined implicitly by the axioms.

As for the axioms, Pasch continues, though some may be suggested by experience, once the set is selected it must be possible to execute all proofs without further reference to experience or to the physical meaning of the concepts. Moreover, the axioms are by no means self-evident truths but just assumptions designed to yield the theorems of any particular geometry. He says in his *Vorlesungen* (2nd ed., p. 90),

> . . . if geometry is to become a genuine deductive science, it is essential that the way in which inferences are made should be altogether independent of the *meaning* of the geometrical concepts, and also of the diagrams; all that need be considered are the relationships between the geometrical concepts asserted by the propositions and definitions. In the course of deduction it is both advisable and useful to bear in mind the meaning of the geometrical concepts used, but this is *in no way essential*; in fact it is

4. *Ann. de Math.*, 9, 1818/19, 1.

precisely when this becomes necessary that a gap occurs in the deduction and (when it is not possible to supply the deficiency by modifying the reasoning) we are forced to admit the inadequacy of the propositions invoked as the means of proof.

Pasch did believe that the concepts and axioms should bear on experience, but logically this was irrelevant.

In his *Vorlesungen* Pasch gave axioms for projective geometry, but many of the axioms or their analogues were equally important for the axiomatization of Euclidean and the non-Euclidean geometries when built up as independent subjects. Thus he was first to give a set of axioms for the order of points on a line (or the concept of betweenness). Such axioms must also be incorporated in a complete set for any one of the metric geometries. We shall see below what the order axioms amount to.

His method of building projective geometry was to add the point, line, and plane at infinity to the proper points, lines, and planes. Then he introduced coordinates (on a geometric basis), using the throw construction of von Staudt and Klein (Chap. 35, sec. 3), and finally the algebraic representation of projective transformations. The non-Euclidean and Euclidean geometries were introduced as special cases on a geometric basis by distinguishing the proper and improper lines and points à la Felix Klein.

A more satisfactory approach to projective geometry was given by Peano.[5] This was followed by the work of Mario Pieri (1860–1904), "I Principii della geometria di posizione";[6] Federigo Enriques (1871–1946, *Lezioni di geometria proiettiva*, 1898); Eliakim Hastings Moore (1862–1932);[7] Friedrich H. Schur (1856–1932);[8] Alfred North Whitehead (1861–1947, *The Axioms of Projective Geometry*);[9] and Oswald Veblen (1880–1960) and John W. Young (1879–1932).[10] The last two men gave a completely independent set. The classic text, Veblen and Young's *Projective Geometry* (2 vols., 1910 and 1918), carries out Klein's organization of geometry by starting with projective geometry on a strict axiomatic basis and then specializing this geometry by choosing different absolute quadrics (Chap. 38, sec. 3) to obtain Euclidean and the several non-Euclidean geometries. Their axioms are general enough to include geometries with a finite number of points, geometries with only rational points, and geometries with complex points.

One more point about many of the axiomatic systems for projective

5. *I Principii di geometria*, Fratelli Bocca, Torino, 1889 = *Opere scelte*, 2, 56–91.

6. *Memorie della Reale Accademia delle Scienze di Torino*, (2), 48, 1899, 1–62.

7. *Amer. Math. Soc. Trans.*, 3, 1902, 142–58.

8. *Math. Ann.*, 55, 1902, 265–92.

9. Cambridge University Press, 1906.

10. *Amer. Jour. of Math.*, 30, 1908, 347–78.

geometry and those we shall look at in a moment for Euclidean geometry
is worth noting. Some of Euclid's axioms are existence axioms (Chap. 4,
sec. 3). To guarantee the logical existence of figures, the Greeks used construction with line and circle. The nineteenth-century foundational work
revised the notion of existence, partly to supply deficiencies in Euclid's
handling of this topic, and partly to broaden the notion of existence so that
Euclidean geometry could include points, lines, and angles not necessarily
constructible with line and circle. We shall see what the new kind of existence
axioms amount to in the systems we are about to examine.

3. The Foundations of Euclidean Geometry

In his *Sui fondamenti della geometria* (1894), Giuseppe Peano gave a set of
axioms for Euclidean geometry. He, too, stressed that the basic elements are
undefined. He laid down the principle that there should be as few undefined
concepts as possible and he used point, segment, and motion. The inclusion
of motion seems somewhat surprising in view of the criticism of Euclid's
use of superposition; however, the basic objection is not to the concept of
motion but to the lack of a proper axiomatic basis if it is to be used. A similar
set was given by Peano's pupil Pieri[11] who adopted point and motion as
undefined concepts. Another set, using line, segment, and congruence of
segments as undefined elements, was given by Giuseppe Veronese (1854–
1917) in his *Fondamenti di geometria* (1891).

The system of axioms for Euclidean geometry that seems simplest in
its concepts and statements, hews closest to Euclid's, and has gained most
favor is due to Hilbert, who did not know the work of the Italians. He gave
the first version in his *Grundlagen der Geometrie* (1899) but revised the set
many times. The following account is taken from the seventh (1930) edition
of this book. In his use of undefined concepts and the fact that their properties
are specified solely by the axioms, Hilbert follows Pasch. No explicit meaning
need be assigned to the undefined concepts. These elements, point, line,
plane, and others, could be replaced, as Hilbert put it, by tables, chairs,
beer mugs, or other objects. Of course, if geometry deals with "things," the
axioms are certainly not self-evident truths but must be regarded as arbitrary
even though, in fact, they are suggested by experience.

Hilbert first lists his undefined concepts. They are point, line, plane,
lie on (a relation between point and line), lie on (a relation between point
and plane), betweenness, congruence of pairs of points, and congruence of
angles. The axiom system treats plane and solid Euclidean geometry in one
set and the axioms are broken up into groups. The first group contains
axioms on existence:

11. *Rivista di Matematica* 4, 1894, 51–90 = Opere scelte 3, 115–57.

I. Axioms of Connection

I_1. To each two points A and B there is one line a which lies on A and B.

I_2. To each two points A and B there is not more than one line which lies on A and B.

I_3. On each line there are at least two points. There exist at least three points which do not lie on one line.

I_4. To any three points A, B, and C which do not lie on one line there is a plane α which lies on [contains] these three points. On each plane there is [at least] one point.

I_5. To any three points A, B, and C not on one line there is not more than one plane containing these three points.

I_6. If two points of a line lie on a plane α then every point of the line lies on α.

I_7. If two planes α and β have a point A in common, then they have at least one more point B in common.

I_8. There are at least four points not lying on the same plane.

The second group of axioms supplies the most serious omission in Euclid's set, namely, axioms about the relative order of points and lines:

II. Axioms of Betweenness

II_1. If a point B lies between points A and C then A, B, and C are three different points on one line and B also lies between C and A.

II_2. To any two points A and C there is at least one point B on the line AC such that C lies between A and B.

II_3. Among any three points on a line there is not more than one which lies between the other two.

Axioms II_2 and II_3 amount to making the line infinite.

Definition. Let A and B be two points of a line a. The pair of points A, B or B, A is called the segment AB. The points between A and B are called points of the segment or points interior to the segment. A and B are called endpoints of the segment. All other points of line a are said to be outside the segment.

II_4. (Pasch's Axiom) Let A, B, and C be the three points not on one line and let a be any line in the plane of A, B, and C but which does not go through (lie on) A, B, or C. If a goes through a point of the segment

AB then it must also go through a point of the segment AC or one of the segment BC.

III. *Axioms of Congruence*

III$_1$. If A, B are two points of a line a and A' is a point of a or another line a', then on a given side (previously defined) of A' on the line a' one can find a point B' such that the segment AB is congruent to $A'B'$. In symbols $AB \equiv A'B'$.

III$_2$. If $A'B'$ and $A''B''$ are congruent to AB then $A'B' \equiv A''B''$.

This axiom limits Euclid's "Things equal to the same thing are equal to each other" to line segments.

III$_3$. Let AB and BC be segments without common interior points on a line a and let $A'B'$ and $B'C'$ be segments without common interior points on a line a'. If $AB \equiv A'B'$ and $BC \equiv B'C'$, then $AC \equiv A'C'$.

This amounts to Euclid's "Equals added to equals gives equals" applied to line segments.

III$_4$. Let $\angle(h, k)$ lie in a plane α and let a line a' lie in a plane α' and a definite side of a' in α' be given. Let h' be a ray of a' which emanates from the point O'. Then in α' there is one and only one ray k' such that $\angle(h, k)$ is congruent to $\angle(h', k')$ and all inner points of $\angle(h', k')$ lie on the given side of a'. Each angle is congruent to itself.

III$_5$. If for two triangles ABC and $A'B'C'$ we have $AB \equiv A'B'$, $AC \equiv A'C'$, and $\angle BAC \equiv \angle B'A'C'$ then $\angle ABC \equiv \angle A'B'C'$.

This last axiom can be used to prove that $\angle ACB = \angle A'C'B'$. One considers the same two triangles and the same hypotheses. However, by taking first $AC \equiv A'C'$ and then $AB \equiv A'B'$, we are entitled to conclude that $\angle ACB \equiv \angle A'C'B'$ because the wording of the axiom applied to the new order of the hypotheses yields this new conclusion.

IV. *The Axiom on Parallels*

Let a be a line and A a point not on a. Then in the plane of a and A there exists at most one line through A which does not meet a.

The existence of at least one line through A which does not meet a can be proved and hence is not needed in this axiom.

V. *Axioms of Continuity*

V$_1$. (Axiom of Archimedes) If AB and CD are any two segments, then there exists on the line AB a number of points $A_1, A_2, \ldots, A_n$ such that the segments $AA_1, A_1A_2, A_2A_3, \ldots, A_{n-1}A_n$ are congruent to CD and such that B lies between A and A_n.

V_2. (Axiom of Linear Completeness) The points of a line form a collection of points which, satisfying axioms I_1, I_2, II, III, and V_1, cannot be extended to a larger collection which continues to satisfy these axioms.

This axiom amounts to requiring enough points on the line so that the points can be put into one-to-one correspondence with the real numbers. Though this fact had been used consciously and unconsciously since the days of coordinate geometry, the logical basis for it had not previously been stated.

With these axioms Hilbert proved some of the basic theorems of Euclidean geometry. Others completed the task of showing that all of Euclidean geometry does follow from the axioms.

The arbitrary character of the axioms of Euclidean geometry, that is, their independence of physical reality, brought to the fore another problem, the consistency of this geometry. As long as Euclidean geometry was regarded as the truth about physical space, any doubt about its consistency seemed pointless. But the new understanding of the undefined concepts and axioms required that the consistency be established. The problem was all the more vital because the consistency of the non-Euclidean geometries had been reduced to that of Euclidean geometry (Chap. 38, sec. 4). Poincaré brought this matter up in 1898[12] and said that we could believe in the consistency of an axiomatically grounded structure if we could give it an arithmetic interpretation. Hilbert proceeded to show that Euclidean geometry is consistent by supplying such an interpretation.

He identifies (in the case of plane geometry) point with the ordered pair of real numbers[13] (a, b) and line with the ratio $(u:v:w)$ in which u and v are not both 0. A point lies on a line if

$$ua + vb + w = 0.$$

Congruence is interpreted algebraically by means of the expressions for translation and rotation of analytic geometry; that is, two figures are congruent if one can be transformed into the other by translation, reflection in the x-axis, and rotation.

After every concept has been interpreted arithmetically and it is clear that the axioms are satisfied by the interpretation, Hilbert's argument is that the theorems must also apply to the interpretation because they are logical consequences of the axioms. If there were a contradiction in Euclidean geometry, then the same would hold of the algebraic formulation of geometry,

12. *Monist*, 9, 1898, p. 38.
13. Strictly he uses a more limited set of real numbers.

which is an extension of arithmetic. Hence *if arithmetic is consistent,* so is Euclidean geometry. The consistency of arithmetic remained open at this time (see Chap. 51).

It is desirable to show that no one of the axioms can be deduced from some or all of the others of a given set, for if it can be deduced there is no need to include it as an axiom. This notion of independence was brought up and discussed by Peano in the 1894 paper already referred to, and even earlier in his *Arithmetices Principia* (1889). Hilbert considered the independence of his axioms. However, it is not possible in his system to show that each axiom is independent of all the others because the meaning of some of them depends upon preceding ones. What Hilbert did succeed in showing was that all the axioms of any one group cannot be deduced from the axioms of the other four groups. His method was to give consistent interpretations or models that satisfy the axioms of four groups but do not satisfy all the axioms of the fifth group.

The proofs of independence have a special bearing on non-Euclidean geometry. To establish the independence of the parallel axiom, Hilbert gave a model that satisfies the other four groups of axioms but does not satisfy the Euclidean parallel axiom. His model uses the points interior to a Euclidean sphere and special transformations which take the boundary of the sphere into itself. Hence the parallel axiom cannot be a consequence of the other four groups because if it were, the model as a part of Euclidean geometry would possess contradictory properties on parallelism. This same proof shows that non-Euclidean geometry is possible because if the Euclidean parallel axiom is independent of the other axioms, a denial of this axiom must also be independent; for if it were a consequence, the full set of Euclidean axioms would contain a contradiction.

Hilbert's system of axioms for Euclidean geometry, which first appeared in 1899, attracted a great deal of attention to the foundations of Euclidean geometry and many men gave alternative versions using different sets of undefined elements or variations in the axioms. Hilbert himself, as we have already noted, kept changing his system until he gave the 1930 version. Among the numerous alternative systems we shall mention just one. Veblen[14] gave a set of axioms based on the undefined concepts of point and order. He showed that each of his axioms is independent of the others, and he also established another property, namely, categoricalness. This notion was first clearly stated and used by Edward V. Huntington (1874–1952) in a paper devoted to the real number system.[15] (He called the notion sufficiency.) A set of axioms $P_1, P_2, \ldots, P_n$ connecting a set of undefined symbols $S_1, S_2, \ldots, S_m$ is said to be categorical if between the elements of any two assemblages, each of which contains undefined symbols and satisfies the

14. *Amer. Math. Soc. Trans.,* 5, 1904, 343–84.
15. *Amer. Math. Soc. Trans.,* 3, 1902, 264–79.

axioms, it is possible to set up a one-to-one correspondence of the undefined concepts which is preserved by the relationships asserted by the axioms; that is, the two systems are isomorphic. In effect categoricalness means that all interpretations of the axiom system differ only in language. This property would not hold if, for example, the parallel axiom were omitted, because then Euclidean and hyperbolic non-Euclidean geometry would be nonisomorphic interpretations of the reduced set of axioms.

Categoricalness implies another property which Veblen called disjunctive and which is now called completeness. A set of axioms is called complete if it is impossible to add another axiom that is independent of the given set and consistent with the given set (without introducing new primitive concepts). Categoricalness implies completeness, for if a set A of axioms were categorical but not complete it would be possible to introduce an axiom S such that S and not-S are consistent with the set A. Then, since the original set A is categorical, there would be isomorphic interpretations of A together with S and A together with not-S. But this would be impossible because the corresponding propositions in the two interpretations must hold, whereas S would apply to one interpretation and not-S to the other.

4. Some Related Foundational Work

The clear delineation of the axioms for Euclidean geometry suggested the corresponding investigations for the several non-Euclidean geometries. One of the nice features of Hilbert's axioms is that the axioms for hyperbolic non-Euclidean geometry are obtained at once by replacing the Euclidean parallel axiom by the Lobatchevsky-Bolyai axiom. All the other axioms in Hilbert's system remain the same.

To obtain axioms for either single or double elliptic geometry, one must not only abandon the Euclidean parallel axiom in favor of an axiom to the effect that any two lines have one point in common (single elliptic) or at least one in common (double elliptic), but one must also change other axioms. The straight line of these geometries is not infinite but has the properties of a circle. Hence one must replace the order axioms of Euclidean geometry by order axioms that describe the order relations of points on a circle. Several such axiom systems have been given. George B. Halsted (1853–1922), in his *Rational Geometry*,[16] and John R. Kline (1891–1955)[17] have given axiomatic bases for double elliptic geometry; and Gerhard Hessenberg (1874–1925)[18] gave a system of axioms for single elliptic geometry.

16. 1904, pp. 212–47.
17. *Annals of Math.*, (2), 18, 1916/17, 31–44.
18. *Math. Ann.*, 61, 1905, 173–84.

Another class of investigations in the foundations of geometry is to consider the consequences of denying or just omitting one or more of a set of axioms. Hilbert in his independence proofs had done this himself, for the essence of such a proof is to construct a model or interpretation that satisfies all of the axioms except the one whose independence is to be established. The most significant example of an axiom that was denied is, of course, the parallel axiom. Interesting results have come from dropping the axiom of Archimedes, which can be stated as in Hilbert's V_1. The resulting geometry is called non-Archimedean; in it there are segments such that the multiple of one by any whole number, however large, need not exceed another. In *Fondamenti di geometria*, Giuseppe Veronese constructed such a geometry. He also showed that the theorems of this geometry approximate as closely as one wishes those of Euclidean geometry.

Max Dehn (1878–1952) also obtained[19] many interesting theorems by omitting the Archimedean axiom. For example, there is a geometry in which the angle sum is two right angles, similar but noncongruent triangles exist, and an infinity of straight lines that are parallel to a given line may be drawn through a given point.

Hilbert pointed out that the axiom of continuity, axiom V_2, need not be used in constructing the theory of areas in the plane. But for space Max Dehn proved[20] the existence of polyhedra having the same volume though not decomposable into mutually congruent parts (even after the addition of congruent polyhedra). Hence in three dimensions the axiom of continuity is needed.

The foundation of Euclidean geometry was approached in an entirely different manner by some mathematicians. Geometry, as we know, had fallen into disfavor because mathematicians found that they had unconsciously accepted facts on an intuitive basis, and their supposed proofs were consequently incomplete. The danger that this would continually recur made them believe that the only sound basis for geometry would be arithmetic. The way to erect such a basis was clear. In fact, Hilbert had given an arithmetic interpretation of Euclidean geometry. What had to be done now, for plane geometry say, was not to interpret point as the pair of numbers (x, y) but to *define* point to *be* the pair of numbers, to define a line as a ratio of three numbers $(u:v:w)$, to define the point (x, y) as being on the line (u, v, w) if and only if $ux + vy + w = 0$, to define circle as the set of all (x, y) satisfying the equation $(x - a)^2 + (y - b)^2 = r^2$, and so on. In other words, one would use the analytic geometry equivalents of purely geometric notions as the definitions of the geometrical concepts and algebraic methods to prove theorems. Since analytic geometry contains in algebraic form the

19. *Math. Ann.*, 53, 1900, 404–39.
20. *Math. Ann.*, 55, 1902, 465–78.

complete counterpart of all that exists in Euclidean geometry, there was no question that the arithmetic foundations could be obtained. Actually the technical work involved had really been done, even for n-dimensional Euclidean geometry, for example by Grassmann in his *Calculus of Extension*; and Grassmann himself proposed that this work serve as the foundation for Euclidean geometry.

5. Some Open Questions

The critical investigation of geometry extended beyond the reconstruction of the foundations. Curves had of course been used freely. The simpler ones, such as the ellipse, had secure geometrical and analytical definitions. But many curves were introduced only through equations and functions. The rigorization of analysis had included not only a broadening of the concept of function but the construction of very peculiar functions, such as continuous functions without derivatives. That the unusual functions were troublesome from the geometric standpoint is readily seen. Thus the curve representing Weierstrass's example of a function which is continuous everywhere but is differentiable nowhere certainly did not fit the usual concept, because the lack of a derivative means that such a curve cannot have a tangent anywhere. The question that arose was, Are the geometrical representations of such functions curves? More generally, what is a curve?

Jordan gave a definition of a curve.[21] It is the set of points represented by the continuous functions $x = f(t)$, $y = g(t)$, for $t_0 \leqslant t \leqslant t_1$. Such a curve is now called a Jordan curve. For some purposes Jordan wished to restrict his curves so that they did not possess multiple points. He therefore required that $f(t) \neq f(t')$ and $g(t) \neq g(t')$ for t and t' in (t_0, t_1) or that to each (x, y) there is one t.

It was in this work that he added the notion of closed curve,[22] which requires merely that $f(t_0) = f(t_1)$ and $g(t_0) = g(t_1)$, and stated the theorem that a closed curve divides the plane into two parts, an inside and an outside. Two points of the same region can always be joined by a polygonal path that does not cut the curve. Two points not in the same region cannot be joined by any polygonal line or continuous curve that does not cut the simple closed curve. The theorem is more powerful than it seems at first sight because a simple closed curve can be quite crinkly in shape. In fact, since the functions $f(t)$ and $g(t)$ need be only continuous, the full variety of complicated continuous functions is involved. Jordan himself and many distinguished mathematicians gave incorrect proofs of the theorem. The first rigorous proof is due to Veblen.[23]

21. *Cours d'analyse*, 1st ed. Vol. 3, 1887, 593; 2nd ed., Vol. 1, 1893, p. 90.
22. First ed., p. 593; 2nd ed., p. 98.
23. *Amer. Math. Soc. Trans.*, 6, 1905, 83–98, and 14, 1913, 65–72.

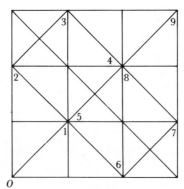

Figure 42.3   O

However, Jordan's definition of a curve, though satisfactory for many applications, was too broad. In 1890 Peano[24] discovered that a curve meeting Jordan's definition can run through all the points of a square at least once. Peano gave a detailed arithmetic description of the correspondence between the points of the interval [0, 1] and the points of the square which, in effect, specifies two functions $x = f(t)$, $y = g(t)$ that are single-valued and continuous for $0 \leq t \leq 1$ and such that x and y take on values belonging to each point of the unit square. However, the correspondence of (x, y) to t is not single-valued, nor is it continuous. A one-to-one continuous correspondence from the t-values to the (x, y)-values is impossible; that is, both $f(t)$ and $g(t)$ cannot be continuous. This was proved by Eugen E. Netto (1846–1919).[25]

The geometrical interpretation of Peano's curve was given by Arthur M. Schoenflies (1853–1928)[26] and E. H. Moore.[27] One maps the line segment [0, 1] into the nine segments shown in Figure 42.3, and then within each subsquare breaks up the segment contained there into the same pattern but making the transition from one subsquare to the next one a continuous one. The process is repeated *ad infinitum*, and the limiting point set covers the original square. Ernesto Cesàro (1859–1906)[28] gave the analytical form of Peano's f and g.

Hilbert[29] gave another example of the continuous mapping of a unit segment onto the square. Divide the unit segment (Fig. 42.4) and the square into four equal parts, thus:

24. *Math. Ann.*, 36, 1890, 157–60 = *Opere scelte*, 1, 110–15.
25. *Jour. für Math.*, 86, 1879, 263–68.
26. *Jahres. der Deut. Math.-Verein.*, 8_2, 1900, 121–25.
27. *Amer. Math. Soc. Trans.*, 1, 1900, 72–90.
28. *Bull. des Sci. Math.*, (2), 21, 1897, 257–66.
29. *Math. Ann.*, 38, 1891, 459–60 = *Ges. Abh.*, 3, 1–2.

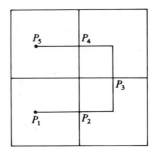

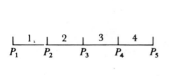

Figure 42.4

Travel through each subsquare so that the path shown corresponds to the unit segment. Now divide the unit square into 16 subsquares numbered as shown in Figure 42.5 and join the centers of the 16 subsquares as shown.

We continue the process of dividing each subsquare into four parts, numbering them so that we can traverse the entire set by a continuous path. The desired curve is the limit of the successive polygonal curves formed at each stage. Since the subsquares and the parts of the unit segment both contract to a point as the subdivision continues, we can see intuitively that each point on the unit segment maps into one point on the square. In fact, if we fix on one point in the unit segment, say $t = 2/3$, then the image of this point is the limit of the successive images of $t = 2/3$ which appear in the successive polygons.

These examples show that the definition of a curve Jordan suggested is not satisfactory because a curve, according to this definition, can fill out a square. The question of what is meant by a curve remained open. Felix Klein remarked in 1898[30] that nothing was more obscure than the notion of a curve. This question was taken up by the topologists (Chap. 50, sec. 2).

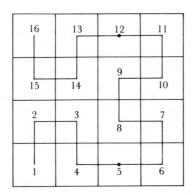

Figure 42.5

30. *Math. Ann.*, 50, 1898, 586.

Beyond the problem of what is meant by a curve, the extension of analysis to functions without derivatives also raised the question of what is meant by the length of a curve. The usual calculus formula

$$L = \int_a^b (1 + y'^2)^{1/2} \, dx,$$

where $y = f(x)$, calls, at the very least, for the existence of the derivative. Hence, the concept no longer applies to the non-differentiable functions. Various efforts to generalize the concept of length of curve were made by Du Bois-Reymond, Peano, Ludwig Scheeffer (1859–85), and Jordan, using either generalized integral definitions or geometric concepts. The most general definition was formulated in terms of the notion of measure, which we shall examine in Chapter 44.

A similar difficulty was noted for the concept of area of a surface. The concept favored in the texts of the nineteenth century was to inscribe in the surface a polyhedron with triangular faces. The limit of the sum of the areas of these triangles when the sides approach 0 was taken to be the area of the surface. Analytically, if the surface is represented by

$$x = \phi(u, v), \qquad y = \psi(u, v), \qquad z = \chi(u, v)$$

then the formula for surface area becomes

$$\int_D \sqrt{A^2 + B^2 + C^2} \, du \, dv,$$

where A, B, and C are the Jacobians of y and z, x and z, and x and y, respectively. Again the question arose of what the definition should be if x, y, and z do not possess derivatives. To complicate the situation, in a letter to Hermite, H. A. Schwarz gave an example[31] in which the choice of the triangles leads to an infinite surface area even for any ordinary cylinder.[32] The theory of surface area was also reconsidered in terms of the notion of measure.

By 1900 no one had proved that every closed plane curve, as defined by Jordan and Peano, encloses an area. Helge von Koch (1870–1924)[33] complicated the area problem by giving an example of a continuous but non-differentiable curve with infinite perimeter which bounds a finite area. One starts with the equilateral triangle ABC (Fig. 42.6) of side $3s$. On the middle third of each side construct an equilateral triangle of side s and delete the base of each triangle. There will be three such triangles. Then on each *outside* segment of length s of the new figure construct on each middle

31. *Ges. Math. Abh.*, 2, pp. 309–11.
32. This example can be found in James Pierpont, *The Theory of Functions of Real Variables*, Dover (reprint), 1959, Vol. 2, p. 26.
33. *Acta Math.*, 30, 1906, 145–76.

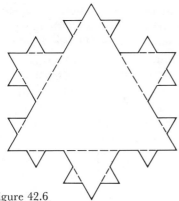

Figure 42.6

third an equilateral triangle of side $s/3$ and delete the base of each triangle. There will be twelve such triangles. Then on the outside segments of the resulting figure, construct a new equilateral triangle of side $s/9$. There will be 48 of these triangles. The perimeters of the successive figures are $9s$, $12s$, $16s$, . . . and these perimeters become infinite. However, the area of the limiting figure is finite. For, by the well-known formula for the area of an equilateral triangle in terms of its side, namely, if the side is b, the area is $(b^2/4)\sqrt{3}$, then the area of the original triangle is $[(3s)^2/4]\sqrt{3}$. The area of the first three triangles added is $3 \cdot (s^2/4)\sqrt{3}$. Since the side of the next triangles added is $s/3$ and there are 12, these areas are $12(s/3)^2\sqrt{3}/4 = (s^2/3)\sqrt{3}$. The sum of the areas is

$$S = \frac{9s^2}{4}\sqrt{3} + \frac{3}{4}s^2\sqrt{3} + \frac{s^2}{3}\sqrt{3} + \frac{4s^2}{27}\sqrt{3} + \cdots.$$

This is an infinite geometric progression (aside from the first term) with common ratio $4/9$. Then

$$S = \frac{9s^2}{4}\sqrt{3} + \frac{(3/4)s^2\sqrt{3}}{1 - 4/9} = \frac{18}{5}s^2\sqrt{3}.$$

The Peano and Hilbert curves also raised the question of what we mean by dimension. The square is two-dimensional in itself, but, as the continuous image of a curve, should be one-dimensional. Moreover, Cantor had shown that the points of a line segment can be put into one-to-one correspondence with the points of a square (Chap. 41, sec. 7). Though this correspondence is not continuous from the line segment to the square or the other way, it did show that dimension is not a matter of multiplicity of points. Nor is it a matter of the number of coordinates needed to fix the position of a point, as Riemann and Helmholtz had thought, because the Peano curve assigns a unique (x, y) to each value of t.

In the light of these difficulties we see that the rigorization of geometry certainly did not answer all the questions that were raised. Many were settled by the topologists and analysts of the next century. The very fact that questions about fundamental concepts continued to arise illustrates once more that mathematics does not grow as a logical structure. Advances into new fields and even the perfection of old ones reveal new and unsuspected defects. Beyond the resolution of the problems involving curves and surfaces we have yet to see whether the ultimate stage in rigor was reached by the foundational work in analysis, the real number system, and basic geometry.

Bibliography

Becker, Oskar: *Grundlagen der Mathematik in geschichtlicher Entwicklung*, Karl Alber, 1954, 199–212.

Enriques, Federigo: "Prinzipien der Geometrie," *Encyk. der Math. Wiss.*, B. G. Teubner, 1907–10, III AB1, 1–129.

Hilbert, David: *Grundlagen der Geometrie*, 7th ed., B. G. Teubner, 1930.

Pasch, M. and M. Dehn: *Vorlesungen über neuere Geometrie*, 2nd ed., Julius Springer, 1926, 185–271.

Peano, Giuseppe: *Opere scelte*, 3 vols., Edizioni Cremonese, 1957–59.

Reichardt, Hans: *C. F. Gauss, Leben und Werke*, Haude und Spenersche, 1960, 111–50.

Schmidt, Arnold: "Zu Hilberts Grundlegung der Geometrie," in Hilbert's *Gesammelte Abhandlungen*, 2, 404–14.

Singh, A. N.: "The Theory and Construction of Non-Differentiable Functions," in E. W. Hobson: *Squaring the Circle and Other Monographs*, Chelsea (reprint), 1953.

43
Mathematics as of 1900

> I have not hesitated in 1900, at the Congress of Mathematicians in Paris, to call the nineteenth century the century of the theory of functions.   VITO VOLTERRA

1. The Chief Features of the Nineteenth-Century Developments

It was true in the nineteenth as in the two preceding centuries that the progress in mathematics brought with it larger changes barely perceptible in the year-to-year developments but vital in themselves and in their effect on future developments. The vast expansion in subject matter and the opening of new fields as well as the extension of older ones are of course apparent. Algebra received a totally new impetus with Galois; geometry again became active and was radically altered with the introduction of non-Euclidean geometry and the revival of projective geometry; number theory flowed over into analytic number theory; and analysis was immeasurably broadened by the introduction of complex function theory and the expansion in ordinary and partial differential equations. From the standpoint of technical development, complex function theory was the most significant of the new creations. But from the standpoint of intellectual importance and ultimate effect on the nature of mathematics, the most consequential development was non-Euclidean geometry. As we shall see, its effects were far more revolutionary than we have thus far pointed out. The circle within which mathematical studies appeared to be enclosed at the beginning of the century was broken at all points, and mathematics exploded into a hundred branches. The flood of new results contradicted sharply the leading opinion at the end of the eighteenth century that the mine of mathematics was exhausted.

Mathematical activity during the nineteenth century expanded in other respects. The number of mathematicians increased enormously as a consequence of the democratization of learning. Though Germany, France, and Britain were the major centers, Italy reappears in the arena, and the United States, with Benjamin Peirce, G. W. Hill, and Josiah Willard Gibbs, enters for the first time. In 1863 the United States founded the National Academy of Sciences. However, unlike the Royal Society of London, the

Academy of Sciences of Paris, and the Academy of Sciences of Berlin, the National Academy has not been a scientific meeting place at which papers have been presented and reviewed. It does publish a journal, the *Proceedings of the Academy*. More mathematical societies were organized (Chap. 26, sec. 6) for the meeting of research men, the presentation of papers, and the sponsorship of journals. By the end of the century the number of journals devoted partly or entirely to mathematical research increased to about 950. In 1897 the practice of holding an international congress every four years was begun.

Accompanying the explosion of mathematical activity was a less healthy development. The many disciplines became autonomous, each featuring its own terminology and methodology. The pursuit of any one imposed the assumption of more specialized and more difficult problems, requiring more and more ingenious ideas, rich inspirations, and less perspicuous proofs. To make progress, mathematicians had to acquire a great deal of background in theory and technical facility. Specialization became apparent in the work of Abel, Jacobi, Galois, Poncelet, and others. Though some stress was laid on interrelationships among the many branches through such notions as groups, linear transformations, and invariance, the overall effect was a segregation into numerous distinct and unrelated divisions. It did seem to Felix Klein in 1893 that the specialization and divergence of the various branches could be overcome by means of the concepts just mentioned, but the hope was vain. Cauchy and Gauss were the last men to know the subject as a whole, though Poincaré and Hilbert were almost universal men.

From the nineteenth century on one finds mathematicians who work only in small corners of mathematics; quite naturally, each rates the importance of his area above all others. His publications are no longer for a large public but for particular colleagues. Most articles no longer contain any indication of their connection with the larger problems of mathematics, are hardly accessible to many mathematicians, and are certainly not palatable to a large circle.

Beyond its achievements in subject matter, the nineteenth century reintroduced rigorous proof. No matter what individual mathematicians may have thought about the soundness of their results, the fact is that from about 200 B.C. to about 1870 almost all of mathematics rested on an empirical and pragmatic basis. The concept of deductive proof from explicit axioms had been lost sight of. It is one of the astonishing revelations of the history of mathematics that this ideal of the subject was, in effect, ignored during the two thousand years in which its content expanded so extensively. Though some earlier efforts to rigorize analysis were made, notably by Lagrange (Chap. 19, sec. 7), the more characteristic note was sounded by Lacroix (Chap. 26, sec. 3). Fourier's work makes a modern analyst's hair stand on end; and as far as Poisson was concerned, the derivative and integral were just shorthand for the difference quotient and a finite sum. The movement to shore up the foundations, initiated by Bolzano and Cauchy, undoubtedly arose from the concern for the rapidly increasing mass of mathematics that rested on the loose foundations of the calculus. The movement was accelerated by Hamilton's discovery of the non-commutative quaternions, which surely challenged the uncritically accepted principles of number. But even more disturbing was the creation of non-Euclidean geometry. Not only did this destroy the very notion of the self-evidence of axioms and their too-superficial acceptance, but the work revealed inadequacies in proofs that had been regarded as the soundest in all of mathematics. The mathematicians realized that they had been gullible and had relied upon intuition.

By 1900 the goal of establishing mathematics rigorously seemed to have been achieved, and the mathematicians were almost smug about this accomplishment. In his address before the Second International Congress in Paris,[1] Poincaré boasted, "Have we at last attained absolute rigor? At each stage of its evolution our forerunners believed they too had attained it. If they were deceived are we not deceived like them? . . . Now in analysis today, if we care to take pains to be rigorous, there are only syllogisms or appeals to the intuition of pure number that can not possibly deceive us. One may say today that absolute rigor has been attained." When one considers the key results in the foundations of the number system and geometry and the erection of analysis on the basis of the number system, one can see reason for gloating. Mathematics now had the foundation that almost all men were happy to accept.

The precise formulation of basic concepts such as the irrational number, continuity, integral, and derivative was not greeted enthusiastically by all mathematicians. Many did not understand the new ε–δ language and regarded the precise definitions as fads, unnecessary for the comprehension of mathematics or even for rigorous proof. These men felt that intuition was good enough, despite the surprises of continuous functions without derivatives, space-filling curves, and curves without length. Emile Picard said, apropos of the rigor in partial differential equations, ". . . true rigor is productive, being distinguished in this from another rigor which is purely formal and tiresome, casting a shadow over the problems it touches."[2]

Despite the fact that geometry too had been rigorized, one consequence of the rigorization movement was that number and analysis took precedence over geometry. The recognition by mathematicians, during and after the creation of non-Euclidean geometry, that they had unconsciously relied upon intuition in accepting the proofs of Euclidean geometry made them fearful that they would continue to do so in all geometric reasoning. Hence they preferred a mathematics built upon number. Many favored going further

1. *Comp. Rendu du Deuxième Congrès Internat. des Math.*, 1900, pub. 1902, pp. 121–22.
2. *Amer. Math. Soc. Bull.*, 11, 1904/05, 417; see also Chap. 40, sec. 7, and Chap. 41, secs. 5 and 9.

and building up all of geometry on number, which could be done through analytic geometry in the manner already described. Thus most mathematicians spoke of the arithmetization of mathematics, though it would have been more accurate to speak of the arithmetization of analysis. Where Plato could say that "God eternally geometrizes," Jacobi, even by the middle of the century, said, "God ever arithmetizes." At the Second International Congress Poincaré asserted, "Today there remain in analysis only integers and finite and infinite systems of integers, interrelated by a net of relations of equality or inequality. Mathematics, as we say, has been arithmetized." Pascal had said, "*Tout ce qui passe la Géométrie nous passe.*"[3] In 1900 the mathematicians preferred to say, "*Tout ce qui passe l'Arithmétique nous passe.*"

The erection of the logical foundations of mathematics, quite apart from whether arithmetic or geometry was the preferred basis, completed another step in the break from metaphysics. The vagueness of the foundations and justifications of mathematical steps had been evaded in the eighteenth and early nineteenth centuries by allusions to metaphysical arguments which, though never made explicit, were mentioned as grounds for accepting the mathematics. The axiomatization of the real numbers and geometry gave mathematics a clear-cut, independent, and self-sufficient basis. The recourse to metaphysics was no longer needed. As Lord Kelvin put it, "Mathematics is the only good metaphysics."

The rigorization of mathematics may have filled a nineteenth-century need, but it also teaches us something about the development of the subject. The newly founded logical structure presumably guaranteed the soundness of mathematics; but the guarantee was somewhat of a sham. Not a theorem of arithmetic, algebra, or Euclidean geometry was changed as a consequence, and the theorems of analysis had only to be more carefully formulated. In fact, all that the new axiomatic structures and rigor did was substantiate what mathematicians knew had to be the case. Indeed the axioms had to yield the existing theorems rather than determine them. All of which means that mathematics rests not on logic but on sound intuitions. Rigor, as Jacques Hadamard pointed out, merely sanctions the conquests of the intuition; or, as Hermann Weyl stated, logic is the hygiene the mathematician practices to keep his ideas healthy and strong.

2. The Axiomatic Movement

The rigorization of mathematics was achieved by axiomatizing the various branches. The essence of an axiomatic development, in accordance with the pattern we have examined in Chapters 41 and 42, is to start with undefined terms whose properties are specified by the axioms; the goal of the work is to

3. "All that transcends geometry transcends our comprehension."

derive the consequences of the axioms. In addition the independence, consistency, and categoricalness of the axioms (notions we have already examined in the two preceding chapters) are to be established for each system.

By the early part of the twentieth century the axiomatic method not only permitted the establishment of the logical foundations of many old and newer branches of mathematics, but also revealed precisely what assumptions underlie each branch and made possible the comparison and clarification of the relationships of various branches. Hilbert was enthusiastic about the values of this method. In discussing the perfect state mathematics had presumably attained by founding each of its branches on sound axiomatic bases, Hilbert remarked,[4]

> Indeed the axiomatic method is and remains the one suitable and indispensable aid to the spirit of every exact investigation no matter in what domain; it is logically unassailable and at the same time fruitful; it guarantees thereby complete freedom of investigation. To proceed axiomatically means in this sense nothing else than to think with knowledge of what one is about. While earlier without the axiomatic method one proceeded naively in that one believed in certain relationships as dogma, the axiomatic approach removes this naiveté and yet permits the advantages of belief.

Again, in the last part of his "Axiomatisches Denken,"[5] he praised the method:

> Everything that can be the object of mathematical thinking, as soon as the erection of a theory is ripe, falls into the axiomatic method and thereby directly into mathematics. By pressing to ever deeper layers of axioms ... we can obtain deeper insights into scientific thinking and learn the unity of our knowledge. Especially by virtue of the axiomatic method mathematics appears called upon to play a leading role in all knowledge.

The opportunity to explore new problems by omitting, negating, or varying in some other manner the axioms of established systems enticed many mathematicians. This activity and the erection of axiomatic bases for the various branches of mathematics are known as the axiomatic movement. It continues to be a favorite activity. In part its great attraction is explained by the fact that after sound axiomatic bases for the major branches have been erected, variations of the kind just described are relatively easy to introduce and explore. However, any new development in mathematics has always attracted a number of men who seek fields wide-open to exploration or are sincerely convinced that the future of mathematics lies in that particular area.

4. *Abh. Math. Seminar der Hamburger Univ.*, 1, 1922, 157–77 = *Ges. Abh.*, 3, 157–77.
5. *Math. Ann.*, 78, 1918, 405–15 = *Ges. Abh.*, 3, 145–56.

3. Mathematics as Man's Creation

From the standpoint of the future development of mathematics, the most significant happening of the century was the acquisition of the correct view of the relationship of mathematics to nature. Though we have not treated the views on mathematics of many of the men whose work we have described, we have said that the Greeks, Descartes, Newton, Euler, and many others believed mathematics to be the accurate description of real phenomena and that they regarded their work as the uncovering of the mathematical design of the universe. Mathematics did deal with abstractions, but these were no more than the ideal forms of physical objects or happenings. Even such concepts as functions and derivatives were demanded by real phenomena and served to describe them.

Beyond what we have already reported that supports this view of mathematics, the position of mathematicians on the number of dimensions that can be considered in geometry shows clearly how closely mathematics had been tied to reality. Thus, in the first book of the *Heaven*, Aristotle says, "The line has magnitude in one way, the plane in two ways, and the solid in three ways, and beyond these there is no other magnitude because the three are all. . . . There is no transfer into another kind, like the transfer from length to area and from area to solid." In another passage he says, ". . . no magnitude can transcend three because there are no more than three dimensions," and adds, "for three is the perfect number." In his *Algebra*, John Wallis regarded a higher-dimensional space as a "monster in nature, less possible than a chimera or a centaure." He says, "Length, Breadth and Thickness take up the whole of space. Nor can Fansie imagine how there should be a Fourth Local Dimension beyond these Three." Cardan, Descartes, Pascal, and Leibniz had also considered the possibility of a fourth dimension and rejected it as absurd. As long as algebra was tied to geometry, the product of more than three quantities was also rejected. Jacques Ozanam pointed out that a product of more than three letters will be a magnitude of "as many dimensions as there are letters, but it would only be imaginary because in nature we do not know of any quantity which has more than three dimensions."

The idea of a mathematical geometry of more than three dimensions was rejected even in the early nineteenth century. Möbius, in his *Der barycentrische Calcul* (1827), pointed out that geometrical figures that could not be superposed in three dimensions because they are mirror images of each other could be superposed in four dimensions. But then he says,[6] "Since, however, such a space cannot be thought about, the superposition is impossible." Kummer in the 1860s mocked the idea of a four-dimensional geometry. The objections that all of these men made to a higher-dimensional

6. *Ges. Werke*, 1, 172.

geometry were sound as long as geometry was identified with the study of physical space.

But gradually and unwittingly mathematicians began to introduce concepts that had little or no direct physical meaning. Of these, negative and complex numbers were the most troublesome. It was because these two types of numbers had no "reality" in nature that they were still suspect at the beginning of the nineteenth century, even though freely utilized by then. The geometrical representation of negative numbers as distances in one direction on a line and of complex numbers as points or vectors in the complex plane, which, as Gauss remarked of the latter, gave them intuitive meaning and so made them admissible, may have delayed the realization that mathematics deals with man-made concepts. But then the introduction of quaternions, non-Euclidean geometry, complex elements in geometry, n-dimensional geometry, bizarre functions, and transfinite numbers forced the recognition of the artificiality of mathematics.

In this connection the impact of non-Euclidean geometry has already been noted (Chap. 36, sec. 8) and the impact of n-dimensional geometry must now be observed. The concept appears innocuously in the analytical work of d'Alembert, Euler, and Lagrange. D'Alembert suggested thinking of time as a fourth dimension in his article "Dimension" in the *Encyclopédie*. Lagrange, in studying the reduction of quadratic forms to standard forms, casually introduces forms in n variables. He, too, used time as a fourth dimension in his *Mécanique analytique* (1788) and in his *Théorie des fonctions analytiques* (1797). He says in the latter work, "Thus we may regard mechanics as a geometry of four dimensions and analytic mechanics as an extension of analytic geometry." Lagrange's work put the three spatial coordinates and the fourth one representing time on the same footing. Further, George Green in his paper of 1828 on potential theory did not hesitate to consider potential problems in n dimensions; he says of the theory, " It is no longer confined, as it was, to the three dimensions of space."

These early involvements in n dimensions were not intended as a study of geometry proper. They were natural generalizations of analytical work that was no longer tied to geometry. In part, this introduction of n-dimensional language was intended only as a convenience and aid to analytical thinking. It was helpful to think of $(x_1, x_2, \ldots, x_n)$ as a point and of an equation in n variables as a hypersurface in n-dimensional space, for by thinking in terms of what these mean in three-dimensional geometry one might secure some insight into the analytical work. In fact Cauchy[7] actually emphasized that the concept of n-dimensional space is useful in many analytical investigations, especially those of number theory.

However, the serious study of n-dimensional geometry, though not

7. *Comp. Rend.*, 24, 1847, 885–87 = *Œuvres*, (1), 10, 292–95.

implying a physical space of n dimensions, was also undertaken in the nineteenth century; the founder of this abstract geometry is Grassmann, in his *Ausdehnungslehre* of 1844. There one finds the concept of n-dimensional geometry in full generality. Grassmann said in a note published in 1845,

> My Calculus of Extension builds the abstract foundation of the theory of space; that is, it is free of all spatial intuition and is a pure mathematical science; only the special application to [physical] space constitutes geometry.
>
> However the theorems of the Calculus of Extension are not merely translations of geometrical results into an abstract language; they have a much more general significance, for while the ordinary geometry remains bound to the three dimensions of [physical] space, the abstract science is free of this limitation.

Grassmann adds that geometry in the usual sense is improperly regarded as a branch of pure mathematics, but that it is really a branch of applied mathematics since it deals with a subject not created by the intellect but given to it. It deals with matter. But, he says, it should be possible to create a purely intellectual subject that would deal with extension as a concept, rather than with the space perceived by the sensations. Thus Grassmann's work is representative of the development asserting that pure thought can build arbitrary constructions that may or may not be physically applicable.

Cayley, independently of Grassmann, also undertook to treat n-dimensional geometry analytically, and, as he says, "without recourse to any metaphysical notions." In the *Cambridge Mathematical Journal* of 1843,[8] Cayley published "Chapters in the Analytical Geometry of N-Dimensions." This work gives analytical results in n variables, which for $n = 3$ state known theorems about surfaces. Though he did nothing especially novel in n-dimensional geometry, the concept is fully grasped there.

By the time that Riemann gave his *Habilitationsvortrag* of 1854, "Über die Hypothesen welche die Geometrie zu Grunde liegen," he had no hesitation in dealing with n-dimensional manifolds, though he was primarily concerned with the geometry of three-dimensional physical space. Those who followed up on this basic paper—Helmholtz, Lie, Christoffel, Beltrami, Lipschitz, Darboux, and others—continued to work in n-dimensional space.

The notion of n-dimensional geometry encountered stiff-necked resistance from some mathematicians even long after it was introduced. Here, as in the case of negative and complex numbers, mathematics was progressing beyond concepts suggested by experience, and mathematicians had yet to grasp that their subject could consider concepts created by the mind and was no longer, if it ever had been, a reading of nature.

8. 4. 119–27 = *Collected Math. Papers*, 1, 55–62,

However after about 1850, the view that mathematics can introduce and deal with rather arbitrary concepts and theories that do not have immediate physical interpretation but may nevertheless be useful, as in the case of quaternions, or satisfy a desire for generality, as in the case of n-dimensional geometry, gained acceptance. Hankel in his *Theorie der complexen Zahlensysteme* (1867, p. 10) defended mathematics as "purely intellectual, a pure theory of forms, which has for its object, not the combination of quantities or of their images, the numbers, but things of thought to which there could correspond effective objects or relations even though such a correspondence is not necessary."

In defense of his creation of transfinite numbers as existing, real definite quantities, Cantor claimed that mathematics is distinguished from other fields by its freedom to create its own concepts without regard to transient reality. He said[9] in 1883, "Mathematics is entirely free in its development and its concepts are restricted only by the necessity of being noncontradictory and coordinated to concepts previously introduced by precise definitions. . . . The essence of mathematics lies in its freedom." He preferred the term "free mathematics" over the usual form, "pure mathematics."

The new view of mathematics extended to the older, physically grounded branches. In his *Universal Algebra* (1898), Alfred North Whitehead says (p. 11),

> . . . Algebra does not depend on Arithmetic for the validity of its laws of transformation. If there were such a dependence it is obvious that as soon as algebraic expressions are arithmetically unintelligible all laws respecting them must lose their validity. But the laws of Algebra, though suggested by Arithmetic, do not depend on it. They depend entirely on the conventions by which it is stated that certain modes of grouping the symbols are to be considered as identical. This assigns certain properties to the marks which form the symbols of Algebra.

Algebra is a logical development independent of meaning. "It is obvious that we can take any marks we like and manipulate them according to any rule we choose to assign" (p. 4). Whitehead does point out that such arbitrary manipulations of symbols can be frivolous and only constructions to which some meaning can be attached or which have some use are significant.

Geometry, too, cut its bonds to physical reality. As Hilbert pointed out in his *Grundlagen* of 1899, geometry speaks of things whose properties are specified in the axioms. Though Hilbert referred only to the strategy by which we must approach mathematics for the purpose of examining its logical structure, he nevertheless supported and encouraged the view that mathematics is quite distinct from the concepts and laws of nature.

9. *Math. Ann.*, 21, 1883, 563–64 = *Ges. Abh.*, 182.

4. *The Loss of Truth*

The introduction and gradual acceptance of concepts that have no immediate counterparts in the real world certainly forced the recognition that mathematics is a human, somewhat arbitrary creation, rather than an idealization of the realities in nature, derived solely from nature. But accompanying this recognition and indeed propelling its acceptance was a more profound discovery—mathematics is not a body of truths about nature. The development that raised the issue of truth was non-Euclidean geometry, though its impact was delayed by the characteristic conservatism and closed-mindedness of all but a few mathematicians. The philosopher David Hume (1711–76) had already pointed out that nature did not conform to fixed patterns and necessary laws; but the dominant view, expressed by Kant, was that the properties of physical space were Euclidean. Even Legendre in his *Eléments de géométrie* of 1794 still believed that the axioms of Euclid were self-evident truths.

With respect to geometry, at least, the view that seems correct today was first expressed by Gauss. Early in the nineteenth century he was convinced that geometry was an empirical science and must be ranked with mechanics, whereas arithmetic and analysis were a priori truths. Gauss wrote to Bessel in 1830,[10]

> According to my deepest conviction the theory of space has an entirely different place in our a priori knowledge than that occupied by pure arithmetic. There is lacking throughout our knowledge of the former the complete conviction of necessity (also of absolute truth) which is characteristic of the latter; we must add in humility, that if number is merely the product of our mind, space has a reality outside our mind whose laws we cannot a priori completely prescribe.

However Gauss seems to have had conflicting views, because he also expressed the opinion that all of mathematics is man-made. In a letter to Bessel of November 21, 1811, in which he speaks of functions of a complex variable he says,[11] "One should never forget that the functions, like all mathematical constructions, are only our own creations, and that when the definition with which one begins ceases to make sense, one should not ask: What is it, but what is it convenient to assume in order that it remain significant?"

Despite Gauss's views on geometry, most mathematicians thought there were basic truths in it. Bolyai thought that the absolute truths in geometry were those axioms and theorems common to Euclidean and hyperbolic geometry. He did not know elliptic geometry, and so in his time still did not

10. *Werke*, 8, 201.
11. *Werke*, 10, 363.

appreciate that many of these common axioms were not common to all geometries.

In his 1854 paper "On the Hypotheses Which Underlie Geometry," Riemann still believed that there were some propositions about space that were a priori, though these did not include the assertion that physical space is truly Euclidean. It was, however, locally Euclidean.

Cayley and Klein remained attached to the reality of Euclidean geometry (see also Chap. 38, sec. 6). In his presidential address to the British Association for the Advancement of Science,[12] Cayley said, ". . . not that the propositions of geometry are only approximately true, but that they remain absolutely true in regard to that Euclidean space which has so long been regarded as being the physical space of our experience." Though they themselves had worked in non-Euclidean geometries, they regarded the latter as novelties that result when new distance functions are introduced in Euclidean geometry. They failed to see that non-Euclidean is as basic and as applicable as Euclidean geometry.

In the 1890s Bertrand Russell took up the question of what properties of space are necessary to and are presupposed by experience. That is, experience would be meaningless if any of these a priori properties were denied. In his *Essay on the Foundations of Geometry* (1897), he agrees that Euclidean geometry is not a priori knowledge. He concludes, rather, that projective geometry is a priori for all geometry, an understandable conclusion in view of the importance of that subject around 1900. He then adds to projective geometry as a priori the axioms common to Euclidean and all the non-Euclidean geometries. The latter facts, the homogeneity of space, finite dimensionality, and a concept of distance make measurement possible. The facts that space is three-dimensional and that our actual space *is* Euclidean he considers to be empirical.

That the metrical geometries can be derived from projective geometry by the introduction of a metric Russell regards as a technical achievement having no philosophical significance. Metrical geometry is a logically subsequent and separate branch of mathematics and is not a priori. With respect to Euclidean and the several basic non-Euclidean geometries, he departs from Cayley and Klein and regards all these geometries as being on an equal footing. Since the only metric spaces that possess the above properties are the Euclidean, hyperbolic, and single and double elliptic, Russell concludes that these are the only possible metrical geometries, and of course Euclidean is the only physically applicable one. The others are of philosophical importance in showing that there can be other geometries. With hindsight it is now possible to see that Russell had replaced the Euclidean bias by a projective bias.

12. *Report of the Brit. Assn. for the Adv. of Sci.*, 1883, 3–37 = *Coll. Math. Papers*, 11, 429–59.

Though mathematicians were slow to recognize the fact, clearly seen by Gauss, that there is no assurance at all to the physical truth of Euclidean geometry, they gradually came around to that view and also to the related conviction of Gauss that the truth of mathematics resides in arithmetic and therefore also in analysis. For example, Kronecker in his essay "Über den Zahlbegriff" (On the Number Concept)[13] maintained the truth of the arithmetical disciplines but denied it to geometry. Gottlob Frege, about whose work we shall say more later, also insisted on the truth of arithmetic.

However, even arithmetic and the analysis built on it soon became suspect. The creation of non-commutative algebras, notably quaternions and matrices, certainly raised the question of how one can be sure that ordinary numbers possess the privileged property of truth about the real world. The attack on the truth of arithmetic came first from Helmholtz. After he had insisted, in a famous essay,[14] that our knowledge of physical space comes only from experience and depends on the existence of rigid bodies to serve, among other purposes, as measuring rods, in his *Zählen und Messen* (Counting and Measuring, 1887) he attacked the truths of arithmetic. He regards as the main problem in arithmetic the meaning or the validity of the objective application of quantity and equality to experience. Arithmetic itself may be just a consistent account of the consequences of the arithmetical operations. It deals with symbols and can be regarded as a game. But these symbols are applied to real objects and to relations among them and give results about real workings of nature. How is this possible? Under what conditions are the numbers and operations applicable to real objects? In particular, what is the objective meaning of the equality of two objects and what character must physical addition have to be treated as arithmetical addition?

Helmholtz points out that the applicability of numbers is neither an accident nor proof of the truths of the laws of numbers. *Some* kinds of experience suggest them and to these they are applicable. To apply numbers to real objects, Helmholtz says, objects must not disappear, or merge with one another, or divide in two. One raindrop added physically to another does not produce two raindrops. Only experience can tell us whether the objects of a physical collection retain their identity so that the collection has a definite number of objects in it. Likewise, knowing when equality between physical quantities can be applied also depends on experience. Any assertion of quantitative equality must satisfy two conditions. If the objects are exchanged they must remain equal. Also, if object *a* equals *c* and object *b* equals *c*, object *a* must equal object *b*. Thus we may speak of the equality of weights and intervals of time, because for these objects equality can be determined. But two pitches may, as far as the ear is concerned, equal an

13. *Jour. für Math.*, 101, 1887, 337–55 = *Werke*, 3, 249–74.
14. *Nachrichten König. Ges. der Wiss. zu Gött.*, 15, 1868, 193–221 = *Wiss. Abh.*, 2, 618–39.

intermediate one and yet the ear might distinguish the original two. Here things equal to the same thing are not equal to each other. One cannot add the values of electrical resistance connected in parallel to obtain the total resistance, nor can one combine in any way the indices of refraction of different media.

By the end of the nineteenth century, the view that all the axioms of mathematics are arbitrary prevailed. Axioms were merely to be the basis for the deduction of consequences. Since the axioms were no longer truths about the concepts involved in them, the physical meaning of these concepts no longer mattered. This meaning could, at best, be a heuristic guide when the axioms bore some relation to reality. Thus even the concepts were severed from the physical world. By 1900 mathematics had broken away from reality; it had clearly and irretrievably lost its claim to the truth about nature, and had become the pursuit of necessary consequences of arbitrary axioms about meaningless things.

The loss of truth and the seeming arbitrariness, the subjective nature of mathematical ideas and results, deeply disturbed many men who considered this a denigration of mathematics. Some therefore adopted a mystical view that sought to grant some reality and objectivity to mathematics. These mathematicians subscribed to the idea that mathematics is a reality in itself, an independent body of truths, and that the objects of mathematics are given to us as are the objects of the real world. Mathematicians merely discover the concepts and their properties. Hermite, in a letter to Stieltjes,[15] said, "I believe that the numbers and functions of analysis are not the arbitrary product of our minds; I believe that they exist outside of us with the same character of necessity as the objects of objective reality; and we find or discover them and study them as do the physicists, chemists and zoologists."

Hilbert said at the International Congress in Bologna in 1928,[16] "How would it be above all with the truth of our knowledge and with the existence and progress of science if there were no truth in mathematics? Indeed in professional writings and public lectures there often appears today a skepticism and despondency about knowledge; this is a certain kind of occultism which I regard as damaging."

Godfrey H. Hardy (1877–1947), an outstanding analyst of the twentieth century, said in 1928,[17] "Mathematical theorems are true or false; their truth or falsity is absolutely independent of our knowledge of them. In *some* sense, mathematical truth is part of objective reality." He expressed the same view in his book *A Mathematician's Apology* (1967 ed., p. 123): "I believe that mathematical reality lies outside us, that our function is to discover or observe

15. *C. Hermite- T. Stieltjes Correspondance*, Gauthier-Villars, 1905, 2, p. 398.
16. *Atti del Congresso*, 1, 1929, 141 = *Grundlagen der Geometrie*, 7th ed., p. 323.
17. *Mind*, 38, 1929, 1–25.

it and that the theorems which we describe grandiloquently as our 'creations,'
are simply the notes of our observations."

5. Mathematics as the Study of Arbitrary Structures

The nineteenth-century mathematicians were primarily concerned with the
study of nature, and physics certainly was the major inspiration for the
mathematical work. The greatest men—Gauss, Riemann, Fourier, Hamilton,
Jacobi, and Poincaré—and the less well-known men—Christoffel, Lipschitz,
Du Bois-Reymond, Beltrami, and hundreds of others—worked directly on
physical problems and on mathematical problems arising out of physical
investigations. Even the men commonly regarded as pure mathematicians,
Weierstrass, for example, worked on physical problems. In fact, more than in
any earlier century, physical problems supplied the suggestions and directions
for mathematical investigations, and highly complex mathematics was
created to master them. Fresnel had remarked that "Nature is not embarrassed by difficulties of analysis" but mathematicians were not deterred and
overcame them. The only major branch that had been pursued for intrinsic
aesthetic satisfaction, at least since Diophantus' work, was the theory of
numbers.

However, in the nineteenth century for the first time, mathematicians
not only carried their work far beyond the needs of science and technology
but raised and answered questions that had no bearing on real problems.
The *raison d'être* of this development might be described as follows. The two-
thousand-year-old conviction that mathematics was the truth about nature
was shattered. But the mathematical theories now recognized to be arbitrary
had nevertheless proved useful in the study of nature. Though the existing
theories historically owed much to suggestions from nature, perhaps new
theories constructed solely by the mind might also prove useful in the
representation of nature. Mathematicians then should feel free to create
arbitrary structures. This idea was seized upon to justify a new freedom in
mathematical research. However, since a few structures already in evidence
by 1900, and many of those created since, seemed so artificial and so far
removed from even potential application, their sponsors began to defend
them as desirable in and for themselves.

The gradual rise and acceptance of the view that mathematics should
embrace arbitrary structures that need have no bearing, immediate or
ultimate, on the study of nature led to a schism that is described today as
pure versus applied mathematics. Such a break from tradition could not but
generate controversy. We can take space to cite only a few of the arguments
on either side.

Fourier had written, in the preface to his *Analytical Theory of Heat*, "The
profound study of nature is the most fertile source of mathematical dis-

coveries. This study offers not only the advantage of a well-determined goal but the advantage of excluding vague questions and useless calculations. It is a means of building analysis itself and of discovering the ideas which matter most and which science must always preserve. The fundamental ideas are those which represent the natural happenings." He also stressed the application of mathematics to socially useful problems.

Though Jacobi had done first-class work in mechanics and astronomy, he took issue with Fourier. He wrote to Legendre on July 2, 1830,[18] "It is true that Fourier is of the opinion that the principal object of mathematics is the public utility and the explanation of natural phenomena; but a scientist like him ought to know that the unique object of science is the honor of the human spirit and on this basis a question of [the theory of] numbers is worth as much as a question about the planetary system. . . ."

Throughout the century, as more men became disturbed by the drift to pure mathematics, voices were raised in protest. Kronecker wrote to Helmholtz, "The wealth of your practical experience with sane and interesting problems will give to mathematics a new direction and a new impetus. . . . One-sided and introspective mathematical speculation leads into sterile fields."

Felix Klein, in his *Mathematical Theory of the Top*, (1897, pp. 1–2) stated, "It is the great need of the present in mathematical science that the pure science and those departments of physical science in which it finds its most important applications should again be brought into the intimate association which proved so fruitful in the works of Lagrange and Gauss." And Emile Picard, speaking in the early part of this century (*La Science moderne et son état actuel*, 1908), warned against the tendency to abstractions and pointless problems.

Somewhat later, Felix Klein spoke out again.[19] Fearing that the freedom to create arbitrary structures was being abused, he emphasized that arbitrary structures are "the death of all science. The axioms of geometry are . . . not arbitrary but sensible statements which are, in general, induced by space perception and are determined as to their precise content by expediency." To justify the non-Euclidean axioms Klein pointed out that visualization can verify the Euclidean parallel axiom only within certain limits. On another occasion he pointed out that "whoever has the privilege of freedom should also bear responsibility." By "responsibility" Klein meant service in the investigation of nature.

Despite the warnings, the trend to abstractions, to generalization of existing results for the sake of generalization, and the pursuit of arbitrarily chosen problems continued. The reasonable need to study an entire class of

18. *Ges. Werke*, 1, 454–55.
19. *Elementary Mathematics from an Advanced Standpoint*, Macmillan, 1939; Dover (reprint), 1945, Vol. 2, p. 187.

problems in order to learn more about concrete cases and to abstract in order to get at the essence of a problem became excuses to tackle generalities and abstractions in and for themselves.

Partly to counter the trend to generalization, Hilbert not only stressed that concrete problems are the lifeblood of mathematics, but took the trouble in 1900 to publish a list of twenty-three outstanding ones (see the bibliography) and to cite them in a talk he gave at the Second International Congress of Mathematicians in Paris. Hilbert's prestige did cause many men to tackle these problems. No honor could be more avidly sought than solving a problem posed by so great a man. But the trend to free creations, abstractions, and generalizations was not stemmed. Mathematics broke away from nature and science to pursue its own course.

6. *The Problem of Consistency*

Mathematics, from a logical standpoint, was by the end of the nineteenth century a collection of structures each built on its own system of axioms. As we have already noted, one of the necessary properties of any such structure is the consistency of its axioms. As long as mathematics was regarded as the truth about nature, the possibility that contradictory theorems could arise did not occur; and indeed the thought would have been regarded as absurd. When the non-Euclidean geometries were created, their seeming variance with reality did raise the question of their consistency. As we have seen, this question was answered by making the consistency of the non-Euclidean geometries depend upon that of Euclidean geometry.

By the 1880s the realization that neither arithmetic nor Euclidean geometry is true made the investigation of the consistency of these branches imperative. Peano and his school began in the 1890s to consider this problem. He believed that clear tests could be devised that would settle it. However, events proved that he was mistaken. Hilbert did succeed in establishing the consistency of Euclidean geometry (Chap. 42, sec. 3) on the assumption that arithmetic is consistent. But the consistency of the latter had not been established, and Hilbert posed this problem as the second in the list he presented at the Second International Congress in 1900; in his "Axiomatisches Denken"[20] he stressed it as the basic problem in the foundations of mathematics. Many other men became aware of the importance of the problem. In 1904 Alfred Pringsheim (1850–1941)[21] emphasized that the truth mathematics seeks is neither more nor less than consistency. We shall examine in Chapter 51 the work on this problem.

20. *Math. Ann.*, 78, 1918, 405–15 = *Ges. Abh.*, 145–56.
21. *Jahres. der Deut. Math.-Verein*, 13, 1904, 381.

7. A Glance Ahead

The pace of mathematical creation has expanded steadily since 1600, and this is certainly true of the twentieth century. Most of the fields pursued in the nineteenth century were further developed in the twentieth. However, the details of the newer work in these fields would be of interest only to specialists. We shall therefore limit our account of twentieth-century work to those fields that first became prominent in this period. Moreover, we shall consider only the beginnings of those fields. Developments of the second and third quarters of this century are too recent to be properly evaluated. We have noted many areas pursued vigorously and enthusiastically in the past, which were taken by their advocates to be the essence of mathematics, but which proved to be passing fancies or to have little consequential impact on the course of mathematics. However confident mathematicians of the last half-century may be that their work is of the utmost importance, the place of their contributions in the history of mathematics cannot be decided at the present time.

Bibliography

Fang, J.: *Hilbert*, Paideia Press, 1970. Sketches of Hilbert's mathematical work.

Hardy, G. H.: *A Mathematician's Apology*, Cambridge University Press, 1940 and 1967.

Helmholtz, H. von: *Counting and Measuring*, D. Van Nostrand, 1930. Translation of *Zählen und Messen = Wissenschaftliche Abhandlungen*, 3, 356–91.

———: "Über den Ursprung Sinn und Bedeutung der geometrischen Sätze"; English translation: "On the Origin and Significance of Geometrical Axioms," in Helmholtz: *Popular Scientific Lectures*, Dover (reprint), 1962, pp. 223–49. Also in James R. Newman: *The World of Mathematics*, Simon and Schuster, 1956, Vol. 1, pp. 647–68. See also Helmholtz's *Wissenschaftliche Abhandlungen*, 2, 640–60.

Hilbert, David: "Sur les problèmes futurs des mathématiques," *Comptes Rendus du Deuxième Congrès International des Mathématiciens*, Gauthier-Villars, 1902, 58–114. Also in German, in *Nachrichten König. Ges. der Wiss. zu Gött.*, 1900, 253–97, and in Hilbert's *Gesammelte Abhandlungen*, 3, 290–329. English translation in *Amer. Math. Soc. Bull.*, 8, 1901/2, 437–79.

Klein, Felix: "Über Arithmetisirung der Mathematik," *Ges. Math. Abh.*, 2, 232–40. English translation in *Amer. Math. Soc. Bull.*, 2, 1895/6, 241–49.

Pierpont, James: "On the Arithmetization of Mathematics," *Amer. Math. Soc. Bull.*, 5, 1898/9, 394–406.

Poincaré, Henri: *The Foundations of Science*, Science Press, 1913. See especially pp. 43–91.

Reid, Constance: *Hilbert*, Springer-Verlag, 1970. A biography.

44
The Theory of Functions of Real Variables

> If Newton and Leibniz had thought that continuous functions do not necessarily have a derivative—and this is the general case—the differential calculus would never have been created. EMILE PICARD

1. *The Origins*

The theory of functions of one or more real variables grew out of the attempt to understand and clarify a number of strange discoveries that had been made in the nineteenth century. Continuous but non-differentiable functions, series of continuous functions whose sum is discontinuous, continuous functions that are not piecewise monotonic, functions possessing bounded derivatives that are not Riemann integrable, curves that are rectifiable but not according to the calculus definition of arc length, and nonintegrable functions that are limits of sequences of integrable functions—all seemed to contradict the expected behavior of functions, derivatives, and integrals. Another motivation for the further study of the behavior of functions came from the research on Fourier series. This theory, as built up by Dirichlet, Riemann, Cantor, Ulisse Dini (1845–1918), Jordan, and other mathematicians of the nineteenth century, was a quite satisfactory instrument for applied mathematics. But the properties of the series, as thus far developed, failed to give a theory that could satisfy the pure mathematicians. Unity, symmetry, and completeness of relation between function and series were still wanting.

The research in the theory of functions emphasized the theory of the integral because it seemed that most of the incongruities could be resolved by broadening that notion. Hence to a large extent this work may be regarded as a direct continuation of the work of Riemann, Darboux, Du Bois-Reymond, Cantor, and others (Chap. 40, sec. 4).

2. The Stieltjes Integral

Actually the first extension of the notion of integral came from a totally different class of problems than those just described. In 1894 Thomas Jan Stieltjes (1856–94) published his "Recherches sur les fractions continues,"[1] a most original paper in which he started from a very particular question and solved it with rare elegance. This work suggested problems of a completely novel nature in the theory of analytic functions and in the theory of functions of a real variable. In particular, in order to represent the limit of a sequence of analytic functions Stieltjes was obliged to introduce a new integral that generalized the Riemann-Darboux concept.

Stieltjes considers a positive distribution of mass along a line, a generalization of the point concept of density which, of course, had already been used. He remarks that such a distribution of mass is given by an increasing function $\phi(x)$ which specifies the total mass in the interval $[0, x]$ for $x > 0$, the discontinuities of ϕ corresponding to masses concentrated at a point. For such a distribution of mass in an interval $[a, b]$ he formulates the Riemann sums

$$\sum_{i=0}^{n} f(\xi_i)(\phi(x_{i+1}) - \phi(x_i))$$

wherein the $x_0, x_1, \ldots, x_n$ are a partition of $[a, b]$ and ξ_i is within $[x_i, x_{i+1}]$. He then showed that when f is continuous in $[a, b]$ and the maximum subinterval of the partitions approaches 0, the sums approach a limit which he denoted by $\int_a^b f(x) \, d\phi(x)$. Though he used this integral in his own work, Stieltjes did not push further the integral notion itself, except that for the interval $(0, \infty)$ he defined

$$\int_0^{\infty} f(x) \, d\phi(x) = \lim_{b \to \infty} \int_0^b f(x) \, d\phi(x).$$

His integral concept was not taken up by mathematicians until much later, when it did find many applications (see Chap. 47, sec. 4).

3. Early Work on Content and Measure

Quite another line of thought led to a different generalization of the notion of integral, the Lebesgue integral. The study of the sets of discontinuities of functions suggested the question of how to measure the extent or "length" of the set of discontinuities, because the extent of these discontinuities determines the integrability of the function. The theory of content and later the theory of

1. *Ann. Fac. Sci. de Toulouse*, 8, 1894, J.1–122, and 9, 1895, A.1–47 = *Œuvres complètes*, 2, 402–559.

measure were introduced to extend the notion of length to sets of points that are not full intervals of the usual straight line.

The notion of content is based on the following idea: Consider a set E of points distributed in some manner over an interval $[a, b]$. To be loose for the moment, suppose that it is possible to enclose or cover these points by small subintervals of $[a, b]$ so that the points of E are either interior to one of the subintervals or at worst an endpoint. We reduce the lengths of these subintervals more and more and add others if necessary to continue to enclose the points of E, while reducing the sum of the lengths of the subintervals. The greatest lower bound of the sum of those subintervals that cover points of E is called the (outer) content of E. This loose formulation is not the definitive notion that was finally adopted, but it may serve to indicate what the men were trying to do.

A notion of (outer) content was given by Du Bois-Reymond in his *Die allgemeine Funktionentheorie* (1882), Axel Harnack (1851–88) in his *Die Elemente der Differential- und Integralrechnung* (1881), Otto Stolz,[2] and Cantor.[3] Stolz and Cantor also extended the notion of content to two and higher-dimensional sets using rectangles, cubes, and so forth in place of intervals.

The use of this notion of content, which was unfortunately not satisfactory in all respects, nevertheless revealed that there were nowhere dense sets (that is, the set lies in an interval but is not dense in any subinterval of that interval) of positive content and that functions with discontinuities on such sets were not integrable in Riemann's sense. Also there were functions with bounded nonintegrable derivatives. But mathematicians of this time, the 1880s, thought that Riemann's notion of the integral could not be extended.

To overcome limitations in the above theory of content and to rigorize the notion of area of a region, Peano (*Applicazioni geometriche del calcolo infinitesimale*, 1887) introduced a fuller and much improved notion of content. He introduced an inner and outer content for regions. Let us consider two dimensions. The inner content is the least upper bound of all polygonal regions contained within the region R and the outer content is the greatest lower bound of all polygonal regions containing the region R. If the inner and outer content are equal, this common value is the area. For a one-dimensional set the idea is similar but uses intervals instead of polygons. Peano pointed out that if $f(x)$ is non-negative in $[a, b]$ then

$$\underline{\int_a^b} f \, dx = C_i(R) \quad \text{and} \quad \overline{\int_a^b} = C_e(R),$$

where the first integral is the least upper bound of the lower Riemann sums of f on $[a, b]$ and the second is the greatest lower bound of the upper Riemann

2. *Math. Ann.*, 23, 1884, 152–56.
3. *Math. Ann.*, 23, 1884, 453–88 = *Ges. Abh.*, 210–46.

sums and $C_i(R)$ and $C_e(R)$ are the inner and outer content of the region R bounded above by the graph of f. Thus f is integrable if and only if R has content in the sense that $C_i(R) = C_e(R)$.

Jordan made the most advanced step in the nineteenth-century theory of content (étendue). He too introduced an inner and outer content,[4] but formulated the concept somewhat more effectively. His definition for a set of points E contained in $[a, b]$ starts with the outer content. One covers E by a finite set of subintervals of $[a, b]$ such that each point of E is interior to or an endpoint of one of these subintervals. The greatest lower bound of the sum of all such sets of subintervals that contain at least one point of E is the outer content of E. The inner content of E is defined to be the least upper bound of the sum of the subintervals that enclose only points of E in $[a, b]$. If the inner and outer content of E are equal, then E has content. The same notion was applied by Jordan to sets in n-dimensional space except that rectangles and the higher-dimensional analogues replace the subintervals. Jordan could now prove what is called the additivity property: The content of the sum of a *finite* number of disjoined sets with content is the sum of the contents of the separate sets. This was not true for the earlier theories of content, except Peano's.

Jordan's interest in content derived from the attempt to clarify the theory of double integrals taken over some plane region E. The definition usually adopted was to divide the plane into squares R_{ij} by lines parallel to the coordinate axes. This partition of the plane induces a partition of E into E_{ij}'s. Then, by definition,

$$\int_E f(x, y) \, dE = \lim_{\Delta x, \Delta y \to 0} \sum_{i,j} f(x_i, y_j) a(R_{ij}),$$

where $a(R_{ij})$ denotes the area of R_{ij} and the sum is over all R_{ij} interior to E and all R_{ij} that contain any points of E but may also contain points outside of E. For the integral to exist it is necessary to show that the R_{ij} that are not entirely interior to E can be neglected or that the sum of the areas of the R_{ij} that contain boundary points of E approaches 0 with the dimensions of the R_{ij}. It had been generally assumed that this was the case and Jordan himself did so in the first edition of his *Cours d'analyse* (Vol. 2, 1883). However the discovery of such peculiar curves as Peano's square-filling curve made the mathematicians more cautious. If E has two-dimensional Jordan content, then one can neglect the R_{ij} that contain the boundary points of E. Jordan was also able to obtain results on the evaluation of double integrals by repeated integration.

The second edition of Jordan's *Cours d'analyse* (Vol. 1, 1893) contains Jordan's treatment of content and its application to integration. Though

4. *Jour. de Math.*, (4), 8, 1892, 69–99 = *Œuvres*, 4, 427–57.

superior to that of his predecessors, Jordan's definition of content was not quite satisfactory. According to it an open bounded set does not necessarily have content and the set of rational points contained in a bounded interval does not have content.

The next step in the theory of content was made by Borel. Borel was led to study the theory, which he called measure, while working on the sets of points on which series representing complex functions converge. His *Leçons sur la théorie des fonctions* (1898) contains his first major work on the subject. Borel discerned the defects in the earlier theories of content and remedied them.

Cantor had shown that every open set U on the line is the union of a *denumerable* family of open intervals, no two having a point in common. In place of approaching U by enclosing it in a finite set of intervals Borel, using Cantor's result, proposed as the measure of a bounded open set U the sum of the lengths of the component intervals. He then defined the measure of the sum of a *countable* number of disjoined measurable sets as the sum of the individual measures and the measure of the set $A - B$, if A and B are measurable and B is contained in A, as the difference of the measures. With these definitions he could attach a measure to sets formed by adding any countable number of disjoined measurable sets and to the difference of any two measurable sets A and B if A contains B. He then considered sets of measure 0 and showed that a set of measure greater than 0 is non-denumerable.

Borel's theory of measure was an improvement over Peano's and Jordan's notions of content, but it was not the final word, nor did he consider its application to integration.

4. The Lebesgue Integral

The generalization of measure and the integral that is now considered definitive was made by Henri Lebesgue (1875–1941), a pupil of Borel and a professor at the Collège de France. Guided by Borel's ideas and also by those of Jordan and Peano, he first presented his ideas on measure and the integral in his thesis, "Intégrale, longueur, aire."[5] His work superseded the nineteenth-century creations and, in particular, improved on Borel's theory of measure.

Lebesgue's theory of integration is based on his notion of measure of sets of points and both ideas apply to sets in n-dimensional space. For illustrative purposes we shall consider the one-dimensional case. Let E be a set of points in $a \le x \le b$. The points of E can be enclosed as interior points in a finite or *countably infinite* set of intervals $d_1, d_2, \ldots$ lying in $[a, b]$. (The endpoints of $[a, b]$ can be endpoints of a d_i.) It can be shown that the set of intervals $\{d_i\}$

5. *Annali di Mat.*, (3), 7, 1902, 231–59.

can be replaced by a set of non-overlapping intervals $\delta_1, \delta_2, \ldots$ such that every point of E is an interior point of one of the intervals or the common endpoint of two adjacent intervals. Let $\sum \delta_n$ denote the sum of the lengths δ_i. The lower bound of $\sum \delta_n$ for all possible sets $\{\delta_i\}$ is called the exterior measure of E and denoted by $m_e(E)$. The interior measure $m_i(E)$ of E is defined to be the exterior measure of the set $C(E)$, that is, the complement of E in $[a, b]$ or the points of $a \leq x \leq b$ not in E.

Now one can prove a number of subsidiary results, including the fact that $m_i(E) \leq m_e(E)$. The set E is defined to be measurable if $m_i(E) = m_e(E)$ and the measure $m(E)$ is taken to be this common value. Lebesgue showed that a union of a countable number of measurable sets that are pairwise disjoined has as its measure the sum of the measures of the component sets. Also all Jordan measurable sets are Lebesgue measurable and the measure is the same. Lebesgue's notion of measure differs from Borel's by the adjunction of a part of a set of measure 0 in the sense of Borel. Lebesgue also called attention to the existence of nonmeasurable sets.

His next significant notion is that of a measurable function. Let E be a bounded measurable set on the x-axis. The function $f(x)$, defined on all points of E, is said to be measurable in E if the set of points of E for which $f(x) > A$ is measurable for every constant A.

Finally we arrive at Lebesgue's notion of an integral. Let $f(x)$ be a bounded and measurable function on the measurable set E contained in $[a, b]$. Let A and B be the greatest lower and least upper bounds of $f(x)$ on E. Divide the interval $[A, B]$ (on the y-axis) into n partial intervals

$$[A, l_1], [l_1, l_2], \cdots, [l_{n-1}, B],$$

with $A = l_0$ and $B = l_n$. Let e_r be the set of points of E for which $l_{r-1} \leq f(x) \leq l_r, r = 1, 2, \ldots, n$. Then $e_1, e_2, \ldots, e_n$ are measurable sets. Form the sums S and s where

$$S = \sum_1^n l_r m(e_r), \qquad s = \sum_1^n l_{r-1} m(e_r).$$

The sums S and s have a greatest lower bound J and a least upper bound I, respectively. Lebesgue showed that for bounded measurable functions $I = J$, and this common value is the Lebesgue integral of $f(x)$ on E. The notation is

$$I = \int_E f(x)\, dx.$$

If E consists of the entire interval $a \leq x \leq b$, then we use the notation $\int_a^b f(x)\, dx$, but the integral is understood in the Lebesgue sense. If $f(x)$ is Lebesgue integrable and the value of the integral is finite, then $f(x)$ is said to be summable, a term introduced by Lebesgue. An $f(x)$ that is Riemann integrable on $[a, b]$ is Lebesgue integrable but not necessarily conversely. If $f(x)$ is integrable in both senses, the values of the two integrals are the same. The generality of the Lebesgue integral derives from the fact that a Lebesgue integrable function need not be continuous almost everywhere (that is, except on a set of measure 0). Thus the Dirichlet function, which is 1 for rational values of x and 0 for irrational values of x in $[a, b]$, is totally discontinuous and though not Riemann integrable is Lebesgue integrable. In this case $\int_a^b f(x)\, dx = 0$.

The notion of the Lebesgue integral can be extended to more general functions, for example unbounded functions. If $f(x)$ is Lebesgue integrable but not bounded in the interval of integration, the integral converges absolutely. Unbounded functions may be Lebesgue integrable but not Riemann integrable and conversely.

For practical purposes the Riemann integral suffices. In fact Lebesgue showed (*Leçons sur l'intégration et la recherche des fonctions primitives*, 1904) that a bounded function is Riemann integrable if and only if the points of discontinuity form a set of measure 0. But for theoretical work the Lebesgue integral affords simplifications. The new theorems rest on the countable additivity of Lebesgue measure as contrasted with the finite additivity of Jordan content.

To illustrate the simplicity of theorems using Lebesgue integration, we have the result given by Lebesgue himself in his thesis. Suppose $u_1(x)$, $u_2(x), \ldots$ are summable functions on a measurable set E and $\sum_1^\infty u_n(x)$ converges to $f(x)$. Then $f(x)$ is measurable. If in addition $s_n(x) = \sum_1^n u_n(x)$ is uniformly bounded ($|s_n(x)| \le B$ for all x in E and all n), then it is a theorem that $f(x)$ is Lebesgue integrable on $[a, b]$ and

$$\int_a^b f(x)\, dx = \lim_{n \to \infty} \int_a^b s_n(x)\, dx.$$

If we were working with Riemann integrals we would need the additional hypothesis that the sum of the series is integrable; this case for the Riemann integral is a theorem due to Cesare Arzelà (1847–1912).[6] Lebesgue made his theorem the cornerstone in the exposition of his theory in his *Leçons sur l'intégration*.

The Lebesgue integral is especially useful in the theory of Fourier series and most important contributions were made in this connection by Lebesgue himself.[7] According to Riemann, the Fourier coefficients a_n and b_n of a bounded and integrable function tend to 0 as n becomes infinite. Lebesgue's generalization states that

$$\lim_{n \to \infty} \int_a^b f(x) \begin{cases} \sin nx \\ \cos nx \end{cases} dx = 0,$$

6. *Atti della Accad. dei Lincei, Rendiconti*, (4), 1, 1885, 321–26, 532–37, 566–69.
7. E.g., *Ann. de l'Ecole Norm. Sup.*, (3), 20, 1903, 453–85.

where $f(x)$ is any function, bounded or not, that is Lebesgue integrable. This fact is now referred to as the Riemann-Lebesgue lemma.

In this same paper of 1903 Lebesgue showed that if f is a bounded function represented by a trigonometric series, that is,

$$f(x) = \frac{a_0}{2} + \sum_1^\infty (a_n \cos nx + b_n \sin nx),$$

then the a_n and b_n are Fourier coefficients. In 1905[8] Lebesgue gave a new sufficient condition for the convergence of the Fourier series to a function $f(x)$ that included all previously known conditions.

Lebesgue also showed (*Leçons sur les séries trigonométriques*, 1906, p. 102) that term-by-term integration of a Fourier series does not depend on the uniform convergence of the series to $f(x)$ itself. What does hold is that

$$\int_{-\pi}^x f(x)\, dx = a_0(x + \pi) + \sum_1^\infty \frac{1}{n} (a_n \sin nx + b_n (\cos n\pi - \cos nx)),$$

where x is any point in $[-\pi, \pi]$, for any $f(x)$ that is Lebesgue integrable, whether or not the original series for $f(x)$ converges. And the new series converges uniformly to the left side of the equation in the interval $[-\pi, \pi]$.

Further, Parseval's theorem that

$$\frac{1}{\pi} \int_{-\pi}^\pi [f(x)]^2\, dx = 2a_0^2 + \sum_1^\infty (a_n^2 + b_n^2)$$

holds for any $f(x)$ whose square is Lebesgue integrable in $[-\pi, \pi]$ (*Leçons*, 1906, p. 100). Then Pierre Fatou (1878–1929) proved[9] that

$$\frac{1}{\pi} \int_{-\pi}^\pi f(x)g(x)\, dx = 2a_0\alpha_0 + \sum_1^\infty (a_n\alpha_n + b_n\beta_n),$$

where a_n and b_n, and α_n and β_n are the Fourier coefficients for $f(x)$ and $g(x)$ whose squares are Lebesgue integrable in $[-\pi, \pi]$. Despite these advances in the theory of Fourier series, there is no known property of an $f(x)$ Lebesgue integrable in $[-\pi, \pi]$ that is necessary and sufficient for the convergence of its Fourier series.

Lebesgue devoted most of his efforts to the connection between the notions of integral and of primitive functions (indefinite integrals). When Riemann introduced his generalization of the integral, the question was posed whether the correspondence between definite integral and primitive function, valid for continuous functions, held in the more general case. But it is possible to give examples of functions f integrable in Riemann's sense and

8. *Math. Ann.*, 61, 1905, 251–80.
9. *Acta Math.*, 30, 1906, 335–400.

such that $\int_a^x f(t)\,dt$ does not have a derivative (not even a right or left derivative) at certain points. Conversely Volterra showed in 1881 [10] that a function $F(x)$ can have a bounded derivative in an interval I that is not integrable in Riemann's sense over the interval. By a subtle analysis Lebesgue showed that if f is integrable in his sense in $[a, b]$, then $F(x) = \int_a^x f(t)\,dt$ has a derivative equal to $f(x)$ almost everywhere, that is, except on a set of measure zero (*Leçons sur l'intégration*). Conversely, if a function g is differentiable in $[a, b]$ and if its derivative $g' = f$ is bounded, then f is Lebesgue integrable and the formula $g(x) - g(a) = \int_a^x f(t)\,dt$ holds. However, as Lebesgue established, the situation is much more complex if g' is not bounded. In this case g' is not necessarily integrable, and the first problem is to characterize the functions g for which g' exists almost everywhere and is integrable. Limiting himself to the case where one of the derived numbers [11] of g is finite everywhere, Lebesgue showed that g is necessarily a function of bounded variation (Chap. 40, sec. 6). Finally Lebesgue established (in the 1904 book) the reciprocal result. A function g of bounded variation admits a derivative almost everywhere and g' is integrable. But one does not necessarily have

(1) $$g(x) - g(a) = \int_a^x g'(t)\,dt;$$

the difference between the two members of this equation is a nonconstant function of bounded variation with derivative zero almost everywhere. As for the functions of bounded variation g for which (1) does hold, these have the following property: The total variation of g in an open set U (that is, the sum of the total variations of g in each of the connected components of U) tends to 0 with the measure of U. These functions were called absolutely continuous by Giuseppe Vitali (1875–1932), who studied them in detail.

Lebesgue's work also advanced the theory of multiple integrals. Under his definition of the double integral, the domain of functions for which the double integral can be evaluated by repeated integration is extended. Lebesgue gave a result in his thesis of 1902, but the better result was given by Guido Fubini (1879–1943) :[12] If $f(x, y)$ is summable over the measurable set G, then

(a) $f(x, y)$ as a function of x and as a function of y is summable for almost all y and x, respectively;

10. *Gior. di Mat.*, 19, 1881, 333–72 = *Opere Mat.*, 1, 16–48.
11. The two right derived numbers of g are the two limits,

$$\limsup_{h \to 0, h > 0} \frac{g(x + h) - g(x)}{h}, \qquad \liminf_{h \to 0, h > 0} \frac{g(x + h) - g(x)}{h}.$$

The two left derived numbers are defined similarly.
12. *Atti della Accad. dei Lincei, Rendiconti*, (5), 16, 1907, 608–14.

(b) the set of points (x_0, y_0) for which either $f(x, y_0)$ or $f(x_0, y)$ is not summable has measure 0;

(c) $$\int \int_G f(x, y) \, dG = \int dy \left(\int f(x, y) \, dx \right) = \int dx \left(\int f(x, y) \, dy \right)$$

where the outer integrals are taken over the set of points y (respectively x) for which $f(x, y)$ as a function of x (as a function of y, respectively) are summable.

Finally, in 1910 [13] Lebesgue arrived at results for multiple integrals that extended those for the derivatives of single integrals. He associated with a function f integrable in every compact region of R^n, the set function (as opposed to functions of numerical variables) $F(E) = \int_E f(x) \, dx$ (x represents n coordinates) defined for each integrable domain E of R^n. This concept generalizes the indefinite integral. He observed that the function F possesses two properties:

(1) It is completely additive; that is, $F(\sum_1^\infty E_n) = \sum_1^\infty F(E_n)$ where the E_n are pairwise disjoined measurable sets.
(2) It is absolutely continuous in the sense that $F(E)$ tends to 0 with the measure of E.

The essential part of this paper of Lebesgue was to show the converse of this proposition, that is, to define a derivative of $F(E)$ at a point P of n-dimensional space. Lebesgue arrived at the following theorem: If $F(E)$ is absolutely continuous and additive, then it possesses a finite derivative almost everywhere, and F is the indefinite integral of that summable function which is the derivative of F where it exists and is finite but is otherwise arbitrary at the remaining points.

The principal tool in the proof is a covering theorem due to Vitali,[14] which remains fundamental in this area of integration. But Lebesgue did not stop there. He indicated the possibility of generalizing the notion of functions of bounded variation by considering functions $F(E)$ where E is a measurable set, the functions being completely additive and such that $\sum_n |F(E_n)|$ remains bounded for every denumerable partition of E into measurable subsets E_n. It would be possible to cite many other theorems of the calculus that have been generalized by Lebesgue's notion of the integral.

Lebesgue's work, one of the great contributions of this century, did win approval but, as usual, not without some resistance. We have already noted (Chap. 40, sec. 7) Hermite's objections to functions without derivatives. He tried to prevent Lebesgue from publishing a "Note on Non-Ruled Surfaces

13. *Ann. de l'Ecole Norm. Sup.*, (3), 27, 1910, 361–450.
14. *Atti Accad. Torino*, 43, 1908, 229–46.

Applicable to the Plane,"[15] in which Lebesgue treated non-differentiable surfaces. Lebesgue said many years later in his *Notice* (p. 13, see bibliography),

> To many mathematicians I became the man of the functions without derivatives, although I never at any time gave myself completely to the study or consideration of such functions. And since the fear and horror which Hermite showed was felt by almost everybody, whenever I tried to take part in a mathematical discussion there would always be an analyst who would say, "This won't interest you; we are discussing functions having derivatives." Or a geometer would say it in his language: "We're discussing surfaces that have tangent planes."

Lebesgue also said (*Notice*, p. 14),

> Darboux had devoted his *Mémoire* of 1875 to integration and to functions without derivatives; he therefore did not experience the same horror as Hermite. Nevertheless I doubt whether he ever entirely forgave my "Note on Applicable Surfaces." He must have thought that those who make themselves dull in this study are wasting their time instead of devoting it to useful research.

5. Generalizations

We have already indicated the advantages of Lebesgue integration in generalizing older results and in formulating neat theorems on series. In subsequent chapters we shall meet additional applications of Lebesgue's ideas. The immediate developments in the theory of functions were many extensions of the notion of integral. Of these we shall merely mention one by Johann Radon (1887–1956), which embraces both Stieltjes's and Lebesgue's integral and is in fact known as the Lebesgue-Stieltjes integral.[16] The generalizations cover not only broader or different notions of integrals on point sets of n-dimensional Euclidean space but on domains of more general spaces such as spaces of functions. The applications of these more general concepts are now found in the theory of probability, spectral theory, ergodic theory, and harmonic analysis (generalized Fourier analysis).

Bibliography

Borel, Emile: *Notice sur les travaux scientifiques de M. Emile Borel*, 2nd ed., Gauthier-Villars, 1921.

Bourbaki, Nicolas: *Eléments d'histoire de mathématiques*, Hermann, 1960, pp. 246–59.

Collingwood, E. F.: "Emile Borel," *Jour. Lon. Math. Soc.*, 34, 1959, 488–512.

15. *Comp. Rend.*, 128, 1899, 1502–05.
16. *Sitzungsber. der Akad. der Wiss. Wien*, 122, Abt. IIa, 1913, 1295–1438.

Fréchet, M.: "La Vie et l'œuvre d'Emile Borel," *L'Enseignement Mathématique*, (2), 11, 1965, 1–94.

Hawkins, T. W., Jr.: *Lebesgue's Theory of Integration: Its Origins and Development*, University of Wisconsin Press, 1970, Chaps. 4–6.

Hildebrandt, T. H.: "On Integrals Related to and Extensions of the Lebesgue Integral," *Amer. Math. Soc. Bull.*, 24, 1918, 113–77.

Jordan, Camille: *Œuvres*, 4 vols., Gauthier-Villars, 1961–64.

Lebesgue, Henri: *Measure and the Integral*, Holden-Day, 1966, pp. 176–94. Translation of the French *La Mesure des grandeurs*.

———: *Notice sur les travaux scientifiques de M. Henri Lebesgue*, Edouard Privat, 1922.

———: *Leçons sur l'intégration et la recherche des fonctions primitives*, Gauthier-Villars, 1904, 2nd ed., 1928.

McShane, E. J.: "Integrals Devised for Special Purposes," *Amer. Math. Soc. Bull.*, 69, 1963, 597–627.

Pesin, Ivan M.: *Classical and Modern Integration Theories*, Academic Press, 1970.

Plancherel, Michel: "Le Développement de la théorie des séries trigonométriques dans le dernier quart de siècle," *L'Enseignement Mathématique*, 24, 1924/25, 19–58.

Riesz, F.: "L'Evolution de la notion d'intégrale depuis Lebesgue," *Annales de l'Institut Fourier*, 1, 1949, 29–42.

45
Integral Equations

Nature is not embarrassed by difficulties of analysis.
AUGUSTIN FRESNEL

1. Introduction

An integral equation is an equation in which an unknown function appears under an integral sign and the problem of solving the equation is to determine that function. As we shall soon see, some problems of mathematical physics lead directly to integral equations, and other problems, which lead first to ordinary or partial differential equations, can be handled more expeditiously by converting them to integral equations. At first, solving integral equations was described as inverting integrals. The term integral equations is due to Du Bois-Reymond.[1]

As in other branches of mathematics, isolated problems involving integral equations occurred long before the subject acquired a distinct status and methodology. Thus Laplace in 1782[2] considered the integral equation for $g(t)$ given by

$$(1) \qquad f(x) = \int_{-\infty}^{\infty} e^{-xt} g(t)\, dt.$$

As equation (1) now stands, it is called the Laplace transform of $g(t)$. Poisson[3] discovered the expression for $g(t)$, namely,

$$g(t) = \frac{1}{2\pi i} \int_{a-i\infty}^{a+i\infty} e^{xt} f(x)\, dx$$

for large enough a. Another of the noteworthy results that really belong to the history of integral equations stems from Fourier's famous 1811 paper on the theory of heat (Chap. 28, sec. 3). Here one finds

$$f(x) = \int_{0}^{\infty} \cos{(xt)} u(t)\, dt$$

1. *Jour. für Math.*, 103, 1888, 228.
2. *Mém. de l'Acad. des Sci.*, Paris, 1782, 1–88, pub. 1785, and 1783, 423–67, pub. 1786 = *Œuvres*, 10, 209–91, p. 236 in particular.
3. *Jour. de l'Ecole Poly.*, 12, 1823, 1–144, 249–403.

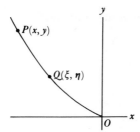

Figure 45.1

and the inversion formula

$$u(t) = \frac{2}{\pi} \int_0^\infty \cos{(xt)} f(x)\ dx.$$

The first conscious direct use and solution of an integral equation go back to Abel. In two of his earliest published papers, the first published in an obscure journal in 1823[4] and the second published in the *Journal für Mathematik*,[5] Abel considered the following mechanics problem: A particle starting at P slides down a smooth curve (Fig. 45.1) to the point O. The curve lies in a vertical plane. The velocity acquired at O is independent of the shape of the curve but the time required to slide from P to O is not. If (ξ, η) are the coordinates of any point Q between P and O and s is the arc OQ, then the velocity of the particle at Q is given by

$$\frac{ds}{dt} = -\sqrt{2g(x - \xi)}$$

where g is the gravitational constant. Hence

$$t = \frac{-1}{\sqrt{2g}} \int_P^Q \frac{ds}{\sqrt{x - \xi}}.$$

Now s can be expressed in terms of ξ. Suppose s is $v(\xi)$. Then the whole time of descent T from P to O is given by

$$T(x) = \frac{1}{\sqrt{2g}} \int_0^x \frac{v'(\xi)\ d\xi}{\sqrt{x - \xi}}.$$

The time T clearly depends upon x for any curve. The problem Abel set was, given T as a function of x, to find $v(\xi)$. If we introduce

$$f(x) = \sqrt{2g}\ T(x)$$

4. *Magazin for Naturwidenskaberne*, 1, 1823 = *Œuvres*, 1, 11–27.
5. *Jour. für Math*, 1, 1826, 153–57 = *Œuvres*, 1, 97–101.

the problem becomes to determine v from the equation

$$f(x) = \int_0^x \frac{v'(\xi)}{\sqrt{x - \xi}} \, d\xi.$$

Abel obtained the solution

$$v(\xi) = \int_0^\xi \frac{f(x) \, dx}{(\xi - x)^{1/2}}.$$

His methods—he gave two—were special and not worth noting.

Actually Abel undertook to solve the more general problem

(2) $$f(x) = \int_a^x \frac{u(\xi) \, d\xi}{(x - \xi)^\lambda}, \qquad 0 < \lambda < 1$$

and obtained

$$u(z) = \frac{\sin \lambda \pi}{\pi} \frac{d}{dz} \int_a^z \frac{f(x) \, dx}{(z - x)^{1 - \lambda}}.$$

Liouville, who worked independently of Abel, solved special integral equations from 1832 on.[6] A more significant step by Liouville[7] was to show how the solution of certain differential equations can be obtained by solving integral equations. The differential equation to be solved is

(3) $$y'' + [\rho^2 - \sigma(x)]y = 0$$

over the interval $a \leq x \leq b$; ρ is a parameter. Let $u(x)$ be the particular solution that satisfies the initial conditions

(4) $$u(a) = 1, \qquad u'(a) = 0.$$

This function will also be a solution of the nonhomogeneous equation

$$y'' + \rho^2 y = \sigma(x)u(x).$$

Then by a basic result on ordinary differential equations,

(5) $$u(x) = \cos \rho(x - a) + \frac{1}{\rho} \int_a^x \sigma(\xi) \sin \rho(x - \xi) u(\xi) \, d\xi.$$

Thus if we can solve this integral equation we shall have obtained that solution of the differential equation (3) that satisfies the initial conditions (4).

Liouville obtained the solution by a method of successive substitutions attributed to Carl G. Neumann, whose work *Untersuchungen über das logarithmische und Newton'sche Potential* (1877) came thirty years later. We shall not describe Liouville's method because it is practically identical with the one given by Volterra, which is to be described shortly.

6. *Jour. de l'Ecole Poly.*, 13, 1832, 1–69.
7. *Jour. de Math.*, 2, 1837, 16–35.

The integral equations treated by Abel and Liouville are of basic types. Abel's is of the form

(6) $$f(x) = \int_a^x K(x, \xi)u(\xi)\, d\xi,$$

and Liouville's of the form

(7) $$u(x) = f(x) + \int_a^x K(x, \xi)u(\xi)\, d\xi.$$

In both of these $f(x)$ and $K(x, \xi)$ are known, and $u(\xi)$ is the function to be determined. The terminology used today, introduced by Hilbert, refers to these equations as the first and second kind, respectively, and $K(x, \xi)$ is called the kernel. As stated, they are also referred to as Volterra's equations, whereas when the upper limit is a fixed number b, they are called Fredholm's equations. Actually the Volterra equations are special cases, respectively, of Fredholm's because one can always take $K(x, \xi) = 0$ for $\xi > x$ and then regard the Volterra equations as Fredholm equations. The special case of the equation of the second kind in which $f(x) \equiv 0$ is called the homogeneous equation.

By the middle of the nineteenth century the chief interest in integral equations centered around the solution of the boundary-value problem associated with the potential equation

(8) $$\Delta u = u_{xx} + u_{yy} = 0.$$

The equation holds in a given plane area that is bounded by some curve C. If the boundary value of u is some function $f(s)$ given as a function of arc length s along C, then a solution of this potential problem can be represented by

$$u(x, y) = \frac{1}{2\pi} \int_C \rho(s) \log \frac{1}{r(s; x, y)}\, ds,$$

wherein $r(s; x, y)$ is the distance from a point s to any point (x, y) in the interior or boundary and $\rho(s)$ is an unknown function satisfying for $s = (x, y)$ on C

(9) $$f(s) = \frac{1}{2\pi} \int_C \rho(t) \log \frac{1}{r(t; x, y)}\, dt.$$

This is an integral equation of the first kind for $\rho(t)$. Alternatively, if one takes as a solution of (8) with the same boundary condition

$$v(x, y) = \frac{1}{2\pi} \int_C \phi(s) \frac{\partial}{\partial n} \left(\log \frac{1}{r(s; x, y)} \right) ds,$$

where $\partial/\partial n$ denotes the normal derivative to the boundary, then $\phi(s)$ must satisfy the integral equation

$$(10) \qquad f(s) = \frac{1}{2}\,\phi(s) + \frac{1}{2\pi}\int_C \phi(t)\,\frac{\partial}{\partial n}\left(\log\frac{1}{r(t;x,y)}\right)dt,$$

an integral equation of the second kind. These equations were solved by Neumann for convex areas in his *Untersuchungen* and later publications.

Another problem of partial differential equations was tackled through integral equations. The equation

$$(11) \qquad\qquad \Delta u + \lambda u = f(x,y)$$

arises in the study of wave motion when the time dependence of the corresponding hyperbolic equation

$$\Delta u - \frac{1}{c^2}\,u_{tt} = f(x,y),$$

usually taken to be $e^{-i\omega t}$, is eliminated. It was known (Chap. 28, sec. 8) that the homogeneous case of (11) subject to boundary conditions has nontrivial solutions only for a discrete set of λ-values, called eigenvalues or characteristic values. Poincaré in 1894[8] considered the inhomogeneous case (11) with complex λ. He was able to produce a function, meromorphic in λ, which represented a unique solution of (11) for any λ which is not an eigenvalue, and whose residues produce eigenfunctions for the homogeneous case, that is, when $f = 0$.

On the basis of these results, Poincaré in 1896[9] considered the equation

$$u(x) + \lambda \int_a^b K(x,y)u(y)\,dy = f(x),$$

which he derived from (11), and affirmed that the solution is a meromorphic function of λ. This result was established by Fredholm in a paper we shall consider shortly.

The conversion of differential equations to integral equations, which is illustrated by the above examples, became a major technique for solving initial- and boundary-value problems of ordinary and partial differential equations and was the strongest impetus for the study of integral equations.

2. *The Beginning of a General Theory*

Vito Volterra (1860–1940), who succeeded Beltrami as professor of mathematical physics at Rome, is the first of the founders of a general theory of

8. *Rendiconti del Circolo Matematico di Palermo*, 8, 1894, 57–155 = *Œuvres*, 9, 123–96.
9. *Acta Math.*, 20, 1896–97, 59–142 = *Œuvres*, 9, 202–72. See also the Hellinger and Toeplitz reference in the bibliography, p. 1354.

integral equations. He wrote papers on the subject from 1884 on and principal ones in 1896 and 1897.[10] Volterra contributed a method of solving integral equations of the second kind,

$$(12) \qquad f(s) = \phi(s) + \int_a^b K(s, t)\phi(t)\, dt$$

wherein $\phi(s)$ is unknown and $K(s, t) = 0$ for $t > s$. Volterra wrote this equation as

$$f(s) = \phi(s) + \int_a^s K(s, t)\phi(t)\, dt.$$

His method of solution was to let

$$f_1(s) = -\int_a^b K(s, t)f(t)\, dt$$

$$(13) \qquad \cdots\cdots\cdots\cdots\cdots\cdots$$

$$f_n(s) = -\int_a^b K(s, t)f_{n-1}(t)\, dt$$

$$\cdots\cdots\cdots\cdots\cdots\cdots$$

and take $\phi(s)$ to be

$$(14) \qquad \phi(s) = f(s) + \sum_{p=1}^{\infty} f_p(s).$$

For his kernel $K(s, t)$ Volterra was able to prove the convergence of (14), and if one substitutes (14) in (12), one can show that it is a solution. This substitution gives

$$\phi(s) = f(s) - \int_a^b K(s, t)f(t)\, dt + \int_a^b \int_a^b K(s, r)K(r, t)f(t)\, dr\, dt + \cdots,$$

which can be written in the form

$$(15) \qquad \phi(s) = f(s) + \int_a^b \overline{K}(s, t)f(t)\, dt,$$

where the kernel $\overline{K}$ (later called the solving kernel or resolvent by Hilbert) is

$$\overline{K}(s, t) = -K(s, t) + \int_a^b K(s, r)K(r, t)\, dr$$

$$- \int_a^b \int_a^b K(s, r)K(r, w)K(w, t)\, dr\, dw + \cdots.$$

10. *Atti della Accad. dei Lincei, Rendiconti*, (5), 5, 1896, 177–85, 289–300; *Atti Accad. Torino*, 31, 1896, 311–23, 400–8, 557–67, 693–708; *Annali di Mat.*, (2), 25, 1897, 139–78; all are in his *Opere matematiche*, 2, 216–313.

Equation (15) is the representation obtained earlier for a particular integral equation by Liouville and credited to Neumann. Volterra also solved integral equations of the first kind $f(s) = \int_a^s K(x, s)\phi(x)\, dx$ by reducing them to equations of the second kind.

In 1896 Volterra observed that an integral equation of the first kind is a limiting form of a system of n linear algebraic equations in n unknowns as n becomes infinite. Erik Ivar Fredholm (1866–1927), professor of mathematics at Stockholm, concerned with solving the Dirichlet problem, took up this idea in 1900[11] and used it to solve integral equations of the second kind, that is, equations of the form (12), without, however, the restriction on $K(s, t)$.

We shall write the equation Fredholm tackled in the form

$$(16) \qquad u(x) = f(x) + \lambda \int_a^b K(x, \xi)u(\xi)\, d\xi,$$

though the parameter λ was not explicit in his work. However, what he did is more intelligible in the light of later work if we exhibit it. To be faithful to Fredholm's formulas, one should set $\lambda = 1$ or regard it as implicitly involved in K.

Fredholm divided the x-interval $[a, b]$ into n equal parts by the points

$$a, x_1 = a + \delta, x_2 = a + 2\delta, \cdots, x_n = a + n\delta = b.$$

He then replaced the definite integral in (16) by the sum

$$(17) \qquad u_n(x) = f(x) + \sum_{j=1}^{n} \lambda K(x, x_j)u_n(x_j)\delta.$$

Now equation (17) is supposed to hold for all values of x in $[a, b]$. Hence it should hold for $x = x_1, x_2, \ldots, x_n$. This gives the system of n equations

$$(18) \quad -\sum_{j=1}^{n} \lambda K(x_i, x_j)u_n(x_j)\delta + u_n(x_i) = f(x_i), \qquad i = 1, 2, \cdots, n.$$

This system is a set of n nonhomogeneous linear equations for determining the n unknowns $u_n(x_1), u_n(x_2), \ldots, u_n(x_n)$.

In the theory of linear equations the following result was known: If the matrix

$$S_n = \begin{Vmatrix} 1 + a_{11} & a_{12} & a_{13} & \cdots & a_{1n} \\ a_{21} & 1 + a_{22} & a_{23} & \cdots & a_{2n} \\ \cdots\cdots\cdots\cdots\cdots\cdots\cdots\cdots\cdots\cdots\cdots\cdots \\ a_{n1} & a_{n2} & a_{n3} & \cdots & 1 + a_{nn} \end{Vmatrix}$$

11. *Acta Math.*, 27, 1903, 365–90.

then the determinant $D(n)$ of S_n has the following expansion:[12]

$$D(n) = 1 + \frac{1}{1!} \sum_{r_1} a_{r_1 r_1} + \frac{1}{2!} \sum_{r_1, r_2} \begin{vmatrix} a_{r_1 r_1} & a_{r_1 r_2} \\ a_{r_2 r_1} & a_{r_2 r_2} \end{vmatrix} + \cdots$$

$$+ \frac{1}{n!} \sum_{r_1, \ldots, r_n} \begin{vmatrix} a_{r_1 r_1} & a_{r_1 r_2} & \cdots & a_{r_1 r_n} \\ a_{r_2 r_1} & a_{r_2 r_2} & \cdots & a_{r_2 r_n} \\ \cdots\cdots\cdots\cdots\cdots\cdots\cdots \\ a_{r_n r_1} & a_{r_n r_2} & \cdots & a_{r_n r_n} \end{vmatrix}$$

where $r_1, r_2, \ldots, r_n$ run independently over all the values from 1 to n. By expanding the determinant of the coefficients in (18), and then letting n become infinite, Fredholm obtained the determinant

$$(19) \quad D(\lambda) = 1 - \lambda \int_a^b K(\xi_1, \xi_1)\, d\xi_1$$

$$+ \frac{\lambda^2}{2!} \int_a^b \int_a^b \begin{vmatrix} K(\xi_1, \xi_1) & K(\xi_1, \xi_2) \\ K(\xi_2, \xi_1) & K(\xi_2, \xi_2) \end{vmatrix} d\xi_1\, d\xi_2 - \cdots.$$

This he called the determinant of (16) or of the kernel K. Likewise, by considering the cofactor of the element in the νth row and μth column of the determinant of the coefficients in (18) and letting n become infinite, Fredholm obtained the function

$$(20) \quad D(x, y, \lambda) = \lambda K(x, y) - \lambda^2 \int_a^b \begin{vmatrix} K(x, y) & K(x, \xi_1) \\ K(\xi_1, y) & K(\xi_1, \xi_1) \end{vmatrix} d\xi_1$$

$$+ \frac{\lambda^3}{2} \int_a^b \int_a^b \begin{vmatrix} K(x, y) & K(x, \xi_1) & K(x, \xi_2) \\ K(\xi_1, y) & K(\xi_1, \xi_1) & K(\xi_1, \xi_2) \\ K(\xi_2, y) & K(\xi_2, \xi_1) & K(\xi_2, \xi_2) \end{vmatrix} d\xi_1\, d\xi_2 - \cdots.$$

Fredholm called $D(x, y, \lambda)$ the first minor of the kernel K because it plays the role analogous to first minors in the case of n linear equations in n unknowns. He also called the zeros of the integral analytic function $D(\lambda)$ the roots of $K(x, y)$. By applying Cramer's rule to the system of linear equations (18) and by letting n become infinite, Fredholm inferred the form of the solution of (16). He then proved that it was correct by direct substitution and could assert the following results: If λ is not one of the roots of K, that is, if $D(\lambda) \neq 0$, (16) has one and only one (continuous) solution, namely,

$$(21) \quad u(x, \lambda) = f(x) + \int_a^b \frac{D(x, y, \lambda)}{D(\lambda)} f(y)\, dy.$$

Further, if λ is a root of $K(x, y)$, then (16) has either no continuous solution or an infinite number of them.

12. A fine exposition can be found in Gerhard Kowalewski, *Integralgleichungen*, Walter de Gruyter, 1930, pp. 101–34.

Fredholm obtained further results involving the relation between the homogeneous equation

$$(22) \qquad u(x) = \lambda \int_a^b K(x, \xi)u(\xi) \, d\xi$$

and the inhomogeneous equation (16). It is almost evident from (21) that when λ is not a root of K the only continuous solution of (22) is $u \equiv 0$. Hence he considered the case when λ is a root of K. Let $\lambda = \lambda_1$ be such a root. Then (22) has the infinite number of solutions

$$c_1 u_1(x) + c_2 u_2(x) + \cdots + c_n u_n(x),$$

where the c_i's are arbitrary constants; the $u_1, u_2, \ldots, u_n$, called principal solutions, are linearly independent; and n depends upon λ_1. The quantity n is called the index of λ_1 [which is not the multiplicity of λ_1 as a zero of $D(\lambda)$]. Fredholm was able to determine the index of any root λ_i and to show that the index can never exceed the multiplicity (which is always finite). The roots of $D(\lambda) = 0$ are called the characteristic values of $K(x, y)$ and the set of roots is called the spectrum. The solutions of (22) corresponding to the characteristic values are called eigenfunctions or characteristic functions.

And now Fredholm was able to establish what has since been called the Fredholm alternative theorem. In the case where λ is a characteristic value of K not only does the integral equation (22) have n independent solutions but the associated or adjoint equation which has the transposed kernel, namely,

$$u(x) = \lambda \int_a^b K(\xi, x)u(\xi) \, d\xi,$$

also has n solutions $\psi_1(x), \ldots, \psi_n(x)$ for the same characteristic value and then the nonhomogeneous equation (16) is solvable if and only if

$$(23) \qquad \int_a^b f(x)\psi_i(x) \, dx = 0, \qquad i = 1, 2, \cdots, n.$$

These last few results parallel very closely the theory of a system of linear algebraic equations, homogeneous and nonhomogeneous.

3. The Work of Hilbert

A lecture by Erik Holmgren (b. 1872) in 1901 on Fredholm's work on integral equations, which had already been published in Sweden, aroused Hilbert's interest in the subject. David Hilbert (1862–1943), the leading mathematician of this century, who had already done superb work on algebraic numbers, algebraic invariants, and the foundations of geometry, now turned his attention to integral equations. He says that an investigation of the subject showed him that it was important for the theory of definite integrals, for the development of arbitrary functions in series (of special functions or trigonometric functions), for the theory of linear differential equations, for potential theory, and for the calculus of variations. He wrote a series of six papers from 1904 to 1910 in the *Nachrichten von der Königlichen Gesellschaft der Wissenschaften zu Göttingen* and reproduced these in his book *Grundzüge einer allgemeinen Theorie der linearen Integralgleichungen* (1912). During the latter part of this work he applied integral equations to problems of mathematical physics.

Fredholm had used the analogy between integral equations and linear algebraic equations, but instead of carrying out the limiting processes for the infinite number of algebraic equations, he boldly wrote down the resulting determinants and showed that they solved the integral equations. Hilbert's first work was to carry out a rigorous passage to the limit on the finite system of linear equations.

He started with the integral equation

$$(24) \qquad f(s) = \phi(s) - \lambda \int_0^1 K(s, t)\phi(t) \, dt,$$

wherein $K(s, t)$ is continuous. The parameter λ is explicit and plays a significant role in the subsequent theory. Like Fredholm, Hilbert divided up the interval $[0, 1]$ into n parts so that p/n or q/n ($p, q = 1, 2, \ldots, n$) denotes a coordinate in the interval $[0, 1]$. Let

$$K_{pq} = K\left(\frac{p}{n}, \frac{q}{n}\right), \qquad f_p = f\left(\frac{p}{n}\right), \qquad \phi_q = \phi\left(\frac{q}{n}\right).$$

Then from (24) we obtain the system of n equations in n unknowns $\phi_1, \ldots, \phi_n$, namely,

$$f_p = \phi_p - \lambda \sum_{q=1}^{n} K_{pq}\phi_q, \qquad p = 1, 2, \cdots, n.$$

After reviewing the theory of solution of a finite system of n linear equations in n unknowns, Hilbert considers equation (24). For the kernel K of (24) the eigenvalues are defined to be zeros of the power series

$$\delta(\lambda) = 1 + \sum_{n=1}^{\infty} (-1)^n d_n \lambda^n,$$

where the coefficients d_n are given by

$$d_n = \frac{1}{n!} \int_0^1 \cdots \int_0^1 |\{K(s_i, s_j)\}| \, ds_1 \cdots ds_n.$$

Here $|\{K(s_i, s_j)\}|$ is the determinant of the n by n matrix $\{K(s_i, s_j)\}$, $i, j = 1, 2, \ldots, n$, and the s_i are values of t in the interval $[0, 1]$. To indicate Hilbert's major result we need the intermediate quantities

$$\Delta_p(x, y) = \frac{1}{p!} \int_0^1 \cdots \int_0^1 \begin{vmatrix} 0 & x(s_1) & \cdots & x(s_p) \\ y(s_1) & K(s_1, s_2) & \cdots & K(s_1, s_p) \\ \vdots & \vdots & \vdots & \vdots \\ y(s_p) & K(s_p, s_1) & \cdots & K(s_p, s_p) \end{vmatrix} ds_1 \cdots ds_p,$$

wherein $x(r)$ and $y(r)$ are arbitrary continuous functions of r on $[0, 1]$, and

$$\Delta(\lambda; x, y) = \sum_{p=1}^{\infty} (-1)^p \Delta_p(x, y) \lambda^{p-1}.$$

Hilbert next defines

$$\Delta^*(\lambda; s, t) = \lambda \Delta(\lambda; x, y) - \delta(\lambda)$$

wherein now $x(r) = K(s, r)$ and $y(r) = K(r, t)$. He then proves that if $\overline{K}$ is defined by

$$\overline{K}(s, t) = \frac{\Delta^*(\lambda; s, t)}{-\delta(\lambda)}$$

for values of λ for which $\delta(\lambda) \neq 0$, then

$$K(s, t) = \overline{K}(s, t) - \lambda \int_0^1 \overline{K}(s, r) K(r, t) \, dr$$

$$= \overline{K}(s, t) - \lambda \int_0^1 K(s, r) \overline{K}(r, t) \, dr.$$

Finally if ϕ is taken to be

(25) $$\phi(r) = f(r) + \lambda \int_0^1 \overline{K}(r, t) f(t) \, dt,$$

then ϕ is a solution of (24). The proofs of various steps in this theory involve a number of limit considerations on expressions which occur in Hilbert's treatment of the finite system of linear equations.

Thus far Hilbert showed that for any continuous (not necessarily symmetric) kernel $K(s, t)$ and for any value of λ such that $\delta(\lambda) \neq 0$, there exists the solving function (resolvent) $\overline{K}(s, t)$, which has the property that (25) solves equation (24).

Now Hilbert assumes $K(s, t)$ to be symmetric, which enables him to use facts about symmetric matrices in the finite case, and shows that the zeros of $\delta(\lambda)$, that is, the eigenvalues of the symmetric kernel, are real. Then the zeros of $\delta(\lambda)$ are ordered according to increasing absolute values (for equal absolute values the positive zero is taken first and any multiplicities are to be counted). The eigenfunctions of (24) are now defined by

$$\phi^k(s) = \left(\frac{\lambda_k}{\Delta^*(\lambda_k; s^*, s^*)}\right)^{1/2} \Delta^*(\lambda_k; s, s^*),$$

where s^* is chosen so that $\Delta^*(\lambda_k; s^*, s^*) \neq 0$ and λ_k is any eigenvalue of $K(s, t)$.

The eigenfunctions associated with the separate eigenvalues can be chosen to be an orthonormal (orthogonal and normalized[13]) set and for each eigenvalue λ_k and for each eigenfunction belonging to λ_k

$$\phi^k(s) = \lambda_k \int_0^1 K(s, t)\phi^k(t) \, dt.$$

With these results Hilbert is able to prove what is called the generalized principal axis theorem for symmetric quadratic forms. First, let

(26) $$\sum_{p=1}^{n} \sum_{q=1}^{n} k_{pq}x_p x_q$$

be an n-dimensional quadratic form in the n variables $x_1, x_2, \ldots, x_n$. This can be written as (Kx, x) where K is the matrix of the k_{pq}, x stands for the vector $(x_1, x_2, \ldots, x_n)$ and (Kx, x) is the inner product (scalar product) of the two vectors Kx and x. Suppose K has the n distinct eigenvalues $\lambda_1, \lambda_2, \ldots, \lambda_n$. Then for any fixed λ_k the equations

(27) $$0 = \phi_p - \lambda_p \sum_{q=1}^{n} k_{pq}\phi_q, \qquad p = 1, 2, \cdots, n$$

have the solution

$$\phi^k = (\phi_1^k, \phi_2^k, \cdots, \phi_n^k),$$

which is a unique solution up to a constant multiple. It is then possible, as Hilbert showed, to write

(28) $$K(x, x) = \sum_{k=1}^{n} \frac{1}{\lambda_k} \frac{(\phi^k, x)^2}{(\phi^k, \phi^k)}$$

wherein the parentheses again denote an inner product of vectors.

Hilbert's generalized principal axis theorem reads as follows: Let $K(s, t)$ be a continuous symmetric function of s and t. Let $\phi^p(s)$ be the normalized eigenfunction belonging to the eigenvalue λ_p of the integral equation

13. Normalization means that $\phi^k(s)$ is modified so that $\int_0^1 (\phi^k)^2 \, ds = 1$.

(24). Then for arbitrary continuous $x(s)$ and $y(s)$, the following relation holds:

$$(29) \quad \int_a^b \int_a^b K(s, t)x(s)y(t)\,ds\,dt = \sum_{p=1}^{\alpha} \frac{1}{\lambda_p} \left(\int_a^b \phi^p(s)x(s)\,ds \right) \left(\int_a^b \phi^p(s)y(s)\,ds \right),$$

where $\alpha = n$ or ∞ depending on the number of eigenvalues and in the latter case the sum converges uniformly and absolutely for all $x(s)$ and $y(s)$ which satisfy

$$\int_a^b x^2(s)\,ds < \infty \quad \text{and} \quad \int_a^b y^2(s)\,ds < \infty.$$

The generalization of (28) to (29) becomes apparent if we first define $\int_a^b u(s)v(s)\,ds$ as the inner product of any two functions $u(s)$ and $v(s)$ and denote it by (u, v). Now replace $y(s)$ in (29) by $x(s)$ and replace the left side of (28) by integration instead of summation.

Hilbert proved next a famous result, later called the Hilbert-Schmidt theorem. If $f(s)$ is such that for some continuous $g(s)$

$$(30) \qquad\qquad f(s) = \int_a^b K(s, t)\dot{g}(t)\,dt$$

then

$$(31) \qquad\qquad f(s) = \sum_{p=1}^{\infty} c_p \phi^p,$$

where the ϕ^p are the orthonormal eigenfunctions of K and

$$(32) \qquad\qquad c_p = \int_a^b \phi^p(s)f(s)\,ds.$$

Thus an "arbitrary" function $f(s)$ can be expressed as an infinite series in the eigenfunctions of K with coefficients c_p that are the "Fourier" coefficients of the expansion.

Hilbert, in the preceding work, had carried out limiting processes that generalized results on finite systems of linear equations and finite quadratic forms to integrals and integral equations. He decided that a treatment of infinite quadratic forms themselves, that is, quadratic forms with infinitely many variables, would "form an essential completion of the well-known theory of quadratic forms with finitely many variables." He therefore took up what may be called purely algebraic questions. He considers the infinite bilinear form

$$K(x, y) = \sum_{p,q=1}^{\infty} k_{pq} x_p y_q,$$

and by passing to the limit of results for bilinear and quadratic forms in $2n$ and n variables respectively, obtains basic results. The details of the work are considerable and we shall only note some of the results. Hilbert first obtains an expression for a resolvent form $\overline{K}(\lambda; x, x)$, which has the peculiar feature that it is the sum of expressions, one for each of a discrete set of values of λ, and of an integral over a set of λ belonging to a continuous range. The discrete set of λ values belongs to the point spectrum of K, and the continuous set to the continuous or band spectrum. This is the first significant use of continuous spectra, which had been observed for partial differential equations by Wilhelm Wirtinger (b. 1865) in 1896.[14]

To get at the key result for quadratic forms, Hilbert introduces the notion of a bounded form. The notation (x, x) denotes the inner (scalar) product of the vector $(x_1, x_2, \ldots, x_n, \ldots)$ with itself and (x, y) has the analogous meaning. Then the form $K(x, y)$ is said to be bounded if $|K(x, y)| \leq M$ for all x and y such that $(x, x) \leq 1$ and $(y, y) \leq 1$. Boundedness implies continuity, which Hilbert defines for a function of infinitely many variables.

Hilbert's key result is the generalization to quadratic forms in infinitely many variables of the more familiar principal axis theorem of analytic geometry. He proves that there exists an orthogonal transformation T such that in the new variables x_i', where $x' = Tx$, K can be reduced to a "sum of squares." That is, every bounded quadratic form $K(x, x) = \sum_{p,q=1}^{\infty} k_{pq} x_p x_q$ can be transformed by a unique orthogonal transformation into the form

$$(33) \qquad K(x, x) = \sum_{i=1}^{\infty} k_i x_i^2 + \int_{(s)} \frac{d\sigma(\mu, \xi)}{\mu},$$

where the k_i are the reciprocal eigenvalues of K. The integral, which we shall not describe further, is over a continuous range of eigenvalues or a continuous spectrum.

To eliminate the continuous spectrum, Hilbert introduces the concept of complete continuity. A function $F(x_1, x_2, \ldots)$ of infinitely many variables is said to be completely continuous at $a = (a_1, a_2, \ldots)$ if

$$\lim_{\substack{\varepsilon_1 \to 0, \\ \varepsilon_2 \to 0, \\ \cdots}} F(a_1 + \varepsilon_1, a_2 + \varepsilon_2, \cdots) = F(a_1, a_2, \cdots)$$

whenever $\varepsilon_1, \varepsilon_2, \ldots$ are allowed to run through any value system $\varepsilon_1^{(h)}, \varepsilon_2^{(h)}, \ldots$ having the limit

$$\lim_{h \to \infty} \varepsilon_1^{(h)} = 0, \lim_{h \to \infty} \varepsilon_2^{(h)} = 0, \cdots.$$

This is a stronger requirement than continuity as previously introduced by Hilbert.

14. *Math. Ann.*, 48, 1897, 365–89.

For a quadratic form $K(x, x)$ to be completely continuous it is sufficient that $\sum_{p,q=1}^{\infty} k_{pq}^2 < \infty$. With the requirement of complete continuity, Hilbert is able to prove that if K is a completely continuous bounded form, then by an orthogonal transformation it can be brought into the form

$$(34) \qquad\qquad K(x, x) = \sum_j k_j x_j^2,$$

where the k_j are reciprocal eigenvalues and the $(x_1, x_2, \ldots)$ satisfy the condition that $\sum_1^{\infty} x_i^2$ is finite.

Now Hilbert applies his theory of quadratic forms in infinitely many variables to integral equations. The results in many instances are not new but are obtained by clearer and simpler methods. Hilbert starts this new work on integral equations by defining the important concept of a complete orthogonal system of functions $\{\phi_p(s)\}$. This is a sequence of functions all defined and continuous on the interval $[a, b]$ with the following properties:

(a) orthogonality:

$$\int_a^b \phi_p(s)\phi_q(s) = \delta_{pq}, \qquad p, q = 1, 2, \cdots.$$

(b) completeness: for every pair of functions u and v defined on $[a, b]$

$$\int_a^b u(s)v(s)\, ds = \sum_{p=1}^{\infty} \int_a^b \phi_p(s)u(s)\, ds \int_a^b \phi_p(s)v(s)\, ds.$$

The value

$$u_p^* = \int_a^b \phi_p(s)u(s)\, ds$$

is called the Fourier coefficient of $u(s)$ with respect to the system $\{\phi_p\}$.

Hilbert shows that a complete orthonormal system can be defined for any finite interval $[a, b]$, for example, by the use of polynomials. Then a generalized Bessel's inequality is proved, and finally the condition

$$\int_a^b u^2(s)\, ds = \sum_{p=1}^{\infty} u_p^{*2}$$

is shown to be equivalent to completeness.

Hilbert now turns to the integral equation

$$(35) \qquad\qquad f(s) = \phi(s) + \int_a^b K(s, t)\phi(t)\, dt.$$

The kernel $K(s, t)$, not necessarily symmetric, is developed in a double "Fourier" series by means of the coefficients

$$a_{pq} = \int_a^b \int_a^b K(s, t)\phi_p(s)\phi_q(t) \, ds \, dt.$$

It follows that

$$\sum_{p,q=1}^{\infty} a_{pq}^2 \leq \int_a^b \int_a^b K^2(s, t) \, ds \, dt.$$

Also, if

$$a_p = \int_a^b \phi_p(s)f(s) \, ds,$$

that is, if the a_p are the "Fourier" coefficients of $f(s)$, then $\sum_{p=1}^{\infty} a_p^2 < \infty$.

Hilbert next converts the above integral equation into a system of infinitely many linear equations in infinitely many unknowns. The idea is to look at solving the integral equation for $\phi(s)$ as a problem of finding the "Fourier" coefficients of $\phi(s)$. Denoting the as yet unknown coefficients by $x_1, x_2, \ldots$, he gets the following linear equations:

$$(36) \qquad x_p + \sum_{q=1}^{\infty} a_{pq}x_q = a_p, \qquad p = 1, 2, \cdots.$$

He proves that if this system has a unique solution, then the integral equation has a unique continuous solution, and when the linear homogeneous system associated with (36) has n linearly independent solutions, then the homogeneous integral equation associated with (35),

$$(37) \qquad 0 = \phi(s) + \int_a^b K(s, t)\phi(t) \, dt,$$

has n linearly independent solutions. In this case the original nonhomogeneous integral equation has a solution if and only if $\psi^{(h)}(s)$, $h = 1, 2, \ldots, n$, which are the n linearly independent solutions of the transposed homogeneous equation

$$\phi(s) + \int_a^b K(t, s)\phi(t) \, dt = 0$$

and which also exist when (37) has n solutions, satisfy the conditions

$$(38) \qquad \int_a^b \psi^{(h)}(s)f(s) \, ds = 0, \qquad h = 1, 2, \cdots, n.$$

Thus the Fredholm alternative theorem is obtained: Either the equation

$$(39) \qquad f(s) = \phi(s) + \int_a^b K(s, t)\phi(t) \, dt$$

has a unique solution for all f or the associated homogeneous equation has n linearly independent solutions. In the latter case (39) has a solution if and only if the orthogonality conditions (38) are satisfied.

Hilbert turns next to the eigenvalue problem

$$(40) \qquad f(s) = \phi(s) - \lambda \int_a^b K(s, t)\phi(t) \, dt$$

where K is now symmetric. The symmetry of K implies that its "Fourier" coefficients determine a quadratic form $K(x, x)$ which is completely continuous. He shows that there exists an orthogonal transformation T whose matrix is $\{l_{pq}\}$ such that

$$K(x', x') = \sum_{p=1}^{\infty} \mu_p {x'_p}^2,$$

where the μ_p are the reciprocal eigenvalues of the quadratic form $K(x, x)$. The eigenfunctions $\{\phi_p(s)\}$ for the kernel $K(s, t)$ are now defined by

$$L_p(K(s)) = \sum_{q=1}^{\infty} l_{pq} \int_0^b K(s, t) \Phi_q(t) \, dt = \mu_p \phi_p(s)$$

where the $\Phi_q(t)$ are a given complete orthonormal set. The $\phi_p(s)$ [as distinct from the $\Phi_q(t)$] are shown to form an orthonormal set and to satisfy

$$\phi_p(s) = \lambda_p \int_a^b K(s, t)\phi_p(t) \, dt$$

where $\lambda_p = 1/\mu_p$. Thus Hilbert shows anew the existence of eigenfunctions for the homogeneous case of (40) and for every finite eigenvalue of the quadratic form $K(x, x)$ associated with the kernel $K(s, t)$ of (40).

Now Hilbert establishes again (Hilbert-Schmidt theorem) that if $f(s)$ is any continuous function for which there is a g so that

$$\int_a^b K(s, t) g(t) \, dt = f(s),$$

then f is representable as a series in the eigenfunctions of K which is uniformly and absolutely convergent (cf. [31]). Hilbert uses this result to show that the homogeneous equation associated with (40) has no nontrivial solutions except at the eigenvalues λ_p. Then the Fredholm alternative theorem takes the form: For $\lambda \neq \lambda_p$ equation (40) has a unique solution. For $\lambda = \lambda_p$, equation (40) has a solution if and only if the n_p conditions

$$\int_a^b \phi_{p+j}(s) f(s) = 0, \qquad j = 1, 2, \cdots, n_p$$

are satisfied where the $\phi_{p+j}(s)$ are the n_p eigenfunctions associated with λ_p. Finally, he proves anew the extension of the principal axis theorem:

$$\int_a^b \int_a^b K(s, t)u(s)u(t)\, ds\, dt = \sum_{p=1}^{\infty} \frac{1}{\lambda_p} \left\{ \int_a^b u(t)\phi_p(t)\, dt \right\}^2,$$

where $u(s)$ is an arbitrary (continuous) function and wherein all ϕ_p associated with any λ_p are included in the summation.

This later work (1906) of Hilbert dispensed with Fredholm's infinite determinants. In it he showed directly the relation between integral equations and the theory of complete orthogonal systems and the expansion of functions in such systems.

Hilbert applied his results on integral equations to a variety of problems in geometry and physics. In particular, in the third of the six papers he solved Riemann's problem of constructing a function holomorphic in a domain bounded by a smooth curve when the real or the imaginary part of the boundary value is given or both are related by a given linear equation.

One of the most noteworthy achievements in Hilbert's work, which appears in the 1904 and 1905 papers, is the formulation of Sturm-Liouville boundary-value problems of differential equations as integral equations. Hilbert's result states that the eigenvalues and eigenfunctions of the differential equation

(41)
$$\frac{d}{dx}\left(p(x)\frac{du}{dx}\right) + q(x)u + \lambda u = 0$$

subject to the boundary conditions

$$u(a) = 0, \qquad u(b) = 0$$

(and even more general boundary conditions) are the eigenvalues and eigenfunctions of

(42)
$$\phi(x) - \lambda \int_a^b G(x, \xi)\phi(\xi)\, d\xi = 0,$$

where $G(x, \xi)$ is the Green's function for (41), that is, a particular solution of

$$\frac{d}{dx}\left(p\frac{du}{dx}\right) + q(x)u = 0$$

which satisfies certain differentiability conditions and whose first partial derivative $\partial G/\partial x$ has a jump singularity at $x = \xi$ equal to $-1/p(\xi)$. Similar results hold for partial differential equations. Thus integral equations are a way of solving ordinary and partial differential equations.

To recapitulate Hilbert's major results, first of all he established the general spectral theory for symmetric kernels K. Only twenty years earlier, it had required great mathematical efforts (Chap. 28, sec. 8) to prove the existence of the lowest oscillating frequency for a membrane. With integral equations, constructive proof of the existence of the whole series of frequencies and of the actual eigenfunctions was obtained under very general conditions on the oscillating medium. These results were first derived, using Fredholm's theory, by Emile Picard.[15] Another noteworthy result due to Hilbert is that the development of a function in the eigenfunctions belonging to an integral equation of the second kind depends on the solvability of the corresponding integral equation of the first kind. In particular, Hilbert discovered that the success of Fredholm's method rested on the notion of complete continuity, which he carried over to bilinear forms and studied intensively. Here he inaugurated the spectral theory of bilinear symmetric forms.

After Hilbert showed how to convert problems of differential equations to integral equations, this approach was used with increasing frequency to solve physical problems. Here the use of a Green's function to convert has been a major tool. Also, Hilbert himself showed,[16] in problems of gas dynamics, that one can go directly to integral equations. This direct recourse to integral equations is possible because the concept of summation proves as fundamental in some physical problems as the concept of rate of change which leads to differential equations is in other problems. Hilbert also emphasized that not ordinary or partial differential equations but integral equations are the necessary and natural starting point for the theory of expansion of functions in series, and that the expansions obtained through differential equations are just special cases of the general theorem in the theory of integral equations.

4. The Immediate Successors of Hilbert

Hilbert's work on integral equations was simplified by Erhard Schmidt (1876–1959), professor at several German universities, who used methods originated by H. A. Schwarz in potential theory. Schmidt's most significant contribution was his generalization in 1907 of the concept of eigenfunction to integral equations with non-symmetric kernels.[17]

Friedrich Riesz (1880–1956), a Hungarian professor of mathematics, in 1907 also took up Hilbert's work.[18] Hilbert had treated integral equations of the form

$$f(s) = \phi(s) + \int_a^b K(s, t)\phi(t) \, dt,$$

15. *Rendiconti del Circolo Matematico di Palermo*, 22, 1906, 241–59.
16. *Math. Ann.*, 72, 1912, 562–77 = *Grundzüge*, Chap. 22.
17. *Math. Ann.*, 63, 1907, 433–76 and 64, 1907, 161–74.
18. *Comp. Rend.*, 144, 1907, 615–19, 734–36, 1409–11.

where f and K are continuous. Riesz sought to extend Hilbert's ideas to more general functions $f(s)$. Toward this end it was necessary to be sure that the "Fourier" coefficients of f could be determined with respect to a given orthonormal sequence of functions $\{\phi_p\}$. He was also interested in discovering under what circumstances a given sequence of numbers $\{a_p\}$ could be the Fourier coefficients of some function f relative to a given orthonormal sequence $\{\phi_p\}$.

Riesz introduced functions whose squares are Lebesgue integrable and obtained the following theorem: Let $\{\phi_p\}$ be an orthonormal sequence of Lebesgue square integrable functions all defined on the interval $[a, b]$. If $\{a_p\}$ is a sequence of real numbers, then the convergence of $\sum_{p=1}^{\infty} a_p^2$ is a necessary and sufficient condition for there to exist a function f such that

$$\int_a^b f(x)\phi_p(x)\, dx = a_p$$

for each ϕ_p and a_p. The function f proves to be Lebesgue square integrable. This theorem establishes a one-to-one correspondence between the set of Lebesgue square integrable functions and the set of square summable sequences through the mediation of any orthonormal sequence of Lebesgue square integrable functions.

With the introduction of Lebesgue integrable functions, Riesz was also able to show that the integral equation of the second kind

$$f(s) = \phi(s) + \int_a^b K(s, t)\phi(t)\, dt$$

can be solved under the relaxed conditions that $f(s)$ and $K(s, t)$ are Lebesgue square integrable. Solutions are unique up to a function whose Lebesgue integral on $[a, b]$ is 0.

In the same year that Riesz published his first papers, Ernst Fischer (1875–1959), a professor at the University of Cologne, introduced the concept of convergence in the mean.[19] A sequence of functions $\{f_n\}$ defined on an interval $[a, b]$ is said to converge in the mean if

$$\lim_{m,n \to \infty} \int_a^b (f_n(x) - f_m(x))^2\, dx = 0,$$

and $\{f_n\}$ is said to converge in the mean to f if

$$\lim_{n \to \infty} \int_a^b (f - f_n)^2\, dx = 0.$$

The integrals are Lebesgue integrals. The function f is uniquely determined to within a function defined over a set of measure 0, that is, a function $g(x) \neq 0$, called a null function, which satisfies the condition $\int_a^b g^2(x)\, dx = 0$.

19. *Comp. Rend.*, 144, 1907, 1022–24.

The set of functions which are Lebesgue square integrable on an interval $[a, b]$ was denoted later by $L^2(a, b)$ or simply by L^2. Fischer's main result is that $L^2(a, b)$ is complete in the mean; that is, if the functions f_n belong to $L^2(a, b)$ and if $\{f_n\}$ converges in the mean, then there is a function f in $L^2(a, b)$ such that $\{f_n\}$ converges in the mean to f. This completeness property is the chief advantage of square summable functions. Fischer then deduced Riesz's above theorem as a corollary, and this result is known as the Riesz-Fischer theorem. Fischer emphasized in a subsequent note[20] that the use of Lebesgue square integrable functions was essential. No smaller set of functions would suffice.

The determination of a function $f(x)$ that belongs to a given set of Fourier coefficients $\{a_n\}$ with respect to a given sequence of orthonormal functions $\{g_n\}$, or the determination of an f satisfying

$$\int_a^b g_n(x) f(x)\, dx = a_n, \qquad n = 1, 2, \cdots,$$

which had arisen in Riesz's 1907 works, is called the moment problem (Lebesgue integration is understood). In 1910[21] Riesz sought to generalize this problem. Because in this new work Riesz used the Hölder inequalities

$$\sum_{i=1}^n |a_i b_i| \leq \left(\sum_{i=1}^n |a_i|^p \right)^{1/p} \left(\sum_{i=1}^n |b_i|^q \right)^{1/q}$$

and

$$\left| \int_M f(x) g(x)\, dx \right| \leq \left(\int_M |f|^p\, dx \right)^{1/p} \left(\int_M |g|^q\, dx \right)^{1/q},$$

where $1/p + 1/q = 1$, and other inequalities, he was obliged to introduce the set L^p of functions f measurable on a set M and for which $|f|^p$ is Lebesgue integrable on M. His first important theorem is that if a function $h(x)$ is such that the product $f(x)h(x)$ is integrable for every f in L^p, then h is in L^q and, conversely, the product of an L^p function and an L^q function is always (Lebesgue) integrable. It is understood that $p > 1$ and $1/p + 1/q = 1$.

Riesz also introduced the concepts of strong and weak convergence. The sequence of functions $\{f_n\}$ is said to converge strongly to f (in the mean of order p) if

$$\lim_{n \to \infty} \int_a^b |f_n(x) - f(x)|^p\, dx = 0.$$

20. *Comp. Rend.*, 144, 1907, 1148–50.
21. *Math. Ann.*, 69, 1910, 449–97.

The sequence $\{f_n\}$ is said to converge weakly to f if

$$\int_a^b |f_n(x)|^p \, dx < M$$

for M independent of n, and if for every x in $[a, b]$

$$\lim_{n \to \infty} \int_a^x (f_n(t) - f(t)) \, dt = 0.$$

Strong convergence implies weak convergence. (The modern definition of weak convergence, if $\{f_n\}$ belongs to L^p and if f belongs to L^p and if

$$\lim_{n \to \infty} \int_a^b (f(x) - f_n(x))g(x) \, dx = 0$$

holds for every g in L^q, then $\{f_n(x)\}$ converges weakly to f, is equivalent to Riesz's.)

In the same 1910 paper Riesz extended the theory of the integral equation

$$\phi(x) - \lambda \int_a^b K(x, t)\phi(t) \, dt = f(x)$$

to the case where the given f and the unknown ϕ are functions in L^p. The results on solution of the eigenvalue problem for this integral equation are analogous to Hilbert's results. What is more striking is that, to carry out his work, Riesz introduced the abstract concept of an operator, formulated for it the Hilbertian concept of complete continuity, and treated abstract operator theory. We shall say more about this abstract approach in the next chapter. Among other results, Riesz proved that the continuous spectrum of a real completely continuous operator in L^2 is empty.

5. *Extensions of the Theory*

The importance Hilbert attached to integral equations made the subject a world-wide fad for a considerable length of time, and an enormous literature, most of it of ephemeral value, was produced. However, some extensions of the subject have proved valuable. We can merely name them.

The theory of integral equations presented above deals with linear integral equations; that is, the unknown function $u(x)$ enters linearly. The theory has been extended to nonlinear integral equations, in which the unknown function enters to the second or higher degree or in some more complicated fashion.

Moreover, our brief sketch has said little about the conditions on the given functions $f(x)$ and $K(x, \xi)$ which lead to the many conclusions. If these functions are not continuous and the discontinuities are not limited, or if the interval $[a, b]$ is replaced by an infinite interval, many of the results are altered or at least new proofs are needed. Thus even the Fourier transform

$$f(x) = \sqrt{\frac{2}{\pi}} \int_0^\infty \cos(x\xi)u(\xi)\,d\xi,$$

which can be regarded as an integral equation of the first kind and has as its solution the inverse transform

$$u(x) = \sqrt{\frac{2}{\pi}} \int_0^\infty \cos(x\xi)f(\xi)\,d\xi,$$

has just two eigenvalues ± 1 and each has an infinite number of eigenfunctions. These cases are now studied under the heading of singular integral equations. Such equations cannot be solved by the methods applicable to the Volterra and Fredholm equations. Moreover, they exhibit a curious property, namely, there are continuous intervals of λ-values or band spectra for which there are solutions. The first significant paper on this subject is due to Hermann Weyl (1885–1955).[22]

The subject of existence theorems for integral equations has also been given a great deal of attention. This work has been devoted to linear and nonlinear integral equations. For example, existence theorems for

$$y(x) = f(x) + \int_a^x K(x, s, y(s))\,ds,$$

which includes as a special case the Volterra equation of the second kind

$$y(x) = f(x) + \int_a^x K(x, s)y(s)\,ds,$$

have been given by many mathematicians.

Historically, the next major development was an outgrowth of the work on integral equations. Hilbert regarded a function as given by its Fourier coefficients. These satisfy the condition that $\sum_1^\infty a_p^2$ is finite. He had also introduced sequences of real numbers $\{x_n\}$ such that $\sum_{n=1}^\infty x_n^2$ is finite. Then Riesz and Fischer showed that there is a one-to-one correspondence between Lebesgue square summable functions and square summable sequences of their Fourier coefficients. The square summable sequences can be regarded as the coordinates of points in an infinite-dimensional space, which is a generalization of n-dimensional Euclidean space. Thus functions can be regarded as points of a space, now called Hilbert space, and the integral $\int_a^b K(x, y)u(x)\,dx$ can be regarded as an operator transforming $u(x)$ into itself or another function. These ideas suggested for the study of integral equations an abstract approach that fitted into an incipient abstract approach to the calculus of variations. This new approach is now known as functional analysis and we shall consider it next.

22. *Math. Ann.*, 66, 1908, 273–324 = *Ges. Abh.*, 1, 1–86.

Bibliography

Bernkopf, M.: "The Development of Function Spaces with Particular Reference to their Origins in Integral Equation Theory," *Archive for History of Exact Sciences*, 3, 1966, 1–96.

Bliss, G. A.: "The Scientific Work of E. H. Moore," *Amer. Math. Soc. Bull.*, 40, 1934, 501–14.

Bocher, M.: *An Introduction to the Study of Integral Equations*, 2nd ed., Cambridge University Press, 1913.

Bourbaki, N.: *Eléments d'histoire des mathématiques*, Hermann, 1960, pp. 230–45.

Davis, Harold T.: *The Present State of Integral Equations*, Indiana University Press, 1926.

Hahn, H.: "Bericht über die Theorie der linearen Integralgleichungen," *Jahres. der Deut. Math.-Verein.*, 20, 1911, 69–117.

Hellinger, E.: *Hilberts Arbeiten über Integralgleichungen und unendliche Gleichungssysteme*, in Hilbert's *Gesam. Abh.*, 3, 94–145, Julius Springer, 1935.

————: "Begründung der Theorie quadratischer Formen von unendlichvielen Veränderlichen," *Jour. für Math.*, 136, 1909, 210–71.

Hellinger, E., and O. Toeplitz: "Integralgleichungen und Gleichungen mit unendlichvielen Unbekannten," *Encyk. der Math. Wiss.*, B. G. Teubner, 1923–27, Vol. 2, Part 3, 2nd half, 1335–1597.

Hilbert, D.: *Grundzüge einer allgemeinen Theorie der linearen Integralgleichungen*, 1912, Chelsea (reprint), 1953.

Reid, Constance: *Hilbert*, Springer-Verlag, 1970.

Volterra, Vito: *Opere matematiche*, 5 vols., Accademia Nazionale dei Lincei, 1954–62.

Weyl, Hermann: *Gesammelte Abhandlungen*, 4 vols, Springer-Verlag, 1968.

46
Functional Analysis

1. The Nature of Functional Analysis

By the late nineteenth century it was apparent that many domains of
mathematics dealt with transformations or operators acting on functions.
Thus even the ordinary differentiation operation and its inverse antidifferentiation act on a function to produce a new function. In problems of the
calculus of variations wherein one deals with integrals such as

$$J = \int_a^b F(x, y, y') \, dx,$$

the integral can be regarded as operating on a class of functions $y(x)$ of
which that one is sought which happens to maximize or minimize the integral.
The area of differential equations offers another class of operators. Thus the
differential operator

$$L = \frac{d^2}{dx^2} + p(x) \frac{d}{dx} + q(x),$$

acting on a class $y(x)$ of functions, converts these to other functions. Of
course to solve the differential equation, one seeks a particular $y(x)$ such
that L acting on that $y(x)$ yields 0 and perhaps satisfies initial or boundary
conditions. As a final example of operators we have integral equations. The
right-hand side of

$$f(x) = \int_a^b K(x, y)u(x) \, dx$$

can be regarded as an operator acting on various $u(x)$ to produce new
functions, though again, as in the case of differential equations, the $u(x)$
that solves the equation is transformed into $f(x)$.

The idea that motivated the creation of functional analysis is that all of these operators could be considered under one abstract formulation of an operator acting on a class of functions. Moreover, these functions could be regarded as elements or points of a space. Then the operator transforms points into points and in this sense is a generalization of ordinary transformations such as rotations. Some of the above operators carry functions into real numbers, rather than functions. Those operators that do yield real or complex numbers are today called functionals and the term operator is more commonly reserved for the transformations that carry functions into functions. Thus the term functional analysis, which was introduced by Paul P. Lévy (1886–1971) when functionals were the key interest, is no longer appropriate. The search for generality and unification is one of the distinctive features of twentieth century mathematics, and functional analysis seeks to achieve these goals.

2. *The Theory of Functionals*

The abstract theory of functionals was initiated by Volterra in work concerned with the calculus of variations. His work on functions of lines (curves), as he called the subject, covers a number of papers.[1] A function of lines was, for Volterra, a real-valued function F whose values depend on all the values of functions $y(x)$ defined on some interval $[a, b]$. The functions themselves were regarded as points of a space for which a neighborhood of a point and the limit of a sequence of points can be defined. For the functionals $F[y(x)]$ Volterra offered definitions of continuity, derivative and differential. However, these definitions were not adequate for the abstract theory of the calculus of variations and were superseded. His definitions were in fact criticized by Hadamard.[2]

Even before Volterra commenced his work, the notion that a collection of functions $y(x)$ all defined on some common interval be regarded as points of a space had already been suggested. Riemann, in his thesis,[3] spoke of a collection of functions forming a connected closed domain (of points of a space). Giulio Ascoli (1843–1896)[4] and Cesare Arzelà[5] sought to extend to sets of functions Cantor's theory of sets of points, and so regarded functions as points of a space. Arzelà also spoke of functions of lines. Hadamard suggested at the First International Congress of Mathematicians in 1897[6]

1. *Atti della Reale Accademia dei Lincei*, (4), 3, 1887, 97–105, 141–46, 153–58 = *Opere matematiche*, 1, 294–314, and others of the same and later years.
2. *Bull. Soc. Math. de France*, 30, 1902, 40–43 = *Œuvres*, 1, 401–4.
3. *Werke*, p. 30.
4. *Memorie della Reale Accademia dei Lincei*, (3), 18, 1883, 521–86.
5. *Atti della Accad. dei Lincei, Rendiconti*, (4), 5, 1889, 342–48.
6. *Verhandlungen des ersten internationalen Mathematiker-Kongresses*, Teubner, 1898, 201–2.

that curves be considered as points of a set. He was thinking of the family of all continuous functions defined over [0, 1], a family that arose in his work on partial differential equations. Emile Borel made the same suggestion[7] for a different purpose, namely, the study of arbitrary functions by means of series.

Hadamard, too, undertook the study of functionals[8] on behalf of the calculus of variations. The term functional is due to him. According to Hadamard, a functional $U[y(t)]$ is linear if when $y(t) = \lambda_1 y_1(t) + \lambda_2 y_2(t)$, wherein λ_1 and λ_2 are constants, then $U[y(t)] = \lambda_1 U[y_1(t)] + \lambda_2 U[y_2(t)]$.

The first major effort to build up an abstract theory of function spaces and functionals was made by Maurice Fréchet (1878–1973), a leading French professor of mathematics, in his doctoral thesis of 1906.[9] In what he called the functional calculus he sought to unify in abstract terms the ideas in the work of Cantor, Volterra, Arzelà, Hadamard, and others.

To gain the largest degree of generality for his function spaces, Fréchet took over the collection of basic notions about sets developed by Cantor, though for Fréchet the points of the sets are functions. He also formulated more generally the notion of a limit of a set of points. This notion was not defined explicitly but was characterized by properties general enough to embrace the kinds of limit found in the concrete theories Fréchet sought to unify. He introduced a class L of spaces, the L denoting that a limit concept exists for each space. Thus if A is a space of class L and the elements are $A_1, A_2, \ldots$ chosen at random, it must be possible to determine whether or not there exists a unique element A, called the limit of the sequence $\{A_n\}$ when it exists, and such that

(a) If $A_i = A$ for every i, then $\lim \{A_i\} = A$.
(b) If A is the limit of $\{A_n\}$ then A is the limit of every infinite subsequence of $\{A_n\}$.

Then Fréchet introduced a number of concepts for any space of class L. Thus the derived set E' of a set E consists of the set of points of E which are limits of sequences belonging to E. E is closed if E' is contained in E. E is perfect if $E' = E$. A point A of E is an interior point of E (in the narrow sense) if A is not the limit of any sequence not in E. A set E is compact if either it has finitely many elements or if every infinite subset of E has at least one limit element. If E is closed and compact it is called extremal. (Fréchet's "compact" is the modern relatively sequentially compact, and his "extremal" is today's sequentially compact.) Fréchet's first important theorem is a generalization of the closed nested interval theorem: If $\{E_n\}$ is a

7. *Verhandlungen*, 204–5.
8. *Comp. Rend.*, 136, 1903, 351–54 = *Œuvres*, 1, 405–8.
9. *Rendiconti del Circolo Matematico di Palermo*, 22, 1906, 1–74.

decreasing sequence, that is, E_{n+1} is contained in E_n, of closed subsets of an extremal set, then the intersection of all the E_n is not empty.

Fréchet now considers functionals (he calls them functional operations). These are real-valued functions defined on a set E. He defines the continuity of a functional thus: A functional U is said to be continuous at an element A of E if $\lim_{n \to \infty} U(A_n) = U(A)$ for all sequences $\{A_n\}$ contained in E and converging to A. He also introduced semicontinuity of functionals, a notion introduced for ordinary functions by René Baire (1874–1932) in 1899.[10] U is upper semicontinuous in E if $U(A) \geq \lim \sup U(A_n)$ for the $\{A_n\}$ just described. It is lower semicontinuous if $U(A) \leq \lim \inf U(A_n)$.[11]

With these definitions Fréchet was able to prove a number of theorems on functionals. Thus every functional continuous on an extremal set E is bounded and attains its maximum and minimum on E. Every upper semicontinuous functional defined on an extremal set E is bounded above and attains its maximum on E.

Fréchet introduced next concepts for sequences and sets of functionals such as uniform convergence, quasi-uniform convergence, compactness, and equicontinuity. For example, the sequence of functionals $\{U_n\}$ converges uniformly to U if given any positive quantity ε, $|U_n(A) - U(A)| < \varepsilon$ for n sufficiently large but independent of A in E. He was then able to prove generalizations to functionals of theorems previously obtained for real functions.

Having treated the class of spaces L, Fréchet defines more specialized spaces such as neighborhood spaces, redefines the concepts introduced for spaces with limit points, and proves theorems analogous to those already described but often with better results because the spaces have more properties.

Finally he introduces metric spaces. In such a space a function playing the role of distance (*écart*) and denoted by (A, B) is defined for each two points A and B of the space and this function satisfies the conditions:

(a) $(A, B) = (B, A) \geq 0$;
(b) $(A, B) = 0$ if and only if $A = B$;
(c) $(A, B) + (B, C) \geq (A, C)$.

Condition C is called the triangle inequality. Such a space, he says, is of class $\mathscr{E}$. For such spaces, too, Fréchet is able to prove a number of theorems about the spaces and the functionals defined on them much like those previously proven for the more general spaces.

Fréchet gave some examples of function spaces. Thus the set of all real functions of a real variable, all continuous on the same interval I with the

10. *Annali di Mat.*, (3), 13, 1899, 1–122.
11. The inferior limit is the smallest limit point of the sequence $U(A_n)$.

écart of any two functions f and g defined to be $\max\limits_{x \in I} |f(x) - g(x)|$, is a space of class $\mathscr{E}$. Today this *écart* is called the maximum norm.

Another example offered by Fréchet is the set of all sequences of real numbers. If $x = (x_1, x_2, \ldots)$ and $y = (y_1, y_2, \ldots)$ are any two sequences, the *écart* of x and y is defined to be

$$(x, y) = \sum_{p=1}^{\infty} \frac{1}{p!} \frac{|x_p - y_p|}{1 + |x_p - y_p|}.$$

This, as Fréchet remarks, is a space of countably infinite dimensions.

Using his spaces $\mathscr{E}$, Fréchet[12] succeeded in giving a definition of continuity, differential, and differentiability of a functional. Though these were not entirely adequate for the calculus of variations, his definition of differential is worth noting because it is the core of what did prove to be satisfactory. He assumes that there exists a linear functional $L(\eta(x))$ such that

$$F[y(x) + \eta(x)] = F(y) + L(\eta) + \varepsilon M(\eta),$$

where $\eta(x)$ is a variation on $y(x)$, $M(\eta)$ is the maximum absolute value of $\eta(x)$ on $[a, b]$, and ε approaches 0 with M. Then $L(\eta)$ is the differential of $F(y)$. He also presupposes the continuity of $F(y)$, which is more than can be satisfied in problems of the calculus of variations.

Charles Albert Fischer (1884–1922)[13] then improved Volterra's definition of the derivative of a functional so that it did cover the functionals used in the calculus of variations; the differential of a functional could then be defined in terms of the derivative.

So far as the basic definitions of properties of functionals needed for the calculus of variations are concerned, the final formulations were given by Elizabeth Le Stourgeon (1881–1971).[14] The key concept, the differential of a functional, is a modification of Fréchet's. The functional $F(y)$ is said to have a differential at $y_0(x)$ if there exists a linear functional $L(\eta)$ such that for all arcs $y_0 + \eta$ in the neighborhood of y_0 the relation

$$F(y_0 + \eta) = F(y_0) + L(\eta) + M(\eta)\varepsilon(\eta)$$

holds where $M(\eta)$ is the maximum absolute value of η and η' on the interval $[a, b]$ and $\varepsilon(\eta)$ is a quantity vanishing with $M(\eta)$. She also defined second differentials.

Both Le Stourgeon and Fischer deduced from their definitions of differentials necessary conditions for the existence of a minimum of a functional that are of a type applicable to the problems of the calculus of variations. For example, a necessary condition that a functional $F(y)$ have

12. *Amer. Math. Soc. Trans.*, 15, 1914, 135–61.
13. *Amer. Jour. of Math.*, 35, 1913, 369–94.
14. *Amer. Math. Soc. Trans.*, 21, 1920, 357–83.

a minimum for $y = y_0$ is that $L(\eta)$ vanish for every $\eta(x)$ that is continuous and has continuous first derivatives on $[a, b]$ and such that $\eta(a) = \eta(b) = 0$. It is possible to deduce the Euler equation from the condition that the first differential vanish, and by using a definition of the second differential of a functional (which several writers already named had given), it is possible to deduce the necessity of the Jacobi condition of the calculus of variations.

The definitive work, at least up to 1925, on the theory of functionals required for the calculus of variations was done by Leonida Tonelli (1885–1946), a professor at the universities of Bologna and Pisa. After writing many papers on the subject from 1911 on, he published his *Fondamenti di calcolo delle variazioni* (2 vols,, 1922, 1924), in which he considers the subject from the standpoint of functionals. The classical theory had relied a great deal on the theory of differential equations. Tonelli's aim was to replace existence theorems for differential equations by existence theorems for minimizing curves of integrals. Throughout his work the concept of the lower semicontinuity of a functional is a fundamental one because the functionals are not continuous.

Tonelli first treats sets of curves and gives theorems that insure the existence of a limiting curve in a class of curves. Ensuing theorems insure that the usual integral, but in the parametric form

$$\int_{t_1}^{t_2} F(t, x(t), y(t), x', y') \, dt,$$

will be lower semicontinuous as a function of $x(t)$ and $y(t)$. (Later he considers the more fundamental nonparametric integrals.) He derives the four classical necessary conditions of the calculus of variations for the standard types of problems. The main emphasis in the second volume is on existence theorems for a great variety of problems deduced on the basis of the notion of semicontinuity. That is, given an integral of the above form, by imposing conditions on it as a functional and on the class of curves to be considered, he proves there is a curve in the class which minimizes the integral. His theorems deal with absolute and relative minima and maxima.

To some extent Tonelli's work pays dividends in differential equations, for his existence theorems imply the existence of the solutions of the differential equations which in the classical approach supply the minimizing curves as solutions. However, his work was limited to the basic types of problems of the calculus of variations. Though the abstract approach was pursued subsequently by many men, the progress made in the application of the theory of functionals to the calculus of variations has not been great.

3. Linear Functional Analysis

The major work in functional analysis sought to provide an abstract theory for integral equations as opposed to the calculus of variations. The properties of functionals needed for the latter field are rather special and do not hold for functionals in general. In addition, the nonlinearity of these functionals created difficulties that were irrelevant to the functionals and operators that embraced integral equations. While concrete extensions of the theory of solution of integral equations were being made by Schmidt, Fischer, and Riesz, these men and others also began work on the corresponding abstract theory.

The first attempt at an abstract theory of linear functionals and operators was made by the American mathematician E. H. Moore starting in 1906.[15] Moore realized that there were features common to the theory of linear equations in a finite number of unknowns, the theory of infinitely many equations in an infinite number of unknowns, and the theory of linear integral equations. He therefore undertook to build an abstract theory, called General Analysis, which would include the above more concrete theories as special cases. His approach was axiomatic. We shall not present the details because Moore's influence was not extensive, nor did he achieve effective methodology. Moreover, his symbolic language was strange and difficult for others to follow.

The first influential step toward an abstract theory of linear functionals and operators was taken in 1907 by Erhard Schmidt[16] and Fréchet.[17] Hilbert, in his work on integral equations, had regarded a function as given by its Fourier coefficients in an expansion with respect to an orthonormal sequence of functions. These coefficients and the values he attached to the x_i in his theory of quadratic forms in infinitely many variables are sequences $\{x_n\}$ such that $\sum_1^\infty x_n^2$ is finite. However, Hilbert did not regard these sequences as coordinates of a point in space, nor did he use geometrical language. This step was taken by Schmidt and Fréchet. By regarding each sequence $\{x_n\}$ as a point, functions were represented as points of an infinite-dimensional space. Schmidt also introduced complex as well as real numbers in the sequences $\{x_n\}$. Such a space has since been called a Hilbert space. Our account follows Schmidt's work.

The elements of Schmidt's function spaces are infinite sequences of complex numbers, $z = \{z_p\}$, with

$$\sum_{p=1}^{\infty} |z_p|^2 < \infty.$$

Schmidt introduced the notation $\|z\|$ for $\{\sum_{p=1}^{\infty} z_p \bar{z}_p\}^{1/2}$; $\|z\|$ was later called the norm of z. Following Hilbert, Schmidt used the notation (z, w) for

15. See, for example, *Atti del IV Congresso Internazionale dei Matematici* (1908), 2, *Reale Accademia dei Lincei*, 1909, 98–114, and *Amer. Math. Soc. Bull.*, 18, 1911/12, 334–62.
16. *Rendiconti del Circolo Matematico di Palermo*, 25, 1908, 53–77.
17. *Nouvelles Annales de Mathématiques*, 8, 1908, 97–116, 289–317.

$\sum_{p=1}^{\infty} z_p w_p$, so that $\|z\| = \sqrt{(z, \bar{z})}$. (The modern practice is to define (z, w) as $\sum z_p \bar{w}_p$.) Two elements z and w of the space are called orthogonal if and only if $(z, \bar{w}) = 0$. Schmidt then proves a generalized Pythagorean theorem, namely, if $z_1, z_2, \ldots, z_n$ are n mutually orthogonal elements of the space, then

$$w = \sum_{p=1}^{n} z_p$$

implies

$$\|w\|^2 = \sum_{p=1}^{n} \|z_p\|^2.$$

It follows that the n mutually orthogonal elements are linearly independent. Schmidt also obtained Bessel's inequality in this general space; that is, if $\{z_n\}$ is an infinite sequence of orthonormal elements, so that $(z_p, \bar{z}_q) = \delta_{pq}$, and if w is any element, then

$$\sum_{p=1}^{\infty} |(w, \bar{z}_p)|^2 \leq \|w\|^2.$$

Also Schwarz's inequality and the triangle inequality are proved for the norm.

A sequence of elements $\{z_n\}$ is said to converge strongly to z if $\|z_n - z\|$ approaches 0 and every strong Cauchy sequence, i.e. every sequence for which $\|z_p - z_q\|$ approaches 0 as p and q approach ∞, is shown to converge to an element z so that the space of sequences is complete. This is a vital property.

Schmidt next introduces the notion of a (strongly) closed subspace. A subset A of his space H is said to be a closed subspace if it is a closed subset in the sense of the convergence just defined and if it is algebraically closed; that is, if w_1 and w_2 are elements of A, then $a_1 w_1 + a_2 w_2$, where a_1 and a_2 are any complex numbers, is also an element of A. Such closed subspaces are shown to exist. One merely takes any sequence $\{z_n\}$ of linearly independent elements and takes all finite linear combinations of elements of $\{z_n\}$. The closure of this collection of elements is an algebraically closed subspace.

Now let A be any fixed closed subspace. Schmidt first proves that if z is any element of the space, there exist unique elements w_1 and w_2 such that $z = w_1 + w_2$ where w_1 is in A and w_2 is orthogonal to A, which means that w_2 is orthogonal to every element in A. (This result is today called the projection theorem; w_1 is the projection of z in A.) Further, $\|w_2\| = \min \|y - z\|$ where y is any element in A and this minimum is assumed only for $y = w_1$. $\|w_2\|$ is called the distance between z and A.

In 1907 both Schmidt and Fréchet remarked that the space of square summable (Lebesgue integrable) functions has a geometry completely analogous to the Hilbert space of sequences. This analogy was elucidated when, some months later, Riesz, making use of the Riesz-Fischer theorem (Chap. 45, sec. 4), which establishes a one-to-one correspondence between Lebesgue measurable square integrable functions and square summable sequences of real numbers, pointed out that in the set L^2 of square summable functions a distance can be defined and one can use this to build a geometry of this space of functions. This notion of distance between any two square summable functions of the space L^2, all defined on an interval $[a, b]$, was in fact also defined by Fréchet[18] as

$$(1) \qquad \sqrt{\int_a^b [f(x) - g(x)]^2 \, dx},$$

where the integral is understood in the Lebesgue sense; and it is also understood that two functions that differ only on a set of measure 0 are considered equal. The square of the distance is also called the mean square deviation of the functions. The scalar product of f and g is defined to be $(f, g) = \int_a^b f(x)g(x) \, dx$. Functions for which $(f, g) = 0$ are called orthogonal. The Schwarz inequality,

$$\int_a^b f(x) g(x) \, dx \le \sqrt{\int_a^b f^2 \, dx} \sqrt{\int_a^b g^2 \, dx},$$

and other properties that hold in the space of square summable sequences, apply to the function space; in particular this class of square summable functions forms a complete space. Thus the space of square summable functions and the space of square summable sequences that are the Fourier coefficients of these functions with respect to a fixed complete orthonormal system of functions can be identified.

So far as abstract function spaces are concerned, we should recall (Chap. 45, sec. 4) the introduction of L^p spaces, $1 < p < \infty$, by Riesz. These spaces are also complete in the metric $d(f_1, f_2) = \left(\int_a^b |f_1 - f_2|^p \, dx \right)^{1/p}$.

Though we shall soon look into additional creations in the area of abstract spaces, the next developments concern functionals and operators. In the 1907 paper just referred to, in which he introduced the metric or *écart* for functions in L^2 space, and in another paper of that year,[19] Fréchet proved that for every continuous linear functional $U(f)$ defined on L^2 there is a unique $u(x)$ in L^2 such that for every f in L^2

$$U(f) = \int_a^b f(x)u(x) \, dx.$$

18. *Comp. Rend.*, 444, 1907, 1414–16.
19. *Amer. Math. Soc. Trans.*, 8, 1907, 433–46.

This result generalizes one obtained by Hadamard in 1903.[20] In 1909 Riesz[21] generalized this result by expressing $U(f)$ as a Stieltjes integral; that is,

$$U(f) = \int_a^b f(x)\, du(x).$$

Riesz himself generalized this result to linear functionals A that satisfy the condition that for all f in L^p

$$A(f) \leq M \left[\int_a^b |f(x)|^p\, dx \right]^{1/p},$$

where M depends only on A. Then there is a function $a(x)$ in L^q, unique up to the addition of a function with zero integral, such that for all f in L^p

$$(2) \qquad U(f) = \int_a^b a(x) f(x)\, dx.$$

This result is called the Riesz representation theorem.

The central part of functional analysis deals with the abstract theory of the operators that occur in differential and integral equations. This theory unites the eigenvalue theory of differential and integral equations and linear transformations operating in n-dimensional space. Such an operator, for example,

$$g(x) = \int_a^b k(x, y) f(y)\, dy$$

for given k, assigns f to g and satisfies some additional conditions. In the notation A for the abstract operator and the notation $g = Af$, linearity means

$$(3) \qquad A(\lambda_1 f_1 + \lambda_2 f_2) = \lambda_1 A f_1 + \lambda_2 A f_2,$$

where λ_i is any real or complex constant. The indefinite integral $g(x) = \int_a^x f(t)\, dt$ and the derivative $f'(x) = Df(x)$ are linear operators for the proper classes of functions. Continuity of the operator A means that if the sequence of functions f_n approaches f in the sense of the limit in the space of functions, then $A f_n$ must approach Af.

The abstract analogue of what one finds in integral equations when the kernel $k(x, y)$ is symmetric is the self-adjointness property of the operator A. If for any two functions

$$(Af_1, f_2) = (f_1, Af_2),$$

20. *Comp. Rend.*, 136, 1903, 351–54 = *Œuvres*, 1, 405–8.
21. *Comp. Rend.*, 149, 1909, 974–77, and *Ann. de l'Ecole Norm. Sup.*, 28, 1911, 33 ff.

where (Af_1, f_2) denotes the inner or scalar product of two functions of the space, then A is said to be self-adjoint. In the case of integral equations, if

$$Af_1 = \int_a^b k(x, y) f(y) \, dy,$$

then

$$(Af_1, f_2) = \int_a^b \int_a^b k(x, y) f_1(y) f_2(x) \, dy \, dx$$

$$(f_1, Af_2) = \int_a^b \int_a^b k(x, y) f_2(y) f_1(x) \, dy \, dx,$$

and $(Af_1, f_2) = (f_1, Af_2)$ if the kernel is symmetric. For arbitrary self-adjoint operators the eigenvalues are real, and the eigenfunctions corresponding to distinct eigenvalues are orthogonal to each other.

A healthy start on the abstract operator theory, which is the core of functional analysis, was made by Riesz in his 1910 *Mathematische Annalen* paper, in which he introduced L^p spaces (Chap. 45, sec. 4). He set out in that paper to generalize the solutions of the integral equation

$$\phi(x) - \lambda \int_a^b K(x, t) \phi(t) \, dt = f(x)$$

to functions in L^p spaces. Riesz thought of the expression

$$\int_a^b K(s, t) \phi(t) \, dt$$

as a transformation acting on the function $\phi(t)$. He called it a functional transformation and denoted it by $T(\phi(t))$. Moreover, since the $\phi(t)$ with which Riesz was concerned were in L^p space, the transformation takes functions into the same or another space. In particular a transformation or operator that takes functions of L^p into functions of L^p is called linear in L^p if it satisfies (3) and if T is bounded; that is, there is a constant M such that for all f in L^p that satisfy

$$\int_a^b |f(x)|^p \, dx \leq 1$$

then

$$\int_a^b |T(f(x))|^p \, dx \leq M^p.$$

Later the least upper bound of such M was called the norm of T and denoted by $\|T\|$.

Riesz also introduced the notion of the adjoint or transposed operator for T. For any g in L^q and T operating in L^p,

$$(4) \qquad \int_a^b T(f(x)) g(x) \, dx$$

defines for fixed g and varying f in L^p a functional on L^p. Hence by the Riesz representation theorem there is a function $\psi(x)$ in L^q, unique up to a function whose integral is 0, such that

$$(5) \qquad \int_a^b T(f(x)) g(x) \, dx = \int_a^b f(x) \psi(x) \, dx.$$

The adjoint or transpose of T, denoted by T^*, is now defined to be that operator in L^q, depending for fixed T only on g, which assigns to g the ψ of equation (5); that is, $T^*(g) = \psi$. (In modern notation T^* satisfies $(Tf, g) = (f, T^*g)$.) T^* is a linear transformation in L^q and $\| T^* \| = \| T \|$.

Riesz now considers the solution of

$$(6) \qquad T(\phi(x)) = f(x),$$

where T is a linear transformation in L^p, f is known, and ϕ is unknown. He proves that (6) has a solution if and only if

$$\left| \int_a^b f(x) g(x) \, dx \right| \leq M \left(\int_a^b | T^*(g(x)) |^q \, dx \right)^{1/q}$$

for all g in L^q. He is thereby led to the notion of the inverse transformation or operator T^{-1} and in the very same thought to T^{*-1}. With the aid of the adjoint operator he proves the existence of the inverses.

In his 1910 paper Riesz introduced the notation

$$(7) \qquad \phi(x) - \lambda K(\phi(x)) = f(x),$$

where K now stands for $\int_a^b K(x, t) * dt$ and where $*$ stands for the function on which K operates. His additional results are limited to L^2 wherein $K = K^*$. To treat the eigenvalue problem of integral equations he introduces Hilbert's notion of complete continuity, but now formulated for abstract operators. An operator K in L^2 is said to be completely continuous if K maps every weakly convergent sequence (Chap. 45, sec. 4) of functions into a strongly convergent one, that is, $\{f_n\}$ converging weakly implies $\{K(f_n)\}$ converging strongly. He does prove that the spectrum of (7) is discrete (that is, there is no continuous spectrum of a symmetric K) and that the eigenfunctions associated with the eigenvalues are orthogonal.

Another approach to abstract spaces, using the notion of a norm, was also initiated by Riesz.[22] However, the general definition of normed spaces

22. *Acta Math.*, 41, 1918, 71–98.

was given during the years 1920 to 1922 by Stefan Banach (1892–1945), Hans Hahn (1879–1934), Eduard Helly (1884–1943), and Norbert Wiener (1894–1964). Though there is much overlap in the work of these men and the question of priority is difficult to settle, it is Banach's work that had the greatest influence. His motivation was the generalization of integral equations.

The essential feature of all this work, Banach's in particular,[23] was to set up a space with a norm but one which is no longer defined in terms of an inner product. Whereas in L^2, $\|f\| = (f,f)^{1/2}$ it is not possible to define the norm of a Banach space in this way because an inner product is no longer available.

Banach starts with a space E of elements denoted by $x, y, z, \ldots$, whereas $a, b, c, \ldots$ denote real numbers. The axioms for his space are divided into three groups. The first group contains thirteen axioms which specify that E is a commutative group under addition, that E is closed under multiplication by a real scalar, and that the familiar associative and distributive relations are to hold among various operations on the real numbers and the elements.

The second group of axioms characterizes a norm on the elements (vectors) of E. The norm is a real-valued function defined on E and denoted by $\|x\|$. For any real number a and any x in E the norm has the following properties:

(a) $\|x\| \geq 0$;
(b) $\|x\| = 0$ if and only if $x = 0$;
(c) $\|ax\| = |a| \cdot \|x\|$;
(d) $\|x + y\| \leq \|x\| + \|y\|$.

The third group contains just a completeness axiom, which states that if $\{x_n\}$ is a Cauchy sequence in the norm, i.e. if $\lim_{n,p \to \infty} \|x_n - x_p\| = 0$, then there is an element x in E such that $\lim_{n \to \infty} \|x_n - x\| = 0$.

A space satisfying these three groups of axioms is called a Banach space or a complete normed vector space. Though a Banach space is more general than a Hilbert space, because the inner product of two elements is not presupposed to define the norm, as a consequence the key notion of two elements being orthogonal is lost in a Banach space that is not also a Hilbert space. The first and third sets of conditions hold also for Hilbert space, but the second set is weaker than conditions on the norm in Hilbert space. Banach spaces include L^p spaces, the space of continuous functions, the space of bounded measurable functions, and other spaces, provided the appropriate norm is used.

With the concept of the norm Banach is able to prove a number of

23. *Fundamenta Mathematicae*, 3, 1922, 133–81.

familiar facts for his spaces. One of the key theorems states: Let $\{x_n\}$ be a set of elements of E such that

$$\sum_{p=1}^{\infty} \|x_p\| < \infty.$$

Then $\sum_{p=1}^{\infty} x_p$ converges in the norm to an element x of E.

After proving a number of theorems, Banach considers operators defined on the space but whose range is in another space E_1, which is also a Banach space. An operator F is said to be continuous at x_0 relative to a set A if $F(x)$ is defined for all x of A; x_0 belongs to A and to the derived set of A; and whenever $\{x_n\}$ is a sequence of A with limit x_0, then $F(x_n)$ approaches $F(x_0)$. He also defines uniform continuity of F relative to a set A and then turns his attention to sequences of operators. The sequence $\{F_n\}$ of operators is said to converge in the norm to F on a set A if for every x in A, $\lim_{n \to \infty} F_n(x) = F(x)$.

An important class of operators introduced by Banach is the set of continuous additive ones. An operator F is additive if for all x and y, $F(x + y) = F(x) + F(y)$. An additive continuous operator proves to have the property that $F(ax) = aF(x)$ for any real number a. If F is additive and continuous at one element (point) of E, it follows that it is continuous everywhere and it is bounded; that is, there is a constant M, depending only on F, so that $\|F(x)\| \le M\|x\|$ for every x of E. Another theorem asserts that if $\{F_n\}$ is a sequence of additive continuous operators and if F is an additive operator such that for each x, $\lim_{n \to \infty} F_n(x) = F(x)$, then F is continuous and there exists an M such that for all n, $\|F_n(x)\| \le M\|x\|$.

In this paper Banach proves theorems on the solution of the abstract formulation of integral equations. If F is a continuous operator with domain and range in the space E and if there exists a number M, $0 < M < 1$, so that for all x' and x'' in E, $\|F(x') - F(x'')\| \le M\|x' - x''\|$, then there exists a unique element x in E that satisfies $F(x) = x$. More important is the theorem: Consider the equation

(8) $$x + hF(x) = y$$

where y is a known function in E, F is an additive continuous operator with domain and range in E, and h is a real number. Let M be the least upper bound of all those numbers M' satisfying $\|F(x)\| \le M'\|x\|$ for all x. Then for every y and every value of h satisfying $|hM| < 1$ there exists a function x satisfying (8) and

$$x = y + \sum_{n=1}^{\infty} (-1)^n h^n F^{(n)}(y),$$

where $F^{(n)}(y) = F(F^{(n-1)}(y))$. This result is one form of the spectral radius theorem, and it is a generalization of Volterra's method of solving integral equations.

Banach in 1929 [24] introduced another important notion in the subject of functional analysis, the notion of the dual or adjoint space of a Banach space. The idea was also introduced independently by Hahn,[25] but Banach's work was more thorough. This dual space is the space of all continuous bounded linear functionals on the given space. The norm for the space of functionals is taken to be the bounds of the functionals, and the space proves to be a complete normed linear space, that is, a Banach space. Actually Banach's work here generalizes Riesz's work on L^p and L^q spaces, where $q = p/(p - 1)$, because the L^q space is equivalent to the dual in Banach's sense of the L^p space. The connection with Banach's work is made evident by the Riesz representation theorem (2). In other words, Banach's dual space has the same relationship to a given Banach space E that L^q has to L^p.

Banach starts with the definition of a continuous linear functional, that is, a continuous real-valued function whose domain is the space E, and proves that each functional is bounded. A key theorem, which generalizes one in Hahn's work, is known today as the Hahn-Banach theorem. Let p be a real-valued functional defined on a complete normed linear space R and let p satisfy for all x and y in R

(a) $$p(x + y) \leq p(x) + p(y);$$

(b) $$p(\lambda x) = \lambda p(x) \quad \text{for} \quad \lambda \geq 0.$$

Then there exists an additive functional f on R that satisfies

$$-p(-x) \leq f(x) \leq p(x)$$

for all x in R. There are a number of other theorems on the set of continuous functionals defined on R.

The work on functionals leads to the notion of adjoint operator. Let R and S be two Banach spaces and let U be a continuous, linear operator with domain in R and range in S. Denote by $R*$ and $S*$ the set of bounded linear functionals defined on R and S respectively. Then U induces a mapping from $S*$ into $R*$ as follows: If g is an element of $S*$, then $g(U(x))$ is well-defined for all x in R. But by the linearity of U and g this is also a linear functional defined on R, that is, $g(U(x))$ is a member of $R*$. Put otherwise, if $U(x) = y$, then $g(y) = f(x)$ where f is a functional of $R*$. The induced mapping $U*$ from $S*$ to $R*$ is called the adjoint of U.

With this notion Banach proves that if $U*$ has a continuous inverse, then $y = U(x)$ is solvable for any y of S. Also if $f = U*(g)$ is solvable for

24. *Studia Mathematica*, 1, 1929, 211–16.
25. *Jour. für Math.*, 157, 1927, 214–29.

every f of R^*, then U^{-1} exists and is continuous on the range of U, where the range of U in S is the set of all y for which there is a g of S^* with the property that $g(y) = 0$ whenever $U^*(g) = 0$. (This last statement is a generalized form of the Fredholm alternative theorem.)

Banach applied his theory of adjoint operators to Riesz operators, which Riesz introduced in his 1918 paper. These are operators U of the form $U = I - \lambda V$ where I is the identity operator and V is a completely continuous operator. The abstract theory can then be applied to the space L^2 of functions defined on $[0, 1]$ and to the operators

$$U_\lambda(x) = x(s) - \lambda \int_0^1 K(s, t)x(t)\, dt,$$

where $\int_0^1 \int_0^1 K^2(s, t)\, ds\, dt < \infty$. The general theory, applied to

(9) $$y(s) = x(s) - \lambda \int_0^1 K(s, t)x(t)\, dt$$

and its associated transposed equation

(10) $$f(s) = g(s) - \lambda \int_0^1 K(t, s)g(t)\, dt,$$

tells us that if λ_0 is an eigenvalue of (9), then λ_0 is an eigenvalue of (10) and conversely. Further, (9) has a finite number of linearly independent eigenfunctions associated with λ_0 and the same is true of (10). Also (9) does not have a solution for all y when $\lambda = \lambda_0$. In fact a necessary and sufficient condition that (9) have a solution is that the n conditions

$$\int_0^1 y(t) g_o^{(p)}(t)\, dt = 0, \qquad p = 1, 2, \cdots, n$$

be satisfied, where $g_o^{(1)}, \ldots, g_o^{(n)}$ is a set of linearly independent solutions of

$$0 = g(s) - \lambda_0 \int_0^1 K(t, s)g(t)\, dt.$$

4. *The Axiomatization of Hilbert Space*

The theory of function spaces and operators seemed in the 1920s to be heading only toward abstraction for abstraction's sake. Even Banach made no use of his work. This state of affairs led Hermann Weyl to remark, "It was not merit but a favor of fortune when, beginning in 1923 . . . the spectral theory of Hilbert space was discovered to be the adequate mathematical instrument of quantum mechanics." Quantum mechanical research showed that the observables of a physical system can be represented by linear symmetric operators in a Hilbert space and the eigenvalues and eigenvectors (eigenfunctions) of the particular operator that represents energy are the energy levels of an electron in an atom and corresponding stationary quantum states of the system. The differences in two eigenvalues give the frequencies of the emitted quantum of light and thus define the radiation spectrum of the substance. In 1926 Erwin Schrödinger presented his quantum theory based on differential equations. He also showed its identity with Werner Heisenberg's theory of infinite matrices (1925), which the latter had applied to quantum theory. But a general theory unifying Hilbert's work and the eigenfunction theory for differential equations was missing.

The use of operators in quantum theory stimulated work on an abstract theory of Hilbert space and operators; this was first undertaken by John von Neumann (1903–57) in 1927. His approach treated both the space of square summable sequences and the space of L^2 functions defined on some common interval.

In two papers[26] von Neumann presented an *axiomatic* approach to Hilbert space and to operators in Hilbert space. Though origins of von Neumann's axioms can be seen in the work of Norbert Wiener, Weyl, and Banach, von Neumann's work was more complete and influential. His major objective was the formulation of a general eigenvalue theory for a large class of operators called Hermitian.

He introduced an L^2 space of complex-valued, measurable, and square integrable functions defined on any measurable set E of the complex plane. He also introduced the analogue, the complex sequence space, that is, the set of all sequences of complex numbers $a_1, a_2, \ldots$ with the property that $\sum_{p=1}^{\infty} |a_p|^2 < \infty$. The Riesz-Fischer theorem had shown that there exists a one-to-one correspondence between the functions of the function space and the sequences of the sequence space, as follows: In the function space choose a complete orthonormal sequence of functions $\{\phi_n\}$. Then if f is an element of the function space, the Fourier coefficients of f in an expansion with respect to the $\{\phi_n\}$ are a sequence of the sequence space; and conversely, if one starts with such a sequence there is a unique (up to a function whose integral is 0) function of the L^2 function space which has this sequence as its Fourier coefficients with respect to the $\{\phi_n\}$.

Further, if one defines the inner product (f, g) in the function space by

$$(f, g) = \int_E f(z)\overline{g(z)}\, dz,$$

where $\overline{g(z)}$ is the complex conjugate of $g(z)$, and in the sequence space for two sequences a and b by

$$(a, b) = \sum_{p=1}^{\infty} a_p \overline{b}_p,$$

then, if f corresponds to a and g to b, $(f, g) = (a, b)$.

26. *Math. Ann.*, 102, 1929/30, 49–131, and 370–427 = *Coll. Works*, 2, 3–143.

A Hermitian operator R in these spaces is defined to be a linear operator which has the property that for all f and g in its domain $(Rf, g) = (f, Rg)$. Likewise, in the sequence space, $(Ra, b) = (a, Rb)$.

Von Neumann's theory prescribes an axiomatic approach to both the function space and the sequence space. He proposed the following axiomatic foundation:

(A) H is a linear vector space. That is, there is an addition and scalar multiplication defined on H so that if f_1 and f_2 are elements of H and a_1 and a_2 are any complex numbers, then $a_1 f_1 + a_2 f_2$ is also an element of H.

(B) There exists on H an inner product or a complex-valued function of any two vectors f and g, denoted by (f, g), with the properties: (a) $(af, g) = a(f, g)$; (b) $(f_1 + f_2, g) = (f_1, g) + (f_2, g)$; (c) $(f, g) = \overline{(g, f)}$; (d) $(f, f) \geq 0$; (e) $(f, f) = 0$ if and only if $f = 0$.

Two elements f and g are called orthogonal if $(f, g) = 0$. The norm of f, denoted by $\|f\|$, is $\sqrt{(f, f)}$. The quantity $\|f - g\|$ defines a metric on the space.

(C) In the metric just defined H is separable, that is, there exists in H a countable set dense in H relative to the metric $\|f - g\|$.

(D) For every positive integral n there exists a set of n linearly independent elements of H.

(E) H is complete. That is, if $\{f_n\}$ is such that $\|f_n - f_m\|$ approaches 0 as m and n approach ∞, then there is an f in H such that $\|f - f_n\|$ approaches 0 as n approaches ∞. (This convergence is equivalent to strong convergence.)

From these axioms a number of simple properties follow: the Schwarz inequality $\|(f, g)\| \leq \|f\| \cdot \|g\|$, the fact that any complete orthonormal set of elements of H must be countable, and Parseval's inequality, that is, $\sum_{p=1}^{\infty} \|(f, \phi_p)\|^2 \leq \|f\|^2$.

Von Neumann then treats linear subspaces of H and projection operators. If M and N are closed subspaces of H, then $M - N$ is defined to be the set of all elements of M that are orthogonal to every element of N. The projection theorem follows: Let M be a closed subspace of H. Then any f of H can be written in one and only one way as $f = g + h$ where g is in M and h is in $H - M$. The projection operator P_M is defined by $P_M(f) = g$; that is, it is the operator defined on all of H which projects the elements f into their components in M.

In his second paper von Neumann introduces two topologies in H, strong and weak. The strong topology is the metric topology defined through the norm. The weak topology, which we shall not make precise, provides a system of neighborhoods associated with weak convergence.

Von Neumann presents a number of results on operators in Hilbert space. The linearity of the operators is presupposed, and integration is generally understood in the sense of Lebesgue-Stieltjes. A linear bounded transformation or operator is one that transforms elements of one Hilbert space into another, satisfies the linearity condition (3), and is bounded; that is, there exists a number M such that for all f of the space on which the operator R acts

$$\|R(f)\| \leq M \|f\|.$$

The smallest possible value of M is called the modulus of R. This last condition is equivalent to the continuity of the operator. Also continuity at a single point together with linearity is sufficient to guarantee continuity at all points and hence boundedness.

There is an adjoint operator R^*; for Hermitian operators $R = R^*$, and R is said to be self-adjoint. If $RR^* = R^*R$ for any operator, then R is said to be normal. If $RR^* = R^*R = I$, where I is the identity operator, then R is the analogue of an orthogonal transformation and is called unitary. For a unitary R, $\|R(f)\| = \|f\|$.

Another of von Neumann's results states that if R is Hermitian on all of the Hilbert space and satisfies the weaker closed condition, namely that if f_n approaches f and $R(f_n)$ approaches g so that $R(f) = g$, then R is bounded. Further, if R is Hermitian, the operator $I - \lambda R$ has an inverse for all complex and real λ exterior to an interval (m, M) of the real axis in which the values of $(R(f), f)$ lie when $\|f\| = 1$. A more fundamental result is that to every linear bounded Hermitian operator R there correspond two operators E_- and E_+ (*Einzeltransformationen*) having the following properties:

(a) $E_- E_- = E_-$, $E_+ E_+ = E_+$, $I = E_- + E_+$.
(b) E_- and E_+ are commutative and commutative with any operator commutative with R.
(c) RE_- and RE_+ are respectively negative and positive (RE_+ is positive if $(RE_+(f), f) \geq 0$).
(d) For all f such that $Rf = 0$, $E_- f = 0$ and $E_+ f = f$.

Von Neumann also established a connection between Hermitian and unitary operators, namely, if U is unitary and R is Hermitian, then $U = e^{iR}$.

Von Neumann subsequently extended his theory to unbounded operators; though his contributions and those of others are of major importance, an account of these developments would take us too far into recent developments.

Applications of functional analysis to the generalized moment problem, statistical mechanics, existence and uniqueness theorems in partial differential equations, and fixed-point theorems have been and are being made. Functional analysis now plays a role in the calculus of variations and in the theory of representations of continuous compact groups. It is also involved in algebra, approximate computations, topology, and real variable theory. Despite this variety of applications, there has been a deplorable absence of new applications to the large problems of classical analysis. This failure has disappointed the founders of functional analysis.

Bibliography

Bernkopf, M.: "The Development of Function Spaces with Particular Reference to their Origins in Integral Equation Theory," *Archive for History of Exact Sciences*, 3, 1966, 1–96.

————: "A History of Infinite Matrices," *Archive for History of Exact Sciences*, 4, 1968, 308–58.

Bourbaki, N.: *Eléments d'histoire des mathématiques*, Hermann, 1960, 230–45.

Dresden, Arnold: "Some Recent Work in the Calculus of Variations," *Amer. Math. Soc. Bull.*, 32, 1926, 475–521.

Fréchet, M.: *Notice sur les travaux scientifiques de M. Maurice Fréchet*, Hermann, 1933.

Hellinger, E. and O. Toeplitz: "Integralgleichungen und Gleichungen mit unendlichvielen Unbekannten," *Encyk. der Math. Wiss.*, B. G. Teubner, 1923–27, Vol. 2, Part III, 2nd half, 1335–1597.

Hildebrandt, T. H.: "Linear Functional Transformations in General Spaces," *Amer. Math. Soc. Bull.*, 37, 1931, 185–212.

Lévy, Paul: "Jacques Hadamard, sa vie et son œuvre," *L'Enseignement Mathématique*, (2), 13, 1967, 1–24.

McShane, E. J.: "Recent Developments in the Calculus of Variations," *Amer. Math. Soc. Semicentennial Publications*, 2, 1938, 69–97.

Neumann, John von: *Collected Works*, Pergamon Press, 1961, Vol. 2.

Sanger, Ralph G.: "Functions of Lines and the Calculus of Variations," *University of Chicago Contributions to the Calculus of Variations for 1931–32*, University of Chicago Press, 1933, 193–293.

Tonelli, L.: "The Calculus of Variations," *Amer. Math. Soc. Bull.*, 31, 1925, 163–72.

Volterra, Vito: *Opere matematiche*, 5 vols., Accademia Nazionale dei Lincei, 1954–62.

47
Divergent Series

> It is indeed a strange vicissitude of our science that those series which early in the century were supposed to be banished once and for all from rigorous mathematics should, at its close, be knocking at the door for readmission.
>
> JAMES PIERPONT

> The series is divergent; therefore we may be able to do something with it. OLIVER HEAVISIDE

1. Introduction

The serious consideration from the late nineteenth century onward of a subject such as divergent series indicates how radically mathematicians have revised their own conception of the nature of mathematics. Whereas in the first part of the nineteenth century they accepted the ban on divergent series on the ground that mathematics was restricted by some inner requirement or the dictates of nature to a fixed class of correct concepts, by the end of the century they recognized their freedom to entertain any ideas that seemed to offer any utility.

We may recall that divergent series were used throughout the eighteenth century, with more or less conscious recognition of their divergence, because they did give useful approximations to functions when only a few terms were used. After the dawn of rigorous mathematics with Cauchy, most mathematicians followed his dictates and rejected divergent series as unsound. However, a few mathematicians (Chap. 40, sec. 7) continued to defend divergent series because they were impressed by their usefulness, either for the computation of functions or as analytical equivalents of the functions from which they were derived. Still others defended them because they were a method of discovery. Thus De Morgan[1] says, "We must admit that many series are such as we cannot at present safely use, except as a means of discovery, the results of which are to be subsequently verified and the most

---

1. *Trans. Camb. Phil. Soc.*, 8, Part II, 1844, 182–203, pub. 1849.

determined rejector of all divergent series doubtless makes this use of them in his closet. . . ."

Astronomers continued to use divergent series even after they were banished because the exigencies of their science required them for purposes of computation. Since the first few terms of such series offered useful numerical approximations, the astronomers ignored the fact that the series as a whole were divergent, whereas the mathematicians, concerned with the behavior not of the first ten or twenty terms but with the character of the entire series, could not base a case for such series on the sole ground of utility.

However, as we have already noted (Chap. 40, sec. 7), both Abel and Cauchy were not without concern that in banishing divergent series they were discarding something useful. Cauchy not only continued to use them (see below) but wrote a paper with the title "Sur l'emploi légitime des séries divergentes,"[2] in which, speaking of the Stirling series for log $\Gamma(x)$ or log $m!$ (Chap. 20, sec. 4), Cauchy points out that the series, though divergent for all values of x, can be used in computing log $\Gamma(x)$ when x is large and positive. In fact he showed that having fixed the number n of terms taken, the absolute error committed by stopping the summation at the nth term is less than the absolute value of the next succeeding term, and the error becomes smaller as x increases. Cauchy tried to understand why the approximation furnished by the series was so good, but failed.

The usefulness of divergent series ultimately convinced mathematicians that there must be some feature which, if culled out, would reveal why they furnished good approximations. As Oliver Heaviside put it in the second volume of *Electromagnetic Theory* (1899), "I must say a few words on the subject of generalized differentiation and divergent series. . . . It is not easy to get up any enthusiasm after it has been artificially cooled by the wet blanket of the rigorists. . . . There will have to be a theory of divergent series, or say a larger theory of functions than the present, including convergent and divergent series in one harmonious whole." When Heaviside made this remark, he was unaware that some steps had already been taken.

The willingness of mathematicians to move on the subject of divergent series was undoubtedly strengthened by another influence that had gradually penetrated the mathematical atmosphere: non-Euclidean geometry and the new algebras. The mathematicians slowly began to appreciate that mathematics is man-made and that Cauchy's definition of convergence could no longer be regarded as a higher necessity imposed by some superhuman power. In the last part of the nineteenth century they succeeded in isolating the essential property of those divergent series that furnished useful approximations to functions. These series were called asymptotic by Poincaré, though during the century they were called semiconvergent, a term introduced by

2. *Comp. Rend.*, 17, 1843, 370–76 = *Œuvres*, (1), 8, 18-25.

Legendre in his *Essai de la théorie des nombres* (1798, p. 13) and used also for oscillating series.

The theory of divergent series has two major themes. The first is the one already briefly described, namely, that some of these series may, for a fixed number of terms, approximate a function better and better as the variable increases. In fact, Legendre, in his *Traité des fonctions elliptiques* (1825–28), had already characterized such series by the property that the error committed by stopping at any one term is of the order of the first term omitted. The second theme in the theory of divergent series is the concept of summability. It is possible to define the sum of a series in entirely new ways that give finite sums to series that are divergent in Cauchy's sense.

2. *The Informal Uses of Divergent Series*

We have had occasion to describe eighteenth-century work in which both convergent and divergent series were employed. In the nineteenth century, both before and after Cauchy banned divergent series, some mathematicians and physicists continued to use them. One new application was the evalua-tion of integrals in series. Of course the authors were not aware at the time that they were finding either complete asymptotic series expansions or first terms of asymptotic series expansions of the integrals.

The asymptotic evaluation of integrals goes back at least to Laplace. In his *Théorie analytique des probabilités* (1812)[3] Laplace obtained by integra-tion by parts the expansion for the error function

$$Erfc(T) = \int_T^\infty e^{-t^2}\, dt = \frac{e^{-T^2}}{2T}\left\{ 1 - \frac{1}{2T^2} + \frac{1 \cdot 3}{(2T^2)^2} - \frac{1 \cdot 3 \cdot 5}{(2T^2)^3} + \cdots \right\}.$$

He remarked that the series is divergent, but he used it to compute $Erfc(T)$ for large values of T.

Laplace also pointed out in the same book[4] that

$$\int \phi(x)\{u(x)\}^s\, dx$$

when s is large depends on the values of $u(x)$ near its stationary points, that is, the values of x for which $u'(x) = 0$. Laplace used this observation to prove that

$$s! \sim s^{s+1/2}e^{-s}\sqrt{2\pi}\left(1 + \frac{1}{12s} + \frac{1}{288s^2} + \cdots \right),$$

a result that can be obtained as well from Stirling's approximation to $\log s!$ (Chap. 20, sec. 4).

3. Third ed., 1820, 88–109 = *Œuvres*, 7, 89–110.
4. *Œuvres*, 7, 128–31.

Laplace had occasion, in his *Théorie analytique des probabilités*,[5] to consider integrals of the form

(1)
$$f(x) = \int_a^b g(t)e^{xh(t)}\,dt,$$

where g may be complex, h and t are real, and x is large and positive; he remarked that the main contribution to the integral comes from the immediate neighborhood of those points in the range of integration in which $h(t)$ attains its absolute maximum. The contribution of which Laplace spoke is the first term of what we would call today an asymptotic series evaluation of the integral. If $h(t)$ has just one maximum at $t = a$, then Laplace's result is

$$f(x) \sim g(a)e^{xh(a)}\sqrt{\frac{-\pi}{2xh''(a)}}$$

as x approaches ∞.

If in place of (1) the integral to be evaluated is

(2)
$$f(x) = \int_a^b g(t)e^{ixh(t)}\,dt$$

when t and x are real and x is large, then $|e^{ixh(t)}|$ is a constant and Laplace's method does not apply. In this case a method adumbrated by Cauchy in his major paper on the propagation of waves,[6] now called the principle of stationary phase, is applicable. The principle states that the most relevant contribution to the integral comes from the immediate neighborhoods of the stationary points of $h(t)$, that is, the points at which $h'(t) = 0$. This principle is intuitively reasonable because the integrand can be thought of as an oscillating current or wave with amplitude $|g(t)|$. If t is time, then the velocity of the wave is proportional to $xh'(t)$, and if $h'(t) \neq 0$, the velocity of the oscillations increases indefinitely as x becomes infinite. The oscillations are then so rapid that during a full period $g(t)$ is approximately constant and $xh(t)$ is approximately linear, so that the integral over a full period vanishes. This reasoning fails at a value of t where $h'(t) = 0$. Thus the stationary points of $h(t)$ are likely to furnish the main contribution to the asymptotic value of $f(x)$. If τ is a value of t at which $h'(t) = 0$ and $h''(\tau) > 0$, then

$$f(x) \sim \sqrt{\frac{2\pi}{xh''(\tau)}}\,g(\tau)e^{ixh(\tau)+i\pi/4}$$

as x becomes infinite.

This principle was used by Stokes in evaluating Airy's integral (see

5. Third ed., 1820, Vol. 1, Part 2, Chap. 1 = *Œuvres*, 7, 89–110.
6. *Mém. de l'Acad. des Sci., Inst. France*, 1, 1827, Note 16 = *Œuvres*, (1), 1, 230.

below) in a paper of 1856[7] and was formulated explicitly by Lord Kelvin.[8] The first satisfactory proof of the principle was given by George N. Watson (1886–1965).[9]

In the first few decades of the nineteenth century, Cauchy and Poisson evaluated many integrals containing a parameter in series of powers of the parameter. In the case of Poisson the integrals arose in geophysical heat conduction and elastic vibration problems, whereas Cauchy was concerned with water waves, optics, and astronomy. Thus Cauchy, working on the diffraction of light,[10] gave divergent series expressions for the Fresnel integrals:

$$\int_0^m \cos\left(\frac{\pi}{2} z^2\right) dz = \frac{1}{2} - N \cos\frac{\pi}{2} m^2 + M \sin\frac{\pi}{2} m^2$$

$$\int_0^m \sin\left(\frac{\pi}{2} z^2\right) dz = \frac{1}{2} - M \cos\frac{\pi}{2} m^2 - N \sin\frac{\pi}{2} m^2,$$

where

$$M = \frac{1}{m\pi} - \frac{1\cdot 3}{m^5\pi^3} + \frac{1\cdot 3\cdot 5\cdot 7}{m^9\pi^5} - \cdots, \qquad N = \frac{1}{m^3\pi^2} - \frac{1\cdot 3\cdot 5}{m^7\pi^4} + \cdots.$$

During the nineteenth century several other methods, such as the method of steepest descent, were invented to evaluate integrals. The full theory of all these methods and the proper understanding of what the approximations amounted to, whether single terms or full series, had to await the creation of the theory of asymptotic series.

Many of the integrals that were expanded by the above-described methods arose first as solutions of differential equations. Another use of divergent series was to solve differential equations directly. This use can be traced back at least as far as Euler's work[11] wherein, concerned with solving the non-uniform vibrating-string problem (Chap. 22, sec. 3), he gave an asymptotic series solution of an ordinary differential equation that is essentially Bessel's equation of order $r/2$, where r is integral.

Jacobi[12] gave the asymptotic form of $J_n(x)$ for large x:

$$J_n(x) \sim \left(\frac{2}{\pi x}\right)^{1/2}\left[\cos\left(x - \frac{n\pi}{2} - \frac{\pi}{4}\right)\left\{1 - \frac{(4n^2 - 1^2)(4n^2 - 3^2)}{2!\,(8x)^2} + \cdots\right\} - \right.$$

$$\left. \sin\left(x - \frac{n\pi}{2} - \frac{\pi}{4}\right)\left\{\frac{4n^2 - 1^2}{1!\,8x} - \frac{(4n^2 - 1^2)(4n^2 - 3^2)(4n^2 - 5^2)}{3!\,(8x)^3} + \cdots\right\}\right].$$

7. *Trans. Camb. Phil. Soc.*, 9, 1856, 166–87 = *Math. and Phys. Papers*, 2, 329–57.
8. *Phil. Mag.*, (5), 23, 1887, 252–55 = *Math. and Phys. Papers*, 4, 303–6.
9. *Proc. Camb. Phil. Soc.*, 19, 1918, 49–55.
10. *Comp. Rend.*, 15, 1842, 554–56 and 573–78 = *Œuvres*, (1), 7, 149–57.
11. *Novi Comm. Acad. Sci. Petrop.*, 9, 1762/3, 246–304, pub. 1764 = *Opera*, (2), 10, 293–343.
12. *Astronom. Nach.*, 28, 1849, 65–94 = *Werke*, 7, 145–74.

A somewhat different use of divergent series in the solution of differential equations was introduced by Liouville. He sought[13] approximate solutions of the differential equation

$$(3) \qquad \frac{d}{dx}\left(p\,\frac{dy}{dx}\right) + (\lambda^2 q_0 + q_1)y = 0,$$

where p, q_0, and q_1 are positive functions of x and λ is a parameter; the solution is to be obtained in $a \le x \le b$. Here, as opposed to boundary-value problems where discrete values of λ are sought, he was interested in obtaining some approximate form of y for large values of λ. To do this he introduced the variables

$$(4) \qquad t = \int_{x_0}^{x}\left(\frac{q_0}{p}\right)^{1/2}dx, \qquad w = (q_0 p)^{1/4}y$$

and obtained

$$(5) \qquad \frac{d^2w}{dt^2} + \lambda^2 w = rw,$$

where

$$r = (q_0 p)^{-1/4}\frac{d^2}{dt^2}(q_0 p)^{1/4} - \frac{q_1}{q_0}.$$

He then used a process that amounts in modern language to the solution by successive approximations of an integral equation of the Volterra type, namely,

$$w(t) = c_1 \cos \lambda t + c_2 \sin \lambda t + \int_{t_0}^{t}\frac{\sin \lambda(t-s)}{\lambda}r(s)w(s)\,ds.$$

Liouville now argued that for sufficiently large values of λ the first approximation to the solutions of (5) should be

$$(6) \qquad w \sim c_1 \cos \lambda t + c_2 \sin \lambda t.$$

If the values of w and t given by (4) are now used to obtain the approximate solution of (3), we have from (6) that

$$(7) \quad y \sim c_1 \frac{1}{(q_0 p)^{1/4}} \cos\left\{\lambda \int_{x_0}^{x}\left(\frac{q_0}{p}\right)^{1/2}dx\right\} + c_2 \frac{1}{(q_0 p)^{1/4}} \sin\left\{\lambda \int_{x_0}^{x}\left(\frac{q_0}{p}\right)^{1/2}dx\right\}.$$

Though Liouville was not aware of it, he had obtained the first term of an asymptotic series solution of (3) for large λ.

The same method was used by Green[14] in the study of the propagation

13. *Jour. de Math.*, 2, 1837, 16–35.
14. *Trans. Camb. Phil. Soc.*, 6, 1837, 457–62 = *Math. Papers*, 225–30.

of waves in a channel. The method has been generalized slightly to equations of the form

$$(8) \qquad\qquad y'' + \lambda^2 q(x, \lambda)y = 0,$$

in which λ is a large positive parameter and x may be real or complex. The solutions are now usually expressed as

$$(9) \qquad\qquad y \sim q^{-1/4} \exp\left(\pm i\lambda \int_0^x q^{1/2}\, dx\right)\left[1 + O\left(\frac{1}{\lambda}\right)\right].$$

The error term $O(1/\lambda)$ implies that the exact solution would contain a term $F(x, \lambda)/\lambda$ where $|F(x, \lambda)|$ is bounded for all x in the domain under consideration and for $\lambda > \lambda_0$. The form of the error term is valid in a restricted domain of the complex x-plane. Liouville and Green did not supply the error term or conditions under which their solutions were valid. The more general and precise approximation (9) is explicit in papers by Gregor Wentzel (1898–),[15] Hendrik A. Kramers (1894–1952),[16] Léon Brillouin (1889–1969),[17] and Harold Jeffreys (1891–1989),[18] and is familiarly known as the WKBJ solution. These men were working with Schrödinger's equation in quantum theory.

In a paper read in 1850,[19] Stokes considered the value of Airy's integral

$$(10) \qquad\qquad W = \int_0^\infty \cos\frac{\pi}{2}\,(w^3 - mw)\, dw$$

for large $|m|$. This integral represents the strength of diffracted light near a caustic. Airy had given a series for W in powers of m which, though convergent for all m, was not useful for calculation when $|m|$ is large. Stokes's method consisted in forming a differential equation of which the integral is a particular solution and solving the differential equation in terms of divergent series that might be useful for calculation (he called such series semiconvergent).

After showing that $U = (\pi/2)^{1/3}\, W$ satisfies the Airy differential equation

$$(11) \qquad\qquad \frac{d^2 U}{dn^2} + \frac{n}{3}\, U = 0, \qquad n = \left(\frac{\pi}{2}\right)^{2/3} m,$$

Stokes proved that for positive n

$$(12) \qquad U = An^{-1/4}\left(R \cos\frac{2}{3}\sqrt{\frac{n^3}{3}} + S \sin\frac{2}{3}\sqrt{\frac{n^3}{3}}\right)$$

$$+ Bn^{-1/4}\left(R \sin\frac{2}{3}\sqrt{\frac{n^3}{3}} - S \cos\frac{2}{3}\sqrt{\frac{n^3}{3}}\right),$$

15. *Zeit. für Physik*, 38, 1926, 518–29.
16. *Zeit. für Physik*, 39, 1926, 828–40.
17. *Comp. Rend.*, 183, 1926, 24–26).
18. *Proc. London Math. Soc.*, (2), 23, 1923, 428–36.
19. *Trans. Camb. Phil. Soc.*, 9, 1856, 166–87 = *Math. and Phys. Papers*, 2, 329–57.

where

$$(13) \qquad R = 1 - \frac{1 \cdot 5 \cdot 7 \cdot 11}{1 \cdot 2 \cdot 16^2 \cdot 3n^3} + \frac{1 \cdot 5 \cdot 7 \cdot 11 \cdot 13 \cdot 17 \cdot 19 \cdot 23}{1 \cdot 2 \cdot 3 \cdot 4 \cdot 16^4 \cdot 3^2 n^6} \cdots,$$

$$(14) \qquad S = \frac{1 \cdot 5}{1 \cdot 16 (3n^3)^{1/2}} - \frac{1 \cdot 5 \cdot 7 \cdot 11 \cdot 13 \cdot 17}{1 \cdot 2 \cdot 3 \cdot 16^3 (3n^3)^{3/2}} + \cdots.$$

The quantities A and B that make U yield the integral were determined by a special argument (wherein Stokes uses the principle of stationary phase). He also gave an analogous result for n negative.

The series (12) and the one for negative n behave like convergent series for some number of terms but are actually divergent. Stokes observed that they can be used for calculation. Given a value of n, one uses the terms from the first up to the one that becomes smallest for the value of n in question. He gave a qualitative argument to show why the series are useful for numerical work.

Stokes encountered a special difficulty with the solutions of (11) for positive and negative n. He was not able to pass from the series for which n is positive to the series for which n is negative by letting n vary through 0 because the series have no meaning for $n = 0$. He therefore tried to pass from positive to negative n through complex values of n, but this did not yield the correct series and constant multipliers.

What Stokes did discover, after some struggle,[20] was that if for a certain range of the amplitude of n a general solution was represented by a certain linear combination of two asymptotic series each of which is a solution, then in a neighboring range of the amplitude of n it was by no means necessary for the same linear combination of the two fundamental asymptotic expansions to represent the same general solution. He found that the constants of the linear combination changed abruptly as certain lines given by amp. n = const. were crossed. These are now called Stokes lines.

Though Stokes had been primarily concerned with the evaluation of integrals, it was clear to him that divergent series could be used generally to solve differential equations. Whereas Euler, Poisson, and others had solved individual equations in such terms, their results appeared to be tricks that produced answers for specific physical problems. Stokes actually gave several examples in the 1856 and 1857 papers.

The above work on the evaluation of integrals and the solution of differential equations by divergent series is a sample of what was done by many mathematicians and physicists.

3. The Formal Theory of Asymptotic Series

The full recognition of the nature of those divergent series that are useful in the representation and calculation of functions and a formal definition

20. *Trans. Camb. Phil. Soc.*, 10, 1857, 106–28 = *Math. and Phys. Papers*, 4, 77–109.

of these series were achieved by Poincaré and Stieltjes independently in 1886. Poincaré called these series asymptotic while Stieltjes continued to use the term semiconvergent. Poincaré[21] took up the subject in order to further the solution of linear differential equations. Impressed by the usefulness of divergent series in astronomy, he sought to determine which were useful and why. He succeeded in isolating and formulating the essential property. A series of the form

$$(15) \qquad a_0 + \frac{a_1}{x} + \frac{a_2}{x^2} + \cdots,$$

where the a_i are independent of x, is said to represent the function $f(x)$ asymptotically for large values of x whenever

$$(16) \qquad \lim_{x \to \infty} x^n \left[f(x) - \left(a_0 + \frac{a_1}{x} + \cdots + \frac{a_n}{x^n} \right) \right] = 0$$

for $n = 0, 1, 2, 3, \ldots$. The series is generally divergent but may in special cases be convergent. The relationship of the series to $f(x)$ is denoted by

$$f(x) \sim a_0 + \frac{a_1}{x} + \frac{a_2}{x^2} + \cdots.$$

Such series are expansions of functions in the neighborhood of $x = \infty$. Poincaré in his 1886 paper considered real x-values. However, the definition holds also for complex x if $x \to \infty$ is replaced by $|x| \to \infty$, though the validity of the representation may then be confined to a sector of the complex plane with vertex at the origin.

The series (15) is asymptotic to $f(x)$ in the neighborhood of $x = \infty$. However, the definition has been generalized, and one speaks of the series

$$a_0 + a_1 x + a_2 x^2 + \cdots$$

as asymptotic to $f(x)$ at $x = 0$ if

$$\lim_{x \to 0} \frac{1}{x^n} \left[f(x) - \sum_{0}^{n-1} a_i x^i \right] = a_n.$$

Though in the case of some asymptotic series one knows what error is committed by stopping at a definite term, no such information about the numerical error is known for general asymptotic series. However, asymptotic series can be used to give rather accurate numerical results for large x by employing only those terms for which the magnitude of the terms decreases as one takes more and more terms. The order of the magnitude of the error at any stage is equal to the magnitude of the first term omitted.

21. *Acta Math.*, 8, 1886, 295–344 = *Œuvres*, 1, 290–332.

Poincaré proved that the sum, difference, product, and quotient of two functions are represented asymptotically by the sum, difference, product, and quotient of their separate asymptotic series, provided that the constant term in the divisor series is not zero. Also, if

$$f(x) \sim a_0 + \frac{a_1}{x} + \frac{a_2}{x^2} + \cdots,$$

then

$$\int_{x_0}^x f(z)\, dz \sim C + a_0 x + a_1 \log x - \frac{a_2}{x} - \frac{1}{2}\frac{a_3}{x^2} - \cdots.$$

The use of integration involves a slight generalization of the original definition, namely,

$$\phi(x) \sim f(x) + g(x)\left(a_0 + \frac{a_1}{x} + \frac{a_2}{x^2} + \cdots\right)$$

if

$$\frac{\phi(x) - f(x)}{g(x)} \sim a_0 + \frac{a_1}{x} + \frac{a_2}{x^2} + \cdots$$

even when $f(x)$ and $g(x)$ do not themselves have asymptotic series representations. As for differentiation, if $f'(x)$ is known to have an asymptotic series expansion, then it can be obtained by differentiating the asymptotic series for $f(x)$.

If a given function has an asymptotic series expansion it is unique, but the converse is not true because, for example, $(1 + x)^{-1}$ and $(1 + e^{-x}) \cdot (1 + x)^{-1}$ have the same asymptotic expansion.

Poincaré applied his theory of asymptotic series to differential equations, and there are many such uses in his treatise on celestial mechanics, *Les Méthodes nouvelles de la mécanique céleste*.[22] The class of equations treated in his 1886 paper is

(17) $P_n(x)y^{(n)} + P_{n-1}(x)y^{(n-1)} + \cdots + P_0(x)y = 0$

where the $P_i(x)$ are polynomials in x. Actually Poincaré treated only the second order case but the method applies to (17).

The only singular points of equation (17) are the zeros of $P_n(x)$ and $x = \infty$. For a regular singular point (*Stelle der Bestimmtheit*) there are convergent expressions for the integrals given by Fuchs (Chap. 29, sec. 5). Consider then an irregular singular point. By a linear transformation this point can be removed to ∞, while the equation keeps its form. If P_n is of the pth degree, the condition that $x = \infty$ shall be a regular singular point

22. Vol. 2, Chap. 8, 1893.

is that the degrees of $P_{n-1}, P_{n-2}, \ldots, P_0$ be at most $p - 1, p - 2, \ldots, p - n$ respectively. For an irregular singular point one or more of these degrees must be greater. Poincaré showed that for a differential equation of the form (17), wherein the degrees of the P_i do not exceed the degree of P_n, there exist n series of the form

$$e^{ax}x^\alpha \left(A_0 + \frac{A_1}{x} + \cdots \right)$$

and the series satisfy the differential equation formally. He also showed that to each such series there corresponds an exact solution in the form of an integral to which the series is asymptotic.

Poincaré's results are included in the following theorem due to Jakob Horn (1867–1946).[23] He treats the equation

(18) $y^{(n)} + a_1(x)y^{(n-1)} + \cdots + a_n(x)y^n = 0,$

where the coefficients are rational functions of x and are assumed to be developable for large positive x in convergent or asymptotic series of the form

$$a_r(x) = x^{rk}\left[a_{r,0} + \frac{a_{r,1}}{x} + \frac{a_{r,2}}{x^2} + \cdots \right], \qquad r = 1, 2, \cdots, n,$$

k being some positive integer or 0. If for the above equation (18) the roots $m_1, m_2, \ldots, m_n$ of the characteristic equation, that is, the algebraic equation

$$m^n + a_{1,0}m^{n-1} + \cdots + a_{n,0} = 0,$$

are distinct, equation (18) possesses n linearly independent solutions $y_1, y_2, \ldots, y_n$ which are developable asymptotically for large positive values of x in the form

$$y_r \sim e^{f_r(x)}x^{\rho_r} \sum_{j=0}^{\infty} \frac{A_{r,j}}{x^j}, \qquad r = 1, 2, \cdots, n,$$

where $f_r(x)$ is a polynomial of degree $k + 1$ in x, the coefficient of whose highest power in x is $m_r/(k + 1)$, while ρ_r and $A_{r,j}$ are constants with $A_{r,0} = 1$. The results of Poincaré and Horn have been extended to various other types of differential equations and generalized by inclusion of the cases where the roots of the characteristic equation are not necessarily distinct.

The existence, form, and range of the asymptotic series solution when the independent variable in (18) is allowed to take on complex values was first taken up by Horn.[24] A general result was given by George David Birkhoff (1884–1944), one of the first great American mathematicians.[25]

23. *Acta Math.*, 24, 1901, 289–308.
24. *Math. Ann.*, 50, 1898, 525–56.
25. *Amer. Math. Soc. Trans.*, 10, 1909, 436–70 = *Coll. Math. Papers*, 1, 201–35.

Birkhoff in this paper did not consider equation (18) but the more general system

$$(19) \qquad \frac{dy_i}{dx} = \sum_{j=1}^{n} a_{ij}(x)y_j, \qquad i = 1, 2, \ldots, n,$$

in which for $|x| > R$ we have for each a_{ij}

$$a_{ij}(x) \sim a_{ij}x^q + a_{ij}^{(1)}x^{q-1} + \cdots + a_{ij}^{(q)} + a_{ij}^{(q+1)} \frac{1}{x} + \cdots$$

and for which the characteristic equation in α

$$|a_{ij} - \delta_{ij}\alpha| = 0$$

has distinct roots. He gave asymptotic series solutions for the y_i which hold in various sectors of the complex plane with vertices at $x = 0$.

Whereas Poincaré's and the other expansions above are in powers of the independent variable, further work on asymptotic series solutions of differential equations turned to the problem first considered by Liouville (sec. 2) wherein a parameter is involved. A general result on this problem was given by Birkhoff.[26] He considered

$$(20) \qquad \frac{d^n z}{dx^n} + \rho a_{n-1}(x, \rho) \frac{d^{n-1}z}{dx^{n-1}} + \cdots + \rho^n a_0(x, \rho)z = 0$$

for large $|\rho|$ and for x in the interval $[a, b]$. The functions $a_i(x, \rho)$ are supposed analytic in the complex parameter ρ at $\rho = \infty$ and have derivatives of all orders in the real variable x. The assumptions on $a_i(x, \rho)$ imply that

$$a_i(x, \rho) = \sum_{j=0}^{\infty} a_{ij}(x)\rho^{-j}$$

and that the roots of $w_1(x), w_2(x), \ldots, w_n(x)$ of the characteristic equation

$$w^n + a_{n-1,0}(x)w^{n-1} + \cdots + a_{00}(x) = 0$$

are distinct for each x. He proves that there are n independent solutions

$$z_1(x, \rho), \cdots, z_n(x, \rho)$$

of (20) that are analytic in ρ in a region S of the ρ plane (determined by the argument of ρ), such that for any integer m and large $|\rho|$

$$(21) \qquad z_i(x, \rho) = u_i(x, \rho) + \exp\left[\rho \int_a^x w_i(t)\, dt\right] E_0 \rho^{-m},$$

26. *Amer. Math. Soc. Trans.*, 9, 1908, 219–31 and 380–82 = *Coll. Math. Papers*, 1, 1–36.

where

$$(22) \qquad u_i(x, \rho) = \exp \left[\rho \int_a^x w_i(t) \, dt \sum_{j=0}^{m-1} u_{ij}(x) \rho^{-j} \right]$$

and E_0 is a function of x, ρ, and m bounded for all x in $[a, b]$ and ρ in S. The $u_{ij}(x)$ are themselves determinable. The result (21), in view of (22), states that z_i is given by a series in $1/\rho$ up to $1/\rho^{m-1}$ plus a remainder term, namely, the second term on the right, which contains $1/\rho^m$. Moreover, since m is arbitrary, one can take as many terms in $1/\rho$ as one pleases into the expression for $u_i(x, \rho)$. Since E_0 is bounded, the remainder term is of higher order in $1/\rho$ than $u_i(x, \rho)$. Then the infinite series, which one can obtain by letting m become infinite, is asymptotic to $z_i(x, \rho)/\exp \left[\rho \int_a^x w_i(t) \, dt \right]$ in Poincaré's sense.

In Birkhoff's theorem the asymptotic series for complex ρ is valid only in a sector S of the complex ρ plane. The Stokes phenomenon enters. That is, the analytic continuation of $z_i(x, \rho)$ across a Stokes line is not given by the analytic continuation of the asymptotic series for $z_i(x, \rho)$.

The use of asymptotic series or of the WKBJ approximation to solutions of differential equations has raised another problem. Suppose we consider the equation

$$(23) \qquad\qquad y'' + \lambda^2 q(x) y = 0$$

where x ranges over $[a, b]$. The WKBJ approximation for large λ, in view of (7), gives two solutions for $x > 0$ and two for $x < 0$. It breaks down at the value of x where $q = 0$. Such a point is called a transition point, turning point, or Stokes point. The exact solutions of (23) are, however, finite at such a point. The problem is to relate the WKBJ solutions on each side of the transition point so that they represent the same exact solution over the interval $[a, b]$ in which the differential equation is being solved. To put the problem more specifically, consider the above equation in which $q(x)$ is real for real x and such that $q(x) = 0$, $q'(x) \neq 0$ for $x = 0$. Suppose also that $q(x)$ is negative for x positive (or the reverse). Given a linear combination of the two WKBJ solutions for $x < 0$ and another for $x > 0$, the question arises as to which of the solutions holding for $x > 0$ should be joined to that for $x < 0$. Connection formulas provide the answer.

The scheme for crossing a zero of $q(x)$ was first given by Lord Rayleigh[27] and extended by Richard Gans (1880–1954),[28] who was familiar with Rayleigh's work. Both of these men were working on the propagation of light in a varying medium.

27. *Proc. Roy. Soc.*, A 86, 1912, 207–26 = *Sci. Papers*, 6, 71–90.
28. *Annalen der Phys.*, (4), 47, 1915, 709–36.

The first systematic treatment of connection formulas was given by Harold Jeffreys[29] independently of Gans's work. Jeffreys considered the equation

(24)
$$\frac{d^2y}{dx^2} + \lambda^2 X(x)y = 0,$$

where x is real, λ is large and real, and $X(x)$ has a simple zero at, say, $x = 0$. He derived formulas connecting the asymptotic series solutions of (24) for $x > 0$ and $x < 0$ by using an approximating equation for (24), namely,

(25)
$$\frac{d^2y}{dx^2} + \lambda^2 xy = 0$$

which replaces $X(x)$ in (24) by a linear function in x. The solutions of (25) are

(26)
$$y_\pm(x) = x^{1/2} J_{\pm 1/3}(\xi)$$

where $\xi = (2/3)\lambda x^{3/2}$. The asymptotic expansions for large x of these solutions can be used to join the asymptotic solutions of (24) on each side of $x = 0$. There are many details on the ranges of x and λ that must be considered in the joining process but we shall not enter into them. Extension of the work on connection formulas for the cases where $X(x)$ in (24) has multiple zeros or several distinct zeros and for more complicated second order and higher-order equations and for complex x and λ has appeared in numerous papers.

The theory of asymptotic series, whether used for the evaluation of integrals or the approximate solution of differential equations, has been extended vastly in recent years. What is especially worth noting is that the mathematical development shows that the eighteenth- and nineteenth-century men, notably Euler, who perceived the great utility of divergent series and maintained that these series could be used as analytical equivalents of the functions they represented, that is, that operations on the series corresponded to operations on the functions, were on the right track. Even though these men failed to isolate the essential rigorous notion, they saw intuitively and on the basis of results that divergent series were intimately related to the functions they represented.

4. Summability

The work on divergent series described thus far has dealt with finding asymptotic series to represent functions either known explicitly or existing implicitly as solutions of ordinary differential equations. Another problem that mathematicians tackled from about 1880 on is essentially the converse of finding asymptotic series. Given a series divergent in Cauchy's sense,

29. *Proc. London Math. Soc.*, (2), 23, 1922–24, 428–36.

can a "sum" be assigned to the series? If the series consists of variable terms this "sum" would be a function for which the divergent series might or might not be an asymptotic expansion. Nevertheless the function might be taken to be the "sum" of the series, and this "sum" might serve some useful purposes, even though the series will certainly not converge to it or may not be usable to calculate approximate values of the function.

To some extent the problem of summing divergent series was actually undertaken before Cauchy introduced his definitions of convergence and divergence. The mathematicians encountered divergent series and sought sums for them much as they did for convergent series, because the distinction between the two types was not sharply drawn and the only question was, What is the appropriate sum? Thus Euler's principle (Chap. 20, sec. 7) that a power series expansion of a function has as its sum the value of the function from which the series is derived gave a sum to the series even for values of x for which the series diverges in Cauchy's sense. Likewise, in his transformation of series (Chap. 20, sec. 4), he converted divergent series to convergent ones without doubting that there should be a sum for practically all series. However, after Cauchy did make the distinction between convergence and divergence, the problem of summing divergent series was broached on a different level. The relatively naive assignments in the eighteenth century of sums to all series were no longer acceptable. The new definitions prescribe what is now called *summability*, to distinguish the notion from convergence in Cauchy's sense.

With hindsight one can see that the notion of summability was really what the eighteenth- and early nineteenth-century men were advancing. This is what Euler's methods of summing just described amount to. In fact in his letter to Goldbach of August 7, 1745, in which he asserted that the sum of a power series is the value of the function from which the series is derived, Euler also asserted that every series must have a sum, but since the word sum implies the usual process of adding, and this process does not lead to the sum in the case of a divergent series such as $1 - 1! + 2! - 3! + \cdots$, we should use the word value for the "sum" of a divergent series.

Poisson too introduced what is now recognized to be a summability notion. Implied in Euler's definition of sum as the value of the function from which the series comes is the idea that

$$(27) \qquad \sum_{n=0}^{\infty} a_n = \lim_{x \to 1-} \sum_{n=0}^{\infty} a_n x^n.$$

(The notation $1-$ means that x approaches 1 from below.) According to (27) the sum of $1 - 1 + 1 - 1 + \cdots$ is

$$\lim_{x \to 1-} (1 - x + x^2 - x^3 + \cdots) = \lim_{x \to 1-} (1 + x)^{-1} = \frac{1}{2}.$$

Poisson[30] was concerned with the series

$$\sin \theta + \sin 2\theta + \sin 3\theta + \cdots$$

which diverges except when θ is a multiple of π. His concept, expressed for the full Fourier series

(28) $$\frac{a_0}{2} + \sum_{n} (a_n \cos n\theta + b_n \sin n\theta),$$

is that one should consider the associated power series

(29) $$\frac{a_0}{2} + \sum_{n} (a_n \cos n\theta + b_n \sin n\theta)r^n$$

and define the sum of (28) to be the limit of the series (29) as r approaches 1 from below. Of course Poisson did not appreciate that he was suggesting a definition of a sum of a divergent series because, as already noted, the distinction between convergence and divergence was in his time not a critical one.

The definition used by Poisson is now called Abel summability because it was also suggested by a theorem due to Abel,[31] which states that if the power series

$$f(x) = \sum_{n=0}^{\infty} a_n x^n$$

has a radius of convergence r and converges for $x = r$, then

(30) $$\lim_{x \to r-} f(x) = \sum_{n=0}^{\infty} a_n r^n.$$

Then the function $f(x)$ defined by the series in $-r < x \leq r$ is continuous on the left at $x = r$. However, if $\sum a_n$ does not converge and the limit (30) does exist for $r = 1$, then one has a definition of sum for the divergent series. This formal definition of summability for series divergent in Cauchy's sense was not introduced until the end of the nineteenth century in a connection soon to be described.

One of the motivations for reconsidering the summation of divergent series, beyond their continued usefulness in astronomical work, was what has been called the boundary-value (*Grenzwert*) problem in the theory of analytic functions. A power series $\sum a_n x^n$ may represent an analytic function in a circle of radius r but not for values of x on the circle. The problem was whether one could find a concept of sum such that the power series might have a sum for $|x| = r$ and such that this sum might even be the value of

30. *Jour. de l'Ecole Poly.*, 11, 1820, 417–89.
31. *Jour. für Math.*, 1, 1826, 311–39 = *Œuvres*, 1, 219–50.

$f(x)$ as $|x|$ approaches r. It was this attempt to extend the range of the power series representation of an analytic function that motivated Frobenius, Hölder, and Ernesto Cesàro. Frobenius[32] showed that if the power series $\sum a_n x^n$ has the interval of convergence $-1 < x < 1$, and if

$$(31) \qquad\qquad s_n = a_0 + a_1 + \cdots + a_n,$$

then

$$\lim_{x \to 1-} \sum_{n=0}^{\infty} a_n x^n = \lim_{n \to \infty} \frac{s_0 + s_1 + \cdots + s_n}{n + 1}$$

when the right-hand limit exists. Thus the power series normally divergent for $x = 1$ can have a sum. Moreover if $f(x)$ is the function represented by the power series, Frobenius's definition of the value of the series at $x = 1$ agrees with $\lim_{x \to 1-} f(x)$.

Divorced from its connection with power series, Frobenius's work suggested a summability definition for divergent series. If $\sum a_n$ is divergent and s_n has the meaning in (31), then one can take as the sum

$$\sum_{n=0}^{\infty} a_n = \lim_{n \to \infty} S_n = \lim_{n \to \infty} \frac{s_0 + s_1 + \cdots + s_n}{n + 1}$$

if this limit exists. Thus for the series $1 - 1 + 1 - 1 + \cdots$, the S_n have the values $1, 1/2, 2/3, 2/4, 3/5, 1/2, 4/7, 1/2, \ldots$ so that $\lim_{n \to \infty} S_n = 1/2$. If $\sum a_n$ converges, then Frobenius's "sum" gives the usual sum. This idea of averaging the partial sums of a series can be found in the older literature. It was used for special types of series by Daniel Bernoulli[33] and Joseph L. Raabe (1801–59).[34]

Shortly after Frobenius published his paper, Hölder[35] produced a generalization. Given the series $\sum a_n$, let

$$s_n^{(0)} = s_n$$

$$s_n^{(1)} = \frac{1}{n + 1} \left(s_0^{(0)} + s_1^{(0)} + \cdots + s_n^{(0)} \right)$$

$$s_n^{(2)} = \frac{1}{n + 1} \left(s_0^{(1)} + s_1^{(1)} + \cdots + s_n^{(1)} \right)$$

$$\cdots \cdots \cdots \cdots \cdots \cdots \cdots$$

$$s_n^{(r)} = \frac{1}{n + 1} \left(s_0^{(r-1)} + s_1^{(r-1)} + \cdots + s_n^{(r-1)} \right).$$

32. *Jour. für Math.*, 89, 1880, 262–64 = *Ges. Abh.*, 2, 8–10.
33. *Comm. Acad. Sci. Petrop.*, 16, 1771, 71–90.
34. *Jour. für Math.*, 15, 1836, 355–64.
35. *Math. Ann.*, 20, 1882, 535–49.

Then the sum s is given by

(32) $$s = \lim_{n \to \infty} s_n^{(r)}$$

if this limit exists for some r. Hölder's definition is now known as summability (H, r).
Hölder gave an example. Consider the series

$$-\frac{1}{(1 + x)^2} = -1 + 2x - 3x^2 + 4x^3 - \cdots.$$

This series diverges when $x = 1$. However for $x = 1$,

$$s_0 = -1, s_1 = 1, s_2 = -2, s_3 = 2, s_4 = -3, \cdots.$$

Then

$$s_0^{(1)} = -1, s_1^{(1)} = 0, s_2^{(1)} = -\frac{2}{3}, s_3^{(1)} = 0, s_4^{(1)} = -\frac{3}{5}, \cdots;$$

$$s_0^{(2)} = -1, s_1^{(2)} = -\frac{1}{2}, s_2^{(2)} = -\frac{5}{9}, s_3^{(2)} = -\frac{5}{12}, s_4^{(2)} = -\frac{34}{75}, \cdots.$$

It is almost apparent that $\lim_{n \to \infty} s_n^{(2)} = -1/4$, and this is the Hölder sum $(H, 2)$. It is also the value that Euler assigned to the series on the basis of his principle that the sum is the value of the function from which the series is derived.

Another of the now-standard definitions of summability was given by Cesàro, a professor at the University of Naples.[36] Let the series be $\sum_{i=0}^{\infty} a_i$ and let s_n be $\sum_{i=0}^{n} a_i$. Then the Cesàro sum is

(33) $$s = \lim_{n \to \infty} \frac{S_n^{(r)}}{D_n^{(r)}}, \qquad r \text{ integral and } \geq 0,$$

where

$$S_n^{(r)} = s_n + r s_{n-1} + \frac{r(r + 1)}{2!} s_{n-2} + \cdots + \frac{r(r + 1) \cdots (r + n - 1)}{n!} s_0$$

and

$$D_n^{(r)} = \frac{(r + 1)(r + 2) \cdots (r + n)}{n!}.$$

The case $r = 1$ includes Frobenius's definition. Cesàro's definition is now referred to as summability (C, r). The methods of Hölder and Cesàro give the same results. That Hölder summability implies Cesàro's was proved by Konrad Knopp (1882–1957) in an unpublished dissertation of 1907; the converse was proved by Walter Schnee (b. 1885).[37]

36. *Bull. des Sci. Math.*, (2), 14, 1890, 114–20.
37. *Math. Ann.*, 67, 1909, 110–25.

An interesting feature of some of the definitions of summability when applied to power series with radius of convergence 1 is that they not only give a sum that agrees with $\lim_{x \to 1^-} f(x)$, where $f(x)$ is the function whose power series representation is involved, but have the further property that they preserve a meaning in regions where $|x| > 1$ and in these regions furnish the analytic continuation of the original power series.

Further progress in finding a "sum" for divergent series received its motivation from a totally different direction, the work of Stieltjes on continued fractions. The fact that continued fractions can be converted into divergent or convergent series and conversely was utilized by Euler.[38] Euler sought (Chap. 20, secs. 4 and 6) a sum for the divergent series

$$(34) \qquad\qquad 1 - 2! + 3! - 4! + 5! - \cdots.$$

In his article on divergent series[39] and in correspondence with Nicholas Bernoulli (1687–1759),[40] Euler first proved that

$$(35) \qquad\qquad x - (1!)x^2 + (2!)x^3 - (3!)x^4 + \cdots$$

formally satisfies the differential equation

$$x^2 \frac{dy}{dx} + y = x,$$

for which he obtained the integral solution

$$(36) \qquad\qquad y = \int_0^\infty \frac{xe^{-t}}{1 + xt} \, dt.$$

Then by using rules he derived for converting convergent series into continued fractions Euler transformed (35) into

$$(37) \qquad\qquad \frac{x}{1+} \frac{x}{1+} \frac{x}{1+} \frac{2x}{1+} \frac{2x}{1+} \frac{3x}{1+} \frac{3x}{1+} \cdots.$$

This work contains two features. On the one hand, Euler obtained an integral that can be taken to be the "sum" of the divergent series (35); the latter is in fact asymptotic to the integral. On the other hand he showed how to convert divergent series into continued fractions. In fact he used the continued fraction for $x = 1$ to calculate a value for the series (34).

There was incidental work of this nature during the latter part of the eighteenth century and a good deal of the nineteenth, of which the most noteworthy is due to Laguerre.[41] He proved, first of all, that the integral

38. *Novi Comm. Acad. Sci. Petrop.*, 5, 1754/5, 205–37, pub. 1760 = *Opera*, (1), 14, 585–617, and *Nova Acta Acad. Sci. Petrop.*, 2, 1784, 36–45, pub. 1788 = *Opera*, (1), 16, 34–43.
39. *Novi Comm. Acad. Sci. Petrop.*, 5, 1754/5, 205–37, pub. 1760 = *Opera*, (1), 14, 585–617.
40. Euler's *Opera Posthuma*, 1, 545–49.
41. *Bull. Soc. Math. de France*, 7, 1879, 72–81 = *Œuvres*, 1, 428–37.

(36) could be expanded into the continued fraction (37). He also treated the divergent series

(38) $$1 + x + 2! \, x^2 + 3! \, x^3 + \cdots.$$

Since

$$m! = \Gamma(m + 1) = \int_0^\infty e^{-z} z^m \, dz,$$

the series can be written as

$$\int_0^\infty e^{-z} \, dz + x \int_0^\infty e^{-z} z \, dz + x^2 \int_0^\infty e^{-z} z^2 \, dz + \cdots.$$

If we formally interchange summation and integration we obtain

$$\int_0^\infty e^{-z} (1 + xz + x^2 z^2 + \cdots) \, dz$$

or

(39) $$f(x) = \int_0^\infty e^{-z} \frac{1}{1 - zx} \, dz.$$

The $f(x)$ thus derived is analytic for all complex x except real and positive values and can be taken to be the "sum" of the series (38).

In his thesis of 1886 Stieltjes took up the study of divergent series.[42] Here Stieltjes introduced the very same definition of a series asymptotic to a function that Poincaré had introduced, but otherwise confined himself to the computational aspects of some special series.

Stieltjes did go on to study continued fraction expansions of divergent series and wrote two celebrated papers of 1894–95 on this subject.[43] This work, which is the beginning of an analytic theory of continued fractions, considered questions of convergence and the connection with definite integrals and divergent series. It was in these papers that he introduced the integral bearing his name.

Stieltjes starts with the continued fraction

(40) $$\frac{1}{a_1 z+} \, \frac{1}{a_2 +} \, \frac{1}{a_3 z+} \, \frac{1}{a_4 z} \, \frac{1}{a_5 z+} \cdots \frac{1}{a_{2n}+} \, \frac{1}{a_{2n+1} z+} \cdots,$$

where the a_n are positive real numbers and z is complex. He then shows that when the series $\sum_{n=1}^\infty a_n$ diverges, the continued fraction (40) converges

42. "Recherches sur quelques séries semi-convergentes," *Ann. de l'Ecole Norm. Sup.*, (3), 3, 1886, 201–58 = *Œuvres complètes*, 2, 2–58.
43. *Ann. Fac. Sci. de Toulouse*, 8, 1894, J. 1–122, and 9, 1895, A. 1–47 = *Œuvres complètes*, 2, 402–559.

to a function $F(z)$ which is analytic in the complex plane except along the negative real axis and at the origin, and

$$(41) \qquad\qquad F(z) = \int_0^\infty \frac{d\phi(u)}{u + z}.$$

When $\sum a_n$ converges, the even and odd partial sums of (40) converge to distinct limits $F_1(z)$ and $F_2(z)$ where

$$F_1(z) = \int_0^\infty \frac{dg_1(u)}{z + u}, \qquad F_2(z) = \int_0^\infty \frac{dg_2(u)}{z + u}.$$

Now it was known that the continued fraction (40) could be formally developed into a series

$$(42) \qquad\qquad \frac{C_0}{z} - \frac{C_1}{z^2} + \frac{C_2}{z^3} - \frac{C_3}{z^4} + \cdots$$

with positive C_i. The correspondence (with some restrictions) is also reciprocal. To every series (42) there corresponds a continued fraction (40) with positive a_n. Stieltjes showed how to determine the C_i from the a_n and in the case where $\sum a_n$ is divergent he showed that the ratio C_n/C_{n-1} increases. If it has a finite limit λ, the series converges for $|z| > \lambda$, but if the ratio increases without limit then the series diverges for all z.

The relation between the series (42) and the continued fraction (40) is more detailed. Although the continued fraction converges if the series does, the converse is not true. When the series (42) diverges one must distinguish two cases, according as $\sum a_n$ is divergent or convergent. In the former case, as we have noted, the continued fraction gives one and only one functional equivalent, which can be taken to be the sum of the divergent series (42). When $\sum a_n$ is convergent two different functions are obtained from the continued fraction, one from the even convergents and the other from the odd convergents. But to the series (42) (now divergent) there corresponds an infinite number of functions each of which has the series as its asymptotic development.

Stieltjes's results have also this significance: they indicate a division of divergent series into at least two classes, those series for which there is properly a single functional equivalent whose expansion is the series and those for which there are at least two functional equivalents whose expansions are the series. The continued fraction is only the intermediary between the series and the integral; that is, given the series one obtains the integral through the continued fraction. Thus, a divergent series belongs to one or more functions, which functions can be taken to be the sum of the series in a new sense of sum.

Stieltjes also posed and solved an inverse problem. To simplify the statement a bit, let us suppose that $\phi(u)$ is differentiable so that the integral (41) can be written as

$$\int_0^\infty \frac{f(u)}{z + u}\, du.$$

To the divergent series (42) and in the case where $\sum a_n$ is divergent there corresponds an integral of this form. The problem is, given the series to find $f(u)$. A formal expansion of the integral shows that

$$(43) \qquad C_n = \int_0^\infty f(u)u^n\, du, \qquad n = 0, 1, 2, \cdots.$$

Hence knowing the C_n one must determine $f(u)$ satisfying the infinite set of equations (43). This is what Stieltjes called the "problem of moments." It does not admit a unique solution, for Stieltjes himself gave a function

$$f(u) = e^{-\sqrt[4]{u}} \sin \sqrt[4]{u}$$

which makes $C_n = 0$ for all n. If the supplementary condition is imposed that $f(u)$ shall be positive between the limits of integration, then only a single $f(u)$ is possible.

The systematic development of the theory of summable series begins with the work of Borel from 1895 on. He first gave definitions that generalize Cesàro's. Then, taking off from Stieltjes's work, he gave an integral definition.[44] If the process used by Laguerre is applied to any series of the form

$$(44) \qquad a_0 + a_1 x + a_2 x^2 + \cdots$$

having a finite radius of convergence (including 0), we are led to the integral

$$(45) \qquad \int_0^\infty e^{-z} F(zx)\, dz$$

where

$$F(u) = 1 + a_1 u + \frac{a_2}{2!} u^2 + \cdots + \frac{a_n}{n!} u^n + \cdots.$$

This integral is the expression on which Borel built his theory of divergent series. It was taken by Borel to be the sum of the series (44). The series $F(u)$ is called the associated series of the original series.

If the original series (44) has a radius of convergence R greater than 0, the associated series represents an entire function. Then the integral $\int_0^\infty e^{-z} F(zx)\, dz$ has a sense if x lies within the circle of convergence, and the values of the integral and series are identical. But the integral may have a sense for values of x outside the circle of convergence, and in this case the

44. *Ann. de l'Ecole Norm. Sup.*, (3), 16, 1899, 9–136.

integral furnishes an analytic continuation of the original series. The series is said by Borel to be *summable* (in the sense just explained) at a point x where the integral has a meaning.

If the original series (44) is divergent $(R = 0)$, the associated series may be convergent or divergent. If it is convergent over only a portion of the plane $u = zx$ we understand by $F(u)$ the value not merely of the associated series but of its analytic continuation. Then the integral $\int_0^\infty e^{-z} F(zx)\, dz$ may have a meaning and is, as we see, obtained from the original divergent series. The determination of the region of x-values in which the original series is summable was undertaken by Borel, both when the original series is convergent $(R > 0)$ and divergent $(R = 0)$.

Borel also introduced the notion of absolute summability. The original series is absolutely summable at a value of x when

$$\int_0^\infty e^{-z} F(zx)\, dz$$

is absolutely convergent and the successive integrals

$$\int_0^\infty e^{-z} \left| \frac{d^\lambda F(zx)}{dz^\lambda} \right| dz, \qquad \lambda = 1, 2, \cdots$$

have a sense. Borel then shows that a divergent series, if absolutely summable, can be manipulated precisely as a convergent series. The series, in other words, represents a function and can be manipulated in place of the function. Thus the sum, difference, and product of two absolutely summable series is absolutely summable and represents the sum, difference, and product respectively of the two functions represented by the separate series. The analogous fact holds for the derivative of an absolutely summable series. Moreover the sum in the above sense agrees with the usual sum in the case of a convergent series, and subtraction of the first k terms reduces the "sum" of the entire series by the sum of these k terms. Borel emphasized that any satisfactory definition of summability must possess these properties, though not all do. He did not require that any two definitions necessarily yield the same sum.

These properties made possible the immediate application of Borel's theory to differential equations. If, in fact,

$$P(x, y, y', \ldots, y^{(n)}) = 0$$

is a differential equation which is holomorphic in x at the origin and is algebraic in y and its derivatives, any absolutely summable series

$$a_0 + a_1 x + a_2 x^2 + \cdots$$

which satisfies the differential equation formally defines an analytic function that is a solution of the equation. For example, the Laguerre series

$$1 + x + 2!\, x^2 + 3!\, x^3 + \cdots$$

satisfies formally

$$x^2 \frac{d^2y}{dx^2} + (x - 1)y = -1,$$

and so the function (cf. (39))

$$f(x) = \int_0^\infty \frac{e^{-z}\, dz}{1 - zx}$$

must be a solution of the equation.

Once the notion of summability gained some acceptance, dozens of mathematicians introduced a variety of new definitions that met some or all of the requirements imposed by Borel and others. Many of the definitions of summability have been extended to double series. Also, a variety of problems has been formulated and many solved that involve the notion of summability. For example, suppose a series is summable by some method. What additional conditions can be imposed on the series so that, granted its summability, it will also be convergent in Cauchy's sense? Such theorems are called Tauberian after Alfred Tauber (1866–1942?). Thus Tauber proved[45] that if $\sum a_n$ is Abelian summable to s and na_n approaches 0 as n becomes infinite, then $\sum a_n$ converges to s.

The concept of summability does then allow us to give a value or sum to a great variety of divergent series. The question of what is accomplished thereby necessarily arises. If a given infinite series were to arise directly in a physical situation, the appropriateness of any definition of sum would depend entirely on whether the sum is physically significant, just as the physical utility of any geometry depends on whether the geometry describes physical space. Cauchy's definition of sum is the one that usually fits because it says basically that the sum is what one gets by continually adding more and more terms in the ordinary sense. But there is no logical reason to prefer this concept of sum to the others that have been introduced. Indeed the representation of functions by series is greatly extended by employing the newer concepts. Thus Leopold Fejér (1880–1959), a student of H. A. Schwarz, showed the value of summability in the theory of Fourier series. In 1904[46] Fejér proved that if in the interval $[-\pi, \pi]$, $f(x)$ is bounded and (Riemann) integrable, or if unbounded the integral $\int_{-\pi}^{\pi} f(x)\, dx$ is absolutely convergent, then at every point in the interval at which $f(x + 0)$ and $f(x - 0)$ exist, the Frobenius sum of the Fourier series

$$\frac{a_0}{2} + \sum_{n=1}^{\infty} (a_n \cos nx + b_n \sin nx)$$

45. *Monatshefte für Mathematik und Physik*, 8, 1897, 273–77.
46. *Math. Ann.*, 58, 1904, 51–69.

is $[f(x + 0) + f(x - 0)]/2$. The conditions on $f(x)$ in this theorem are weaker than in previous theorems on the convergence of Fourier series to $f(x)$ (cf. Chap. 40, sec. 6).

Fejér's fundamental result was the beginning of an extensive series of fruitful investigations on the summability of series. We have had numerous occasions to view the need to represent functions by infinite series. Thus in meeting the initial condition in the solution of initial- and boundary-value problems of partial differential equations, it is usually necessary to represent the given initial $f(x)$ in terms of the eigenfunctions that are obtained from the application of the boundary conditions to the ordinary differential equations that result from the method of separation of variables. These eigenfunctions can be Bessel functions, Legendre functions, or any one of a number of types of special functions. Whereas the convergence in the Cauchy sense of such a series of eigenfunctions to the given $f(x)$ may not obtain, the series may indeed be summable to $f(x)$ in one or another of the senses of summability and the initial condition is thereby satisfied. These applications of summability represent a great success for the concept.

The construction and acceptance of the theory of divergent series is another striking example of the way in which mathematics has grown. It shows, first of all, that when a concept or technique proves to be useful even though the logic of it is confused or even nonexistent, persistent research will uncover a logical justification, which is truly an afterthought. It also demonstrates how far mathematicians have come to recognize that mathematics is man-made. The definitions of summability are not the natural notion of continually adding more and more terms, the notion which Cauchy merely rigorized; they are artificial. But they serve mathematical purposes, including even the mathematical solution of physical problems; and these are now sufficient grounds for admitting them into the domain of legitimate mathematics.

Bibliography

Borel, Emile: *Notice sur les travaux scientifiques de M. Emile Borel*, 2nd ed., Gauthier-Villars, 1921.

————: *Leçons sur les séries divergentes*, Gauthier-Villars, 1901.

Burkhardt, H.: "Trigonometrische Reihe und Integrale," *Encyk. der Math. Wiss.*, B. G. Teubner, 1904–16, II A12, 819–1354.

————: "Über den Gebrauch divergenter Reihen in der Zeit von 1750–1860," *Math. Ann.*, 70, 1911, 169–206.

Carmichael, Robert D.: "General Aspects of the Theory of Summable Series," *Amer. Math. Soc. Bull.*, 25, 1918/19, 97–131.

Collingwood E. F.: "Emile Borel," *Jour. Lon. Math. Soc.*, 34, 1959, 488–512.

Ford, W. B.: "A Conspectus of the Modern Theories of Divergent Series," *Amer. Math. Soc. Bull.*, 25, 1918/19, 1–15.

Hardy, G. H.: *Divergent Series*, Oxford University Press, 1949. See the historical notes at the ends of the chapters.

Hurwitz, W. A.: "A Report on Topics in the Theory of Divergent Series," *Amer. Math. Soc. Bull.*, 28, 1922, 17–36.

Knopp, K.: "Neuere Untersuchungen in der Theorie der divergenten Reihen," *Jahres. der Deut. Math.-Verein.*, 32, 1923, 43–67.

Langer, Rudolf E.: "The Asymptotic Solution of Ordinary Linear Differential Equations of the Second Order," *Amer. Math. Soc. Bull.*, 40, 1934, 545–82.

McHugh, J. A. M.: "An Historical Survey of Ordinary Linear Differential Equations with a Large Parameter and Turning Points," *Archive for History of Exact Sciences*, 7, 1971, 277–324.

Moore, C. N.: "Applications of the Theory of Summability to Developments in Orthogonal Functions," *Amer. Math. Soc. Bull.*, 25, 1918/19, 258–76.

Plancherel, Michel: "Le Développement de la théorie des séries trigonométriques dans le dernier quart de siècle," *L'Enseignement Mathématique*, 24, 1924/25, 19–58.

Pringsheim, A.: "Irrationalzahlen und Konvergenz unendlicher Prozesse," *Encyk. der Math. Wiss.*, B. G. Teubner, 1898–1904, IA3, 47–146.

Reiff, R.: *Geschichte der unendlichen Reihen*, H. Lauppsche Buchhandlung, 1889, Martin Sandig (reprint), 1969.

Smail, L. L.: *History and Synopsis of the Theory of Summable Infinite Processes*, University of Oregon Press, 1925.

Van Vleck, E. B.: "Selected Topics in the Theory of Divergent Series and Continued Fractions," *The Boston Colloquium of the Amer. Math. Soc.*, 1903, Macmillan, 1905, 75–187.

48
Tensor Analysis and Differential Geometry

> Either therefore the reality which underlies space must form a discrete manifold or we must seek the ground of its metric in relations outside it, in the binding forces which act on it. This leads us into the domain of another science, that of physics, into which the object of our work does not allow us to go today.   BERNHARD RIEMANN

1. The Origins of Tensor Analysis

Tensor analysis is often described as a totally new branch of mathematics, created *ab initio* either to meet some specific objective or just to delight mathematicians. It is actually no more than a variation on an old theme, namely, the study of differential invariants associated primarily with a Riemannian geometry. These invariants, we may recall (Chap. 37, sec. 5), are expressions that retain their form and value under any change in the coordinate system because they represent geometrical or physical properties.

The study of differential invariants had been launched by Riemann, Beltrami, Christoffel, and Lipschitz. The new approach was initiated by Gregorio Ricci-Curbastro (1853–1925), a professor of mathematics at the University of Padua. He was influenced by Luigi Bianchi, whose work had followed Christoffel's. Ricci sought to expedite the search for geometrical properties and the expression of physical laws in a form invariant under change of coordinates. He did his major work on the subject during the years 1887–96, though he and an Italian school continued to work on it for twenty or more years after 1896. In the main period Ricci worked out his approach and a comprehensive notation for the subject, which he called the absolute differential calculus. In an article published in 1892[1] Ricci gave the first systematic account of his method and applied it to some problems in differential geometry and physics.

1. *Bull. des Sci. Math.*, (2), 16, 1892, 167–89 = *Opere*, 1, 288–310.

Nine years later Ricci and his famous pupil Tullio Levi-Civita (1873–1941) collaborated on a comprehensive paper, "Methods of the Absolute Differential Calculus and Their Applications."[2] The work of Ricci and Levi-Civita gave a more definitive formulation of this calculus. The subject became known as tensor analysis after Einstein gave it this name in 1916. In view of the many changes in notation made by Ricci and later by Levi-Civita and Ricci, we shall use the notation which has by now become rather standard.

2. The Notion of a Tensor

To get at the notion of a tensor as introduced by Ricci, let us consider the function $A(x^1, x^2, \ldots, x^n)$. By A_i we shall mean $\partial A/\partial x^i$. Then the expression

$$(1) \qquad \sum A_j \, dx^j$$

is a differential invariant under transformations of the form

$$(2) \qquad x^i = f_i(y^1, y^2, \ldots, y^n).$$

It is assumed that the functions f_i possess all necessary derivatives and that the transformation is reversible, so that

$$(3) \qquad y^i = g_i(x^1, x^2, \ldots, x^n).$$

Under the transformation (2) the expression (1) becomes

$$(4) \qquad \sum \bar{A}_j(y^1, y^2, \cdots, y^n) \, dy^j.$$

However, $\bar{A}_j$ does not equal A_j. Rather

$$(5) \qquad \bar{A}_j = \frac{\partial \bar{A}}{\partial y^j} = A_1 \frac{\partial x^1}{\partial y^j} + A_2 \frac{\partial x^2}{\partial y^j} + \cdots + A_n \frac{\partial x^n}{\partial y^j}$$

wherein it is understood that the x^i in the A_i are replaced by their values in terms of the y^i. Thus the $\bar{A}_j$ can be related to the A_j by the specific law of transformation (5) wherein the first derivatives of the transformation are involved.

Ricci's idea was that instead of concentrating on the invariant differential form (1), it would be sufficient and more expeditious to treat the set

$$A_1, A_2, \cdots, A_n$$

and to call this set of components a tensor provided that, under a change of coordinates, the new set of components

$$\bar{A}_1, \bar{A}_2, \cdots, \bar{A}_n$$

2. *Math. Ann.*, 54, 1901, 125–201 = Ricci, *Opere*, 2, 185–271.

are related to the original set by the law of transformation (5). It is this
explicit emphasis on the set or system of functions and the law of transformation that marks Ricci's approach to the subject of differential invariants. The set of A_j, which happen to be the components of the gradient of
the scalar function A, is an example of a covariant tensor of rank 1. The
notion of a set or system of functions characterizing an invariant quantity
was in itself not new because vectors were already well known in Ricci's
time; these are represented by their components in a coordinate system and
are also subject to a law of transformation if the vector is to remain, as it
should be, invariant under a change of coordinates. However, the new
systems that Ricci introduced were far more general and the emphasis on
the law of transformation was also new.

As another example of the point of view introduced by Ricci, let us
consider the expression for the element of distance. This is given by

$$(6) \qquad\qquad ds^2 = \sum_{i,j=1}^{n} g_{ij}\, dx^i\, dx^j.$$

Under change of coordinates the value of the distance ds must remain the
same on geometrical grounds. However, if we perform the transformation
(2) and write the new expression in the form

$$(7) \qquad\qquad \overline{ds^2} = \sum_{i,j=1}^{n} G_{ij}\, dy^i\, dy^j,$$

then $g_{ij}(x^1, \ldots, x^n)$ will not equal $G_{ij}(y^i, \ldots, y^n)$ (when the values of the y^i
represent the same point as the x^i do). What does hold is that

$$(8) \qquad\qquad G_{kl} = \sum_{i,j=1}^{n} g_{ij}\, \frac{\partial x^i}{\partial y^k}\frac{\partial x^j}{\partial y^l},$$

when the x^i in the g_{ij} are replaced by their values in terms of the y^i. To see
that (8) holds we have but to replace dx^i in (6) by

$$dx^i = \sum_{k=1}^{n} \frac{\partial x^i}{\partial y^k}\, dy^k$$

and dx^j in (6) by

$$dx^j = \sum_{l=1}^{n} \frac{\partial x^j}{\partial y^l}\, dy^l$$

and extract the coefficient of $dy^k\, dy^l$. Thus, though G_{kl} is not at all g_{kl}, we
do know how to obtain G_{kl} from the g_{kl}. The set of n^2 coefficients g_{ik} of the

fundamental quadratic form are a tensor, a covariant tensor of rank 2, and the law of transformation is that given by (8).

Ricci also introduced contravariant tensors. Let us consider the inverse of the transformation hitherto considered. If this inverse is

(9) $$y^j = g_j(x^1, x^2, \cdots, x^n),$$

then

(10) $$dy^j = \sum_{k=1}^{n} \frac{\partial y^j}{\partial x^k} \, dx^k.$$

If we now regard the dx^k as a set of quantities constituting a tensor, then we see that, of course, $dy^j \neq dx^j$, but we can obtain the dy^j from the dx^j by the law of transformation illustrated in (10). The set of elements dx^k is called a contravariant tensor of rank 1, the term contravariant pointing to the presence of the $\partial y^j / \partial x^k$ in the transformation, as opposed to the derivatives $\partial x^i / \partial y^j$ which appear in (8) and (5). Thus the very differentials of the transformation variables form a contravariant tensor of rank 1.

Correspondingly, we can have a tensor of rank 2 that transforms contravariantly in both indices. If the set of functions $A^{kl}(x^1, x^2, \ldots, x^n)$, $k, l = 1, 2, \ldots, n$, transform under (9) so that

(11) $$\bar{A}^{ij} = \sum_{k,l=1}^{n} \frac{\partial y^i}{\partial x^k} \frac{\partial y^j}{\partial x^l} A^{kl},$$

then this set is a contravariant tensor of rank 2. Moreover, we can have what are called mixed tensors, which transform covariantly in some indices and contravariantly in others. For example, the set A^k_{ij}, $i, j, k = 1, 2, \ldots, n$, denotes a mixed tensor wherein, following Ricci, the lower indices are the ones in which it transforms covariantly and the upper index is the one in which it transforms contravariantly. The tensor whose elements are A^k_{ij} is called a tensor of rank 3. We can have covariant, contravariant, and mixed tensors of rank r. An n-dimensional tensor of rank r will have n^r components. Equation (21) in Chapter 37 shows that Riemann's four-index symbol (rk, ih) is a covariant tensor of rank 4. A covariant tensor of rank 1 is a vector. To it Levi-Civita associated the contravariant vector which is defined as follows: If the set of λ_i are the components of a covariant vector, then the set

$$\lambda^i = \sum_{k=1}^{n} g^{ik} \lambda_k$$

is the associated contravariant vector; g^{ik} is a quotient whose numerator is the cofactor of g_{ik} in the determinant of the g_{ik} and whose denominator is g, the value of the determinant.

There are operations on tensors. Thus if we have two tensors of the same kind, that is, having the same number of covariant and the same number of contravariant indices, we may add them by adding the components with the identical indices. Thus

$$A_i^j + B_i^j = C_i^j.$$

One must and can show that the C_i^j constitute a tensor covariant in the index i and contravariant in the index j.

One may multiply any two tensors whose indices run from 1 to n. An example may suffice to show the idea. Thus

$$A_i^h B_j^k = C_{ij}^{hk},$$

and one can show that the tensor with the n^4 components C_{ij}^{hk} is covariant in the lower indices and contravariant in the upper ones. There is no operation of division of tensors.

The operation of contraction is illustrated by the following example: Given the tensor A_{ir}^{hs}, we define the quantity

$$B_i^h = \sum_{r=1}^{n} A_{ir}^{hr},$$

wherein on the right side we add the components. We could show that the set of quantities B_i^h is a tensor of rank 2, covariant in the index i and contravariant in the index h.

To sum up, a tensor is a set of functions (components), fixed relative to one frame of reference or coordinate system, that transform under change of coordinates in accordance with certain laws. Each component in one coordinate system is a linear homogeneous function of the components in another coordinate system. If the components of one tensor equal those of another when both are expressed in one coordinate system, they will be equal in all coordinate systems. In particular, if the components vanish in one system they vanish in all. Equality of tensors is then an invariant with respect to change of reference system. The physical, geometrical, or even purely mathematical significance which a tensor possesses in one coordinate system is preserved by the transformation so that it obtains again in the second coordinate system. This property is vital in the theory of relativity, wherein each observer has his own coordinate system. Since the true physical laws are those that hold for all observers, to reflect this independence of coordinate system these laws are expressed as tensors.

With the tensor concept in hand, one can re-express many of the concepts of Riemannian geometry in tensor form. Perhaps the most important is the curvature of the space. Riemann's concept of curvature (Chap. 37, sec. 3) can be formulated as a tensor in many ways. The modern expressions use the summation convention introduced by Einstein, namely, that if an index is repeated in a product of two symbols, then summation is understood. Thus

$$g^{ij}\lambda_j = \sum_{j=1}^{n} g^{ij}\lambda_j.$$

In this notation the curvature tensor is (cf. [20] of Chap. 37)

$$R_{\lambda\mu\rho\sigma} = \frac{\partial}{\partial x^\rho}[\mu\sigma, \lambda] - \frac{\partial}{\partial x^\sigma}[\mu\rho, \lambda] + \{\mu\rho, \varepsilon\}[\lambda\sigma, \varepsilon] - \{\mu\sigma, \varepsilon\}[\lambda\rho, \varepsilon]$$

or equivalently

$$R^i_{jlk} = \frac{\partial}{\partial x^l}\{jk, i\} - \frac{\partial}{\partial x^k}\{jl, i\} - [\{sk, i\}\{jl, s\} - \{sl, i\}\{jk, s\}]$$

wherein the brackets denote Christoffel symbols of the first kind and the braces, symbols of the second kind. Either form is now called the Riemann-Christoffel curvature tensor. By reason of certain relationships (which we shall not describe) among the components, the number of distinct components of this tensor is $n^2(n^2 - 1)/12$. For $n = 4$, which is the case in the general theory of relativity, the number of distinct components is 20. In a two-dimensional Riemannian space there is just one distinct component, which can be taken to be R_{1212}. In this case Gauss's total curvature K proves to be

$$K = \frac{R_{1212}}{g},$$

where g is the determinant of the g_{ij} or $g_{11}g_{22} - g_{12}^2$. If all components vanish the space is Euclidean.

From the Riemann-Christoffel tensor Ricci obtained by contraction what is now called the Ricci tensor. The components R_{jl} are $\sum_{k=1}^{n} R^k_{jlk}$. This tensor for $n = 4$ was used by Einstein[3] to express the curvature of his space-time Riemannian geometry.

3. Covariant Differentiation

Ricci also introduced into tensor analysis[4] an operation that he and Levi-Civita later called covariant differentiation. This operation had already appeared in Christoffel's and Lipschitz's work.[5] Christoffel had given a method (Chap. 37, sec. 4) whereby from differential invariants involving

3. *Zeit. für Math. und Phys.*, 62, 1914, 225–61.
4. *Atti della Accad. dei Lincei, Rendiconti*, (4), 3, 1887, 15–18 = *Opere*, 1, 199–203.
5. *Jour. für Math.*, 70, 1869, 46–70 and 241–45, and 71–102.

the derivatives of the fundamental form for ds^2 and of functions $\phi(x_1, x_2, \ldots, x_n)$ one could derive invariants involving higher derivatives. Ricci appreciated the importance of this method for his tensor analysis and adopted it.

Whereas Christoffel and Lipschitz treated covariant differentiation of the entire form, Ricci, in accordance with his emphasis on the components of a tensor, worked with these. Thus if $A_i(x^1, x^2, \ldots, x^n)$ is a covariant component of a vector or tensor of rank 1, the covariant derivative of A_i is not simply the derivative with respect to x^l, but the tensor of second rank

$$(12) \qquad A_{i,l} = \frac{\partial A_i}{\partial x^l} - \sum_{j=1}^{n} \{il, j\} A_j,$$

where the braces indicate the Christoffel symbol of the second kind. Likewise, if A_{ik} is a component of a covariant tensor of rank 2, then its covariant derivative with respect to x^l is given by

$$(13) \qquad A_{ik,l} = \frac{\partial A_{ik}}{\partial x^l} - \sum_{j=1}^{n} \{il, j\} A_{jk} - \sum_{j=1}^{n} \{kl, j\} A_{ij}.$$

For a contravariant tensor with components A^i of rank 1 the covariant derivative $A^i_{,l}$ is given by

$$A^i_{,l} = \frac{\partial A^i}{\partial x^l} + \sum_{j=1}^{n} A^j \{jl, i\},$$

and this is a mixed tensor of the second order. For the mixed tensor with components A^h_i the covariant derivative is

$$A^h_{i,l} = \frac{\partial A^h_i}{\partial x^l} - \sum_{j=1}^{n} A^h_j \{il, j\} + \sum_{j=1}^{n} A^j_i \{jl, h\}.$$

The covariant derivative of a scalar invariant ϕ is the covariant vector whose components are given by $\phi_{,i} = \partial\phi/\partial x_i$. This vector is called the gradient of the scalar invariant.

From the purely mathematical standpoint, the covariant derivative of a tensor is a tensor of one rank higher in the covariant indices. This fact is important because it enables one to treat such derivatives within the framework of tensor analysis. It also has geometrical meaning. Suppose we have a constant vector field in the plane, that is, a set of vectors one at each point but all having the same magnitude and direction. Then the components of any vector, expressed with respect to a rectangular coordinate system, are also constant. However, the components of these vectors with respect to the polar coordinate system, one component along the radius vector and the other perpendicular to the radius vector, change from point to point, because the directions in which the components are taken change from point to point.

If one takes the derivatives with respect to the coordinates, r and θ say, of these components, then the rate of change expressed by these derivatives reflects the change in the components caused by the coordinate system and not by any change in the vectors themselves. The coordinate systems used in Riemannian geometry are curvilinear. The effect of the curvilinearity of the coordinates is given by the Christoffel symbol of the second kind (here denoted by braces). The full covariant derivative of a tensor gives the actual rate of change of the underlying physical or geometrical quantity represented by the original tensor as well as the change due to the underlying coordinate system.

In Euclidean spaces, where the ds^2 can always be reduced to a sum of squares with constant coefficients, the covariant derivative reduces to the ordinary derivative because the Christoffel symbols are 0. Also the covariant derivative of each g_{ij} in a Riemannian metric is 0. This last fact was proved by Ricci[6] and is called Ricci's lemma.

The concept of covariant differentiation enables us to express readily for tensors generalizations of notions already known in vector analysis but now treatable in Riemannian geometry. Thus if the $A_i(x^1, x^2, \ldots, x^n)$ are the components of an n-dimensional vector A, then

(14) $$\theta = \sum_{i,l=1}^{n} g^{il} A_{i,l},$$

where g^{il} has been defined above, is a differential invariant. When the fundamental metric is a rectangular Cartesian coordinate system (in Euclidean space), the constants $g^{il} = 0$ except when $i = l$, and in this case the covariant and ordinary derivatives are identical. Then (14) becomes

$$\theta = \sum_{i=1}^{n} \frac{\partial A_i}{\partial x^i},$$

and this is the n-dimensional analogue in Euclidean space of what is called the divergence in three dimensions. Hence (14) is also called the divergence of the tensor with components A_i. One can also show, by using (14), that if A is a scalar point function then the divergence of the gradient of A is given by

(15) $$\Delta_2 A = \frac{1}{\sqrt{g}} \sum_{l=1}^{n} \frac{\partial}{\partial x^l} (\sqrt{g}\, A^l),$$

where

$$A^l = \sum_{i=1}^{n} g^{il} \frac{\partial A}{\partial x^i} = \sum_{i=1}^{n} g^{il} A_i.$$

6. *Atti della Accad. dei Lincei, Rendiconti*, (4), 5, 1889, 112–18 = *Opere*, 1, 268–75.

This expression for $\Delta_2 A$ is also Beltrami's expression for $\Delta_2 A$ in a Riemannian geometry (Chap. 37, sec. 5).

Though Ricci and Levi-Civita devoted much of their 1901 paper to developing the technique of tensor analysis, they were primarily concerned to find differential invariants. They posed the following general problem: Given a positive differential quadratic form ϕ and an arbitrary number of associated functions S, to determine all the absolute differential invariants that one can form from the coefficients of ϕ, the functions S, and the derivatives of the coefficients and functions up to a definite order m. They gave a complete solution. It is sufficient to find the *algebraic* invariants of the system consisting of the fundamental differential quadratic form ϕ, the covariant derivatives of any associated functions S up to order m, and, for $m > 1$, a certain quadrilinear form G_4 whose coefficients are the Riemann expressions (ih, jk) and its covariant derivatives up to order $m - 2$.

They conclude their paper by showing how some partial differential equations and physical laws can be expressed in tensor form so as to render them independent of the coordinate system. This was Ricci's avowed goal. Thus tensor analysis was used to express the mathematical invariance of physical laws many years before Einstein used it for this purpose.

4. Parallel Displacement

From 1901 to 1915, research on tensor analysis was limited to a very small group of mathematicians. However, Einstein's work changed the picture. Albert Einstein (1879–1955), while engaged as an engineer in the Swiss patent office, greatly stirred the scientific world by the announcement of his restricted or special theory of relativity.[7] In 1914 Einstein accepted a call to the Prussian Academy of Science in Berlin, as successor to the celebrated physical chemist Jacobus Van't Hoff (1852–1911). Two years later he announced his general theory of relativity.[8]

Einstein's revolutionary views on the relativity of physical phenomena aroused intense interest among physicists, philosophers, and mathematicians throughout the world. Mathematicians especially were excited by the nature of the geometry Einstein found it expedient to use in the creation of his theories.

The exposition of the restricted theory, involving the properties of four-dimensional pseudo-Euclidean manifolds (space-time), is best made with the aid of vectors and tensors, but the exposition of the general theory, involving properties of four-dimensional Riemannian manifolds (space-time), demands the use of the special calculus of tensors associated with such manifolds.

7. *Annalen der Phys.*, 17, 1905, 891–921; an English translation can be found in the Dover edition of A. Einstein, *The Principle of Relativity* (1951).
8. *Annalen der Physik*, 49, 1916, 769–822.

Fortunately, the calculus had already been developed but had not as yet attracted particular notice from physicists.

Actually Einstein's work on the restricted theory did not use Riemannian geometry or tensor analysis.[9] But the restricted theory did not involve the action of gravitation. Einstein then began to work on the problem of dispensing with the force of gravitation and accounting for its effect by imposing a structure on his space-time geometry so that objects would automatically move along the same paths as those derived by assuming the action of the gravitational force. In 1911 he made public a theory that accounted in this manner for a gravitational force that had a constant direction throughout space, knowing of course that this theory was unrealistic. Up to this time Einstein had used only the simplest mathematical tools and had even been suspicious of the need for "higher mathematics," which he thought was often introduced to dumbfound the reader. However, to make progress on his problem he discussed it in Prague with a colleague, the mathematician Georg Pick, who called his attention to the mathematical theory of Ricci and Levi-Civita. In Zurich Einstein found a friend, Marcel Grossmann (1878–1936), who helped him learn the theory; and with this as a basis, he succeeded in formulating the general theory of relativity.

To represent his four-dimensional world of three space coordinates and a fourth coordinate denoting time, Einstein used the Riemannian metric

$$(16) \qquad ds^2 = \sum_{i,j=1}^{4} g_{ij}\, dx_i\, dx_j,$$

wherein x_4 represents time. The g_{ij} were chosen so as to reflect the presence of matter in various regions. Moreover, since the theory is concerned with the determination of lengths, time, mass, and other physical quantities by different observers who are moving in arbitrary fashion with respect to each other, the "points" of space-time are represented by different coordinate systems, one attached to each observer. The relation of one coordinate system to another is given by a transformation

$$x_i = \phi_i(y_1, y_2, \ldots, y_4), \qquad i = 1, \ldots, 4.$$

The laws of nature are those relationships or expressions that are the same for all observers. Hence they are invariants in the mathematical sense.

From the standpoint of mathematics, the importance of Einstein's work was, as already noted, the enlargement of interest in tensor analysis and Riemannian geometry. The first innovation in tensor analysis following upon the theory of relativity is due to Levi-Civita. In 1917, improving on an idea

9. The metric is $ds^2 = dx^2 + dy^2 + dz^2 - c^2\, dt^2$. This is a space of constant curvature. Any section by a plane $t = $ const. is Euclidean.

of Ricci, he introduced[10] the notion of parallel displacement or parallel transfer of a vector. The idea was also created independently by Gerhard Hessenberg in the same year.[11] In 1906 Brouwer had introduced it for surfaces of constant curvature. The objective in the notion of parallel displacement is to define what is meant by parallel vectors in a Riemannian space. The difficulty in doing so may be seen by considering the surface of a sphere which, considered as a space in itself and with distance on the surface given by arcs of great circles, is a Riemannian space. If a vector, say one starting on a circle of latitude and pointing north (the vector will be tangent to the sphere), is moved by having the initial point follow the circle and is kept parallel to itself in the Euclidean three-space, then when it is halfway around the circle it is no longer tangent to the sphere and so is not in that space. To obtain a notion of parallelism of vectors suitable for a Riemannian space, the Euclidean notion is generalized, but some of the familiar properties are lost in the process.

The geometrical idea used by Levi-Civita to define parallel transfer or displacement is readily understood for a surface. Consider a curve C on the surface, and let a vector with one endpoint on C be moved parallel to itself in the following sense: At each point of C there is a tangent plane. The envelope of this family of planes is a developable surface, and when this surface is flattened out onto a plane, the vectors parallel along C are truly parallel in the Euclidean plane.

Levi-Civita generalized this idea to fit n-dimensional Riemannian spaces. It is true in the Euclidean plane that when a vector is carried parallel to itself and its initial point follows a straight line—a geodesic in the plane— the vector always makes the same angle with the line. Accordingly, parallelism in a Riemannian space is defined thus: When a vector in the space moves so that its initial point follows a geodesic, then the vector must continue to make the same angle with the geodesic (tangent to the geodesic). In particular, a tangent to a geodesic remains parallel to itself as it moves along a geodesic. By definition the vector continues to have the same magnitude. It is understood that the vector remains in the Riemannian space, even if that space is imbedded in a Euclidean space. The definition of parallel transfer also requires that the angle between two vectors be preserved as each is moved parallel to itself along the same curve C. In the general case of parallel transfer around an arbitrary closed curve C, the initial vector and the final vector will usually not have the same (Euclidean) direction. The deviation in direction will depend on the path C. Thus consider a vector starting at a point P on a circle of latitude on a sphere and tangent to the sphere along a meridian. When carried by *parallel transfer* around the circle it will end up at

10. *Rendiconti del Circolo Matematico di Palermo*, 42, 1917, 173–205.
11. *Math. Ann.*, 78, 1918, 187–217.

P tangent to the surface; but if ϕ is the co-latitude of P, it will make an angle $2\pi - 2\pi \cos \phi$ with the original vector.

If one uses the general definition of parallel displacement along a curve of a Riemannian space, he obtains an analytic condition. The differential equation satisfied by the components X^α of a contravariant vector under parallel transfer along a curve is (summation understood)

$$(17) \qquad \frac{dX^\alpha}{dt} + \{\beta\gamma, \alpha\}X^\beta \frac{du^\gamma}{dt} = 0, \qquad \alpha = 1, 2, \cdots, n,$$

wherein it is presupposed that the $u^i(t)$, $i = 1, 2, \ldots, n$, define a curve. For a covariant vector X_α the condition is

$$(18) \qquad \frac{dX_\alpha}{dt} - \{\alpha l, j\}X_j \frac{du^l}{dt} = 0, \qquad \alpha = 1, 2, \cdots, n.$$

These equations can be used to define parallel transfer along any curve C. The solution determined uniquely by the values of the components at a definite P is a vector having values at each point of C and by definition parallel to the initial vector at P. Equation (18) states that the covariant derivative of X_α is 0.

Once the notion of parallel displacement is introduced, one can describe the curvature of a space in terms of it, specifically in terms of the change in an infinitesimal vector upon parallel displacement by infinitesimal steps. Parallelism is at the basis of the notion of curvature even in Euclidean space, because the curvature of an infinitesimal arc depends on the change in direction of the tangent vector over the arc.

5. *Generalizations of Riemannian Geometry*

The successful use of Riemannian geometry in the theory of relativity revived interest in that subject. However, Einstein's work raised an even broader question. He had incorporated the gravitational effect of mass in space by using the proper functions for the g_{ij}. As a consequence the geodesics of his space-time proved to be precisely the paths of objects moving freely as, for example, the earth does around the sun. Unlike the situation in Newtonian mechanics, no gravitational force was required to account for the path. The elimination of gravity suggested another problem, namely, to account for attraction and repulsion of electric charges in terms of the metric of space. Such an accomplishment would supply a unified theory of gravitation and electromagnetism. This work led to generalizations of Riemannian geometry known collectively as non-Riemannian geometries.

In Riemannian geometry the ds^2 ties together the points of the space. It specifies how points are related to each other by prescribing the distance between points. In the non-Riemannian geometries the connection between points is specified in ways that do not necessarily rely upon a metric. The variety of these geometries is great, and each has a development as extensive as Riemannian geometry itself. Hence we shall give only some examples of the basic ideas in these geometries.

This area of work was initiated primarily by Hermann Weyl[12] and the class of geometries he introduced is known as the geometry of affinely connected spaces. In Riemannian geometry the proof that the covariant derivative of a tensor is itself a tensor depends only on relations of the form

$$(19) \qquad \overline{\{ik, h\}} = \{ab, j\} \frac{\partial x^a}{\partial y^i} \frac{\partial x^b}{\partial y^k} \frac{\partial y^h}{\partial x^j} + \frac{\partial^2 x^j}{\partial y^i \partial y^k} \frac{\partial y^h}{\partial x^j},$$

wherein the left side is the transform of $\{ab, j\}$ under a change of coordinates from the x^i to the y^i. These relations are satisfied by the Christoffel symbols, and the symbols themselves are defined in terms of the coefficients of the fundamental form. Consider instead functions L^i_{jk} and $\bar{L}^i_{jk}$ of the x^i and y^i, respectively, which satisfy the same relations (19) but are specified only as given functions unrelated to the fundamental quadratic form. A set of functions L^i_{jk} associated with a space V_n and having the transformation property (19) is said to constitute an affine connection. The functions are called the coefficients of affine connection and the space V_n is said to be affinely connected or to be an affine space. Riemannian geometry is the special case in which the coefficients of affine connection are the Christoffel symbols of the second kind and are derived from the fundamental tensor of the space. Given the L-functions, concepts such as covariant differentiation, curvature, and other notions analogous to those in Riemannian geometry can be introduced. However, one cannot speak of the magnitude of a vector in this new geometry.

In an affinely connected space a curve such that its tangents are parallel with respect to the curve (in the sense of parallel displacement for that space) is called a path of the space. Paths are thus a generalization of the geodesics of a Riemannian space. All affinely connected spaces that have the same L's have the same paths. Thus the geometry of affinely connected spaces dispenses with the Riemannian metric. Weyl did derive Maxwell's equations from the properties of the space, but the theory as a whole did not accord with other established facts.

Another non-Riemannian geometry, due to Luther P. Eisenhart (1876–1965) and Veblen,[13] and called the geometry of paths, proceeds somewhat differently. It starts with n^3 given functions $\Gamma^i_{\lambda\mu}$ of $x^1, \ldots, x^n$. The system of n differential equations

$$(20) \qquad \frac{d^2 x^i}{ds^2} + \sum_{\lambda\mu} \Gamma^i_{\lambda\mu} \frac{dx^\lambda}{ds} \frac{dx^\mu}{ds} = 0, \qquad i = 1, 2, \cdots, n,$$

12. *Mathematische Zeitschrift*, 2, 1918, 384–411 = *Ges. Abh.*, 2, 1–28.
13. *Proceedings of the National Academy of Sciences*, 8, 1922, 19–23.

with $\Gamma^i_{\lambda\mu} = \Gamma^i_{\mu\lambda}$, defines a family of curves called paths. These are the geodesics of the geometry. (In Riemannian geometry equations (20) are precisely those for geodesics.) Given the geodesics in the sense just described, one can build up the geometry of paths in a manner quite analogous to that pursued in Riemannian geometry.

A different generalization of Riemannian geometry is due to Paul Finsler (1894–) in his excellent thesis of 1918 at Göttingen.[14] The Riemannian ds^2 is replaced by a more general function $F(x, dx)$ of the coordinates and differentials. There are restrictions on F to insure the possibility of minimizing the integral $\int F(x, (dx/dt))\, dt$ and obtaining thereby the geodesics. the geodesics.

The attempts to generalize the concept of Riemannian geometry so as to incorporate electromagnetic as well as gravitational phenomena have thus far failed. However, mathematicians have continued to work on the abstract geometries.

14. The thesis was ultimately published. Uber Kurven und Flachen in allgemeinen Raumen, Birkhauser Verlag, Basel, 1951.

Bibliography

Cartan, E.: "Les récentes généralisations de la notion d'espace," *Bull. des Sci. Math.*, 48, 1924, 294–320.

Pierpont, James: "Some Modern Views of Space," *Amer. Math. Soc. Bull.*, 32, 1926, 225–58.

Ricci-Curbastro, G.: *Opere*, 2 vols., Edizioni Cremonese, 1956–57.

Ricci-Curbastro, G., and T. Levi-Civita: "Méthodes de calcul différentiel absolu et leurs applications," *Math. Ann.*, 54, 1901, 125–201.

Thomas, T. Y.: "Recent Trends in Geometry," *Amer. Math. Soc. Semicentennial Publications* II, 1938, 98–135.

Weatherburn, C. E.: *The Development of Multidimensional Differential Geometry*, Australian and New Zealand Association for the Advancement of Science, 21, 1933, 12–28.

Weitzenbock, R.: "Neuere Arbeiten der algebraischen Invariantentheorie. Differentialinvarianten," *Encyk. der Math. Wiss.*, B. G. Teubner, 1902–27, III, Part III, E1, 1–71.

Weyl, H.: *Mathematische Analyse des Raumproblems* (1923), Chelsea (reprint), 1964.

49
The Emergence of Abstract Algebra

> Perhaps I may without immodesty lay claim to the appella-
> tion of Mathematical Adam, as I believe that I have given
> more names (passed into general circulation) of the creatures
> of the mathematical reason than all the other mathematicians
> of the age combined.   J. J. SYLVESTER

1. The Nineteenth-Century Background

In abstract algebra, as in the case of most twentieth-century developments, the basic concepts and goals were fixed in the nineteenth century. The fact that algebra can deal with collections of objects that are not necessarily real or complex numbers was demonstrated in a dozen nineteenth-century crea- tions. Vectors, quaternions, matrices, forms such as $ax^2 + bxy + cy^2$, hyper- numbers of various sorts, transformations, and substitutions or permutations are examples of objects that were combined under operations and laws of operation peculiar to the respective collections. Even the work on algebraic numbers, though it dealt with classes of complex numbers, brought to the fore the variety of algebras because it demonstrated that only some properties are applicable to these classes as opposed to the entire complex number system.

These various classes of objects were distinguished in accordance with the properties that the operations on them possessed; and we have seen that such notions as group, ring, ideal, and field, and subordinate notions such as subgroup, invariant subgroup, and extension field were introduced to identify the sets of properties. However, nearly all of the nineteenth-century work on these various types of algebras dealt with the concrete systems mentioned above. It was only in the last decades of the nineteenth century that the mathematicians appreciated that they could move up to a new level of efficiency by integrating many separate algebras through abstraction of their common content. Thus permutation groups, the groups of classes of forms treated by Gauss, hypernumbers under addition, and transformation groups could all be treated in one swoop by speaking of a set of elements or things subject to an operation whose nature is specified only by certain abstract properties, the foremost of these being that the operation applied to two elements of the set produces a third element of the set. The same advantages could be achieved for the various collections that formed rings and fields. Though the idea of working with abstract collections preceded the axiomatics of Pasch, Peano, and Hilbert, the latter development undoubtedly accelerated the acceptance of the abstract approach to algebras.

Thus arose abstract algebra as the conscious study of entire classes of algebras, which individually were not only concrete but which served purposes in specific areas as substitution groups did in the theory of equations. The advantage of obtaining results that might be useful in many specific areas by considering abstract versions was soon lost sight of, and the study of abstract structures and the derivation of their properties became an end in itself.

Abstract algebra has been one of the favored fields of the twentieth century and is now a vast area. We shall present only the beginnings of the subject and indicate the almost unlimited opportunities for continued research. The great difficulty in discussing what has been done in this field is terminology. Aside from the usual difficulties, that different authors use different terms and that terms change meaning from one period to another, abstract algebra is marked and marred by the introduction of hundreds of new terms. Every minor variation in concept is distinguished by a new and often imposing sounding term. A complete dictionary of the terms used would fill a large book.

2. Abstract Group Theory

The first abstract structure to be introduced and treated was the group. A great many of the basic ideas of abstract group theory can be found implicitly and explicitly at least as far back as 1800. It is a favorite activity of historians, now that the abstract theory is in existence, to trace how many of the abstract ideas were foreshadowed by the concrete work of Gauss, Abel, Galois, Cauchy, and dozens of other men. We shall not devote space to this re-examination of the past. The only significant point that bears mentioning is that, once the abstract notion was acquired, it was relatively easy for the founders of abstract group theory to obtain ideas and theorems by rephrasing this past work.

Before examining the development of the abstract group concept, it may be well to know where the men were heading. The abstract definition of a group usually used today calls for a collection of elements, finite or infinite in number, and an operation which, when applied to any two elements in the collection, results in an element of the collection (the closure property). The operation is associative; there is an element, e, say, such that for any element a of the group, $ae = ea = a$; and to each element a there exists an inverse element a' such that $aa' = a'a = e$. When the operation is commutative the group is called commutative or Abelian and the operation is called addition and denoted by $+$. The element e is then denoted by 0 and called the zero element. If the operation is not commutative it is called multiplication and the element e is denoted by 1 and called the identity.

The abstract group concept and the properties that should be attached to it were slow in coming to light. We may recall (Chap. 31, sec. 6) that Cayley had proposed the abstract group in 1849, but the merit of the notion was not recognized at the time. In 1858 Dedekind,[1] far in advance of his time, gave an abstract definition of finite groups which he derived from permutation groups. Again in 1877[2] he observed that his modules of algebraic numbers, which call for $\alpha + \beta$ and $\alpha - \beta$ belonging when α and β do, can be generalized so that the elements are no longer algebraic numbers and the operation can be general but must have an inverse and be commutative. Thus he suggested an abstract finite commutative group. Dedekind's understanding of the value of abstraction is noteworthy. He saw clearly in his work on algebraic number theory the value of structures such as ideals and fields. He is the effective founder of abstract algebra.

Kronecker,[3] taking off from work on Kummer's ideal numbers, also gave what amounts to the abstract definition of a finite Abelian group similar to Cayley's 1849 concept. He specifies abstract elements, an abstract operation, the closure property, the associative and commutative properties, and the existence of a unique inverse element of each element. He then proves some theorems. Among the various powers of any element θ, there is one which equals the unit element, 1. If ν is the smallest exponent for which θ^ν equals the unit element, then to each divisor μ of ν there is an element ϕ for which $\phi^\mu = 1$. If θ^ρ and ϕ^σ both equal 1 and ρ and σ are the smallest numbers for which this holds and are relatively prime, then $(\theta\phi)^{\rho\sigma} = 1$. Kronecker also gave the first proof of what is now called a basis theorem. There exists a fundamental finite system of elements $\theta_1, \theta_2, \theta_3, \ldots$ such that the products

$$\theta_1^{h_1}\theta_2^{h_2}\theta_3^{h_3}\cdots, \qquad h_i = 1, 2, 3, \cdots, n_i,$$

represent all elements of the group just once. The lowest possible values $n_1, n_2, n_3, \ldots$ that correspond to $\theta_1, \theta_2, \theta_3, \ldots$ (that is, for which $\theta_i^{n_i} = 1$) are such that each is divisible by the following one and the product $n_1 n_2 n_3 \cdots$ equals the number n of elements in the group. Moreover all the prime factors of n are in n_1.

In 1878 Cayley wrote four more papers on finite abstract groups.[4] In

1. *Werke*, 3, 439–46.
2. *Bull. des Sci. Math.*, (2), 1, 1877, 17–41, p. 41 in particular = *Werke*, 3, 262–96.
3. *Monatsber. Berliner Akad.*, 1870, 881–89 = *Werke*, 1, 271–82.
4. *Math. Ann.*, 13, 1878, 561–65; *Proc. London Math. Soc.*, 9, 1878, 126–33; *Amer. Jour. of Math.*, 1, 1878, 50–52 and 174–76; all in Vol. 10 of his *Collected Math. Papers*.

these, as in his 1849 and 1854 papers, he stressed that a group can be considered as a general concept and need not be limited to substitution groups, though, he points out, every (finite) group can be represented as a substitution group. These papers of Cayley had more influence than his earlier ones because the time was ripe for an abstraction that embraced more than substitution groups.

In a joint paper, Frobenius and Ludwig Stickelberger (1850–1936)[5] made the advance of recognizing that the abstract group concept embraces congruences and Gauss's composition of forms as well as the substitution groups of Galois. They mention groups of infinite order.

Though Eugen Netto, in his book *Substitutionentheorie und ihre Anwendung auf die Algebra* (1882), confined his treatment to substitution groups, his wording of his concepts and theorems recognized the abstractness of the concepts. Beyond putting together results established by his predecessors, Netto treated the concepts of isomorphism and homomorphism. The former means a one-to-one correspondence between two groups such that if $a \cdot b = c$, where a, b, and c are elements of the first group, then $a' \cdot b' = c'$, where a', b', and c' are the corresponding elements of the second group. A homomorphism is a many-to-one correspondence in which again $a \cdot b = c$ implies $a' \cdot b' = c'$.

By 1880 new ideas on groups came into the picture. Klein, influenced by Jordan's work on permutation groups, had shown in his Erlanger Programm (Chap. 38, sec. 5) that infinite transformation groups, that is, groups with infinitely many elements, could be used to classify geometries. These groups, moreover, are continuous in the sense that arbitrarily small transformations are included in any group or, alternatively stated, the parameters in the transformations can take on all real values. Thus in the transformations that express rotation of axes the angle θ can take on all real values. Klein and Poincaré in their work on automorphic functions had utilized another kind of infinite group, the discrete or noncontinuous group (Chap. 29, sec. 6).

Sophus Lie, who had worked with Klein around 1870, took up the notion of continuous transformation groups, but for other purposes than the classification of geometries. He had observed that most of the ordinary differential equations that had been integrated by older methods were invariant under classes of continuous transformation groups, and he thought he could throw light on the solution of differential equations and classify them.

In 1874 Lie introduced his general theory of transformation groups.[6] Such a group is represented by

$$(1) \qquad x_i' = f_i(x_1, x_2, \cdots, x_n, a_1, \cdots, a_n), \qquad i = 1, 2, \cdots, n,$$

5. *Jour. für. Math.*, 86, 1879, 217–62 = Frobenius, *Ges. Abh.*, 1, 545–90.
6. *Nachrichten König. Ges. der Wiss. zu Gött.*, 1874, 529–42 = *Ges. Abh.*, 5, 1–8.

where the f_i are analytic in the x_i and a_i. The a_i are parameters as opposed to the x_i, which are variables, and $(x_1, x_2, \ldots, x_n)$ stands for a point in n-dimensional space. Both the parameters and the variables may take on real or complex values. Thus in one dimension the class of transformations

$$x' = \frac{ax + b}{cx + d},$$

where a, b, c, and d take on all real values, is a continuous group. The groups represented by (1) are called finite, the word finite referring to the number of parameters. The number of transformations is, of course, infinite. The one-dimensional case is a three-parameter group because only the ratios of a, b, and c to d matter. In the general case the product of two transformations

$$x_i' = f_i(x_i, \cdots, x_n, a_1, \cdots, a_n)$$
$$x_i'' = f_i(x_i', \cdots, x_n', b_1, \cdots, b_n)$$

is

$$x_i''' = f_i(x_i'', \cdots, x_n'', c_1, \cdots, c_n),$$

where the c_i are functions of the a_i and b_i. In the case of one variable Lie spoke of the group as a simply extended manifold; for n variables he spoke of an arbitrarily extended manifold.

In a paper of 1883 on continuous groups, published in an obscure Norwegian journal,[7] Lie also introduced infinite continuous transformation groups. These are not defined by equations such as (1) but by means of differential equations. The resulting transformations do not depend upon a finite number of continuous parameters but on arbitrary functions. There is no abstract group concept corresponding to these infinite continuous groups, and though much work has been done on them, we shall not pursue it here.

It is perhaps of interest that Klein and Lie at the beginning of their work defined a group of transformations as one possessing only the closure property. The other properties, such as the existence of an inverse to each transformation, were established by using the properties of the transformations, or, as in the case of the associative law, were used as obvious properties of transformations. Lie recognized during the course of his work that one should postulate as part of the definition of a group the existence of an inverse to each element.

By 1880 four main types of groups were known. These are the discontinuous groups of finite order, exemplified by substitution groups; the infinite discontinuous (or discrete) groups, such as occur in the theory of automorphic functions; the finite continuous groups of Lie exemplified by the transformation groups of Klein and the more general analytic transformations of Lie; and the infinite continuous groups of Lie defined by differential equations.

7. *Ges. Abh.*, 5, 314–60.

With the work of Walther von Dyck (1856–1934), the three main roots of group theory—the theory of equations, number theory, and infinite transformation groups—were all subsumed under the abstract group concept. Dyck was influenced by Cayley and was a student of Felix Klein. In 1882 and 1883[8] he published papers on abstract groups that included discrete and continuous groups. His definition of a group calls for a set of elements and an operation that satisfy the closure property, the associative but not the commutative property, and the existence of an inverse element of each element.

Dyck worked more explicitly with the notion of the generators of a group, which is implicit in Kronecker's basis theorem and explicit in Netto's work on substitution groups. The generators are a fixed subset of independent elements of a group such that every member of the group can be expressed as the product of powers of the generators and their inverses. When there are no restrictions on the generators the group is called a free group. If A_1, A_2, ... are the generators, then an expression of the form

$$A_1^{\mu_1} A_2^{\mu_2} \cdots,$$

where the μ_i are positive or negative integers, is called a word. There may be relations among the generators, and these would be of the form

$$F_i(A_i) = 1;$$

that is, a word or combination of words equals the identity element of the group. Dyck then shows that the presence of relations implies an invariant subgroup and a factor group $\bar{G}$ of the free group G. In his 1883 paper he applied the abstract group theory to permutation groups, finite rotation groups (symmetries of polyhedra), number-theoretic groups, and transformation groups.

Sets of independent postulates for an abstract group were given by Huntington,[9] E. H. Moore,[10] and Leonard E. Dickson (1874–1954).[11] These as well as other postulate systems are minor variations of each other.

Having arrived at the abstract notion of a group, the mathematicians turned to proving theorems about abstract groups that were suggested by known results for concrete cases. Thus Frobenius[12] proved Sylow's theorem (Chap. 31, sec. 6) for finite abstract groups. Every finite group whose order, that is, the number of elements, is divisible by the νth power of a prime p always contains a subgroup of order p^ν.

Beyond searching concrete groups for properties that may hold for abstract groups, many men introduced concepts directly for abstract groups.

8. *Math. Ann.*, 20, 1882, 1–44 and 22, 1883, 70–118.
9. *Amer. Math. Soc. Bull.*, 8, 1902, 296–300 and 388–91, and *Amer. Math. Soc. Trans.*, 6, 1905, 181–97.
10. *Amer. Math. Soc. Trans.*, 3, 1902, 485–92, and 6, 1905, 179–80.
11. *Amer. Math. Soc. Trans.*, 6, 1905, 198–204.
12. *Jour. für Math.*, 100, 1887, 179–81 = *Ges. Abh.*, 2, 301–3.

Dedekind[13] and George A. Miller (1863–1951)[14] treated non-Abelian groups in which every subgroup is a normal (invariant) subgroup. Dedekind in his 1897 paper and Miller[15] introduced the notions of commutator and commutator subgroup. If s and t are any two elements of a group G, the element $s^{-1}t^{-1}st$ is called the commutator of s and t. Both Dedekind and Miller used this notion to prove theorems. For example, the set of all commutators of all (ordered) pairs of elements of a group G generates an invariant subgroup of G. The automorphisms of a group, that is, the one-to-one transformations of the members of a group into themselves under which if $a \cdot b = c$ then $a' \cdot b' = c'$, were studied on an abstract basis by Hölder[16] and E. H. Moore.[17]

The further development of abstract group theory has pursued many directions. One of these taken over from substitution groups by Hölder in the 1893 paper is to find all the groups of a given order, a problem Cayley had also mentioned in his 1878 papers.[18] The general problem has defied solution. Hence particular orders have been investigated, such as p^2q^2 where p and q are prime. A related problem has been the enumeration of intransitive and primitive and imprimitive groups of various degrees (the number of letters in a substitution group).

Another direction of research has been the determination of composite or solvable groups and simple groups, that is, those which have no invariant subgroups (other than the identity). This problem of course derives from Galois theory. Hölder, after introducing the abstract notion of a factor group,[19] treated simple groups[20] and composite groups.[21] Among results are the fact that a cyclic group of prime order is simple and so is the alternating group of all even permutations on n letters for $n \geq 5$. Many other finite simple groups have been found.

As for solvable groups, Frobenius devoted several papers to the problem. He found, for example,[22] that all groups whose order is not divisible by the square of a prime are solvable.[23] The problem of investigating which groups

13. *Math. Ann.*, 48, 1897, 548–61 = *Werke*, 2, 87–102.
14. *Amer. Math. Soc. Bull.*, 4, 1898, 510–15 = *Coll. Works*, 1, 266–69.
15. *Amer. Math. Soc. Bull.*, 4, 1898, 135–39 = *Coll. Works*, 1, 254–57.
16. *Math. Ann.*, 43, 1893, 301–412.
17. *Amer. Math. Soc. Bull.*, 1, 1895, 61–66, and 2, 1896, 33–43.
18. *Coll. Math. Papers*, 10, 403.
19. *Math. Ann.*, 34, 1889, 26–56.
20. *Math. Ann.*, 40, 1892, 55–88, and 43, 1893, 301–412.
21. *Math. Ann.*, 46, 1895, 321–422.
22. *Sitzungsber. Akad. Wiss. zu Berlin*, 1893, 337–45, and 1895, 1027–44 = *Ges. Abh.*, 2, 565–73, 677–94.
23. A recent result of major importance was obtained by Walter Feit (1930–) and John G. Thompson (*Pacific Jour. of Math.*, 13, Part 2, 1963, 775–1029). All finite groups of odd order are solvable. The suggestion that this might be the case was made by Burnside in 1906.

are solvable is part of the broader problem of determining the structure of a given group.

Dyck in his papers of 1882 and 1883 had introduced the abstract idea of a group defined by generators and relations among the generators. Given a group defined in terms of a finite number of generators and relations, the identity or word problem, as formulated by Max Dehn,[24] is the task of determining whether any "word" or product of elements is equal to the unit element. Any set of relations may be given because at worst the trivial group consisting only of the identity satisfies them. To decide whether a group given by generators and relations is trivial is not trivial. In fact there is no effective procedure. For one defining relation Wilhelm Magnus (1907–) showed[25] that the word problem is solvable. But the general problem is not.[26]

Another famous unsolved problem of ordinary group theory is Burnside's problem. Any finite group has the properties that it is finitely generated and every element has finite order. In 1902[27] William Burnside (1852–1927) asked whether the converse was true; that is, if a group G is finitely generated and if every element has finite order, is G finite? This problem has attracted a great deal of attention and only specializations of it have received solutions. Still another problem, the isomorphism problem, is to determine when two groups, each defined by generators and relations, are isomorphic.

One of the surprising turns in group theory is that shortly after the abstract theory had been launched, the mathematicians turned to representations by more concrete algebras in order to obtain results for the abstract groups. Cayley had pointed out in his 1854 paper that any finite abstract group can be represented by a group of permutations. We have also noted (Chap. 31, sec. 6) that Jordan in 1878 introduced the representation of substitution groups by linear transformations. These transformations or their matrices have proved to be the most effective representation of abstract groups and are called linear representations.

A matrix representation of a group G is a homomorphic correspondence of the elements g of G to a set of non-singular square matrices $A(g)$ of fixed order and with complex elements. The homomorphism implies that

$$A(g_i g_j) = A(g_i)A(g_j)$$

for all g_i and g_j of G. There are many matrix representations of a group G because the order (number of rows or columns) can be altered, and even for a given order the correspondence can be altered. Also one can add two

24. *Math. Ann.*, 71, 1911, 116–44.
25. *Math. Ann.*, 106, 1932, 295–307.
26. This was proved in 1955 by P. S. Novikov. See *American Math. Soc. Translations* (2), 9, 1958, 1–122.
27. *Quart. Jour. of Math.*, 33, 230–38.

representations. If for each element g of G, A_g is the corresponding matrix of one representation of order m and B_g is the corresponding matrix of order n, then

$$\begin{pmatrix} A_g & 0 \\ 0 & B_g \end{pmatrix}$$

is another representation, which is called the sum of the separate representations. Likewise if

$$E_g = \begin{pmatrix} B_g & C_g \\ 0 & D_g \end{pmatrix}$$

is another representation, when B_g and D_g are non-singular matrices of orders m and n respectively, the B_g and D_g are also representations, and of lower order than E_g. E_g is called a graduated representation; it and any representation equivalent to it $(F_g^{-1}E_gF_g$ is an equivalent representation if F is non-singular and of the order of E_g) is called reducible. A representation not equivalent to a graduated one is called irreducible. The basic idea of an irreducible representation consisting of a set of linear transformations in n variables is that it is a homomorphic or isomorphic representation in which it is impossible to choose $m < n$ linear functions of the variables which are transformed among themselves by every operation of the group they represent. A representation that is equivalent to the sum of irreducible representations is called completely reducible.

Every finite group has a particular representation called regular. Suppose the elements are labeled $g_1, g_2, \ldots, g_n$. Let a be any one of the g's. We consider an n by n matrix. Suppose $ag_i = g_j$. Then we place a 1 in the (i, j) place of the matrix. This is done for all g_i and the fixed a. All the other elements of the matrix are taken to be 0. The matrix so obtained corresponds to a. There is such a matrix for each g of the group and this set of matrices is a left regular representation. Likewise by forming the products $g_i a$ we get a right regular representation. By reordering the g's of the group we can get other regular representations. The notion of a regular representation was introduced by Charles S. Peirce in 1879.[28]

The representation of substitution groups by linear transformations of the form

$$x_i' = \sum_j a_{ij}x_j, \qquad i = 1, 2, \cdots, n,$$

initiated by Jordan was broadened to the study of representations of all finite abstract groups by Frobenius, Burnside, Theodor Molien (1861–1941), and Issai Schur (1875–1941) in the latter part of the nineteenth century and

28. *Amer. Jour. of Math.*, 4, 1881, 221–25.

the first part of the twentieth. Frobenius[29] introduced for finite groups the notions of reducible and completely reducible representations and showed that a regular representation contains all the irreducible representations. In other papers published from 1897 to 1910, some in conjunction with Schur, he proved many other results, including the fact that there are only a few irreducible representations, out of which all the others are composed.

Burnside[30] gave another major result, a necessary and sufficient condition on the coefficients of a group of linear transformations in n variables in order that the group be reducible. The fact that any finite group of linear transformations is completely reducible was first proved by Heinrich Maschke (1853–1908).[31] Representation theory for finite groups has led to important theorems for abstract groups. In the second quarter of this century, representation theory was extended to continuous groups, but this development will not be pursued here.

Aiding in the study of group representations is the notion of group character. This notion, which can be traced back to the work of Gauss, Dirichlet, and Heinrich Weber (see note 35), was formulated abstractly by Dedekind for Abelian groups in the third edition of Dirichlet's *Vorlesungen über Zahlentheorie* (1879). A character of a group is a function $x(s)$ defined on all the elements s such that it is not zero for any s and $x(ss') = x(s)x(s')$. Two characters are distinct if $x(s) \neq x'(s)$ for at least one s of the group.

This notion was generalized to all finite groups by Frobenius. After having formulated a rather complex definition,[32] he gave a simpler definition,[33] which is now standard. The character function is the trace (sum of the main diagonal elements) of the matrices of an irreducible representation of the group. The same concept was applied later by Frobenius and others to infinite groups.

Group characters furnish in particular a determination of the minimum number of variables in terms of which a given finite group can be represented as a linear transformation group. For commutative groups they permit a determination of all subgroups.

Displaying the usual exuberance for the current vogue, many mathematicians of the late nineteenth and early twentieth centuries thought that all mathematics worth remembering would ultimately be comprised in the theory of groups. Klein in particular, though he did not like the formalism of *abstract* group theory, favored the group concept because he thought it would unify mathematics. Poincaré was equally enthusiastic. He said,[34] ". . . the

29. *Sitzungsber. Akad. Wiss. zu Berlin*, 1897, 994–1015 = *Ges. Abh.*, 3, 82–103.
30. *Proc. London Math. Soc.*, (2), 3, 1905, 430–34.
31. *Math. Ann.*, 52, 1899, 363–68.
32. *Sitzungsber. Akad. Wiss. zu Berlin*, 1896, 985–1021 = *Ges. Abh.*, 3, 1–37.
33. *Sitzungsber. Akad. Wiss. zu Berlin*, 1897, 994–1015 = *Ges. Abh.*, 3, 82–103.
34. *Acta Math.*, 38, 1921, 145.

theory of groups is, so to say, the whole of mathematics divested of its matter and reduced to pure form."

3. The Abstract Theory of Fields

The concept of a field R generated by n quantities $a_1, a_2, \ldots, a_n$, that is, the set of all quantities formed by adding, subtracting, multiplying, and dividing these quantities (except division by 0) and the concept of an extension field formed by adjoining a new element λ not in R, are in Galois's work. His fields were the domains of rationality of the coefficients of an equation and extensions were made by adjunction of a root. The concept also has a quite different origin in Dedekind's and Kronecker's work on algebraic numbers (Chap. 34, sec. 3), and in fact the word "field" (Körper) is due to Dedekind.

The abstract theory of fields was initiated by Heinrich Weber, who had already espoused the abstract viewpoint for groups. In 1893[35] he gave an abstract formulation of Galois theory wherein he introduced (commutative) fields as extensions of groups. A field, as Weber specifies, is a collection of elements subject to two operations, called addition and multiplication, which satisfy the closure condition, the associative and commutative laws, and the distributive law. Moreover each element must have a unique inverse under each operation, except for division by 0. Weber stressed group and field as the two major concepts of algebra. Somewhat later, Dickson[36] and Huntington[37] gave independent postulates for a field.

To the fields that were known in the nineteenth century, the rational, real and complex numbers, algebraic number fields, and fields of rational functions in one or several variables, Kurt Hensel added another type, p-adic fields, which initiated new work in algebraic numbers (*Theorie der algebraischen Zahlen*, 1908). Hensel observed, first of all, that any ordinary integer D can be expressed in one and only one way as a sum of powers of a prime p. That is,

$$D = d_0 + d_1 p + \cdots + d_k p^k,$$

in which d_i is some integer from 0 to $p - 1$. For example,

$$14 = 2 + 3 + 3^2$$
$$216 = 2 \cdot 3^3 + 2 \cdot 3^4.$$

Similarly any rational number r (not 0) can be written in the form

$$r = \frac{a}{b} p^n,$$

35. *Math. Ann.*, 43, 1893, 521–49. For groups see *Math. Ann.*, 20, 1882, 301–29.
36. *Amer. Math. Soc. Trans.*, 4, 1903, 13–20, and 6, 1905, 198–204.
37. *Amer. Math. Soc. Trans.*, 4, 1903, 31–37, and 6, 1905, 181–97.

where a and b are integers not divisible by p and n is 0 or a positive or negative integer. Hensel generalized on these observations and introduced p-adic numbers. These are expressions of the form

$$(2) \qquad\qquad \sum_{i=-\rho}^{\infty} c_i p^i,$$

where p is a prime and the coefficients, the c_i's, are ordinary rational numbers reduced to their lowest form whose denominator is not divisible by p. Such expressions need not in general have values as ordinary numbers. However, by definition they are mathematical entities.

Hensel defined the four basic operations with these numbers and showed that they are a field. A subset of the p-adic numbers can be put into one-to-one correspondence with the ordinary rational numbers, and in fact this subset is isomorphic to the rational numbers in the full sense of an isomorphism between two fields. In the field of p-adic numbers, Hensel defined units, integral p-adic numbers, and other notions analogous to those of the ordinary rational numbers.

By introducing polynomials whose coefficients are p-adic numbers, Hensel was able to speak of p-adic roots of polynomial equations and extend to these roots all of the concepts of algebraic number fields. Thus there are p-adic integral algebraic numbers and more general p-adic algebraic numbers, and one can form fields of p-adic algebraic numbers that are extensions of the "rational" p-adic numbers defined by (2). In fact, all of the ordinary theory of algebraic numbers is carried over to p-adic numbers. Surprisingly perhaps, the theory of p-adic algebraic numbers leads to results on ordinary algebraic numbers. It has also been useful in treating quadratic forms and has led to the notion of valuation fields.

The growing variety of fields motivated Ernst Steinitz (1871–1928), who was very much influenced by Weber's work, to undertake a comprehensive study of abstract fields; this he did in his fundamental paper, *Algebraischen Theorie der Körper*.[38] All fields, according to Steinitz, can be divided into two principal types. Let K be a field and consider all subfields of K (for example, the rational numbers are a subfield of the real numbers). The elements common to all the subfields are also a subfield, called the prime field P of K. Two types of prime fields are possible. The unit element e is contained in P and, therefore, so are

$$e, 2e, \cdots, ne, \cdots.$$

These elements are either all different or there exists an ordinary integer p such that $pe = 0$. In the first case P must contain all fractions ne/me, and

38. *Jour. für Math.*, 137, 1910, 167–309.

since these elements form a field, P must be isomorphic to the field of rational numbers and K is said to have characteristic 0.

If, on the other hand, $pe = 0$, it is readily shown that the smallest such p must be a prime and the field must be isomorphic to the field of integral residues modulo p, that is, $0, 1, \ldots, p - 1$. Then K is said to be a field of characteristic p. Any subfield of K has the same characteristic. Then $pa = pea = 0$; that is, all expressions in K can be reduced modulo p.

From the prime field P in either of the types just described, the original field K can be obtained by the process of adjunction. The method is to take an element a in K but not in P and to form all rational functions $R(a)$ of a with coefficients in P and then, if necessary, to take b not in $R(a)$ and do the same with b, and to continue the process as long as necessary.

If one starts with an arbitrary field K one can make various types of adjunctions. A simple adjunction is obtained by adjoining a single element x. The enlarged field must contain all expressions of the form

$$(3) \qquad\qquad a_0 + a_1 x + \cdots + a_n x^n,$$

where the a_i are elements of K. If these expressions are all different, then the extended field is the field $K(x)$ of all rational functions of x with coefficients in K. Such an adjunction is called a transcendental adjunction and $K(x)$ is called a transcendental extension. If some of the expressions (3) are equal, one can show that there must exist a relation (using α for x)

$$f(\alpha) = \alpha^m + b_1 \alpha^{m-1} + \cdots + b_m = 0$$

with the b_i in K and with $f(x)$ irreducible in K. Then the expressions

$$C_1 \alpha^{m-1} + \cdots + C_m$$

with the C_i in K constitute a field $K(\alpha)$ formed by the adjunction of α to K. This field is called a simple algebraic extension of K. In $K(\alpha), f(x)$ has a root, and conversely, if an arbitrary irreducible $f(x)$ in K is chosen, then one can construct a $K(\alpha)$ in which $f(x)$ has a root.

A fundamental result due to Steinitz is that every field can be obtained from its prime field by first making a series of (possibly infinite) transcendental adjunctions and then a series of algebraic adjunctions to the transcendental field. A field K' is said to be an algebraic extension of K if it can be obtained by successive simple algebraic adjunctions. If the number of adjunctions is finite, K' is said to be of finite rank.

Not every field can be enlarged by algebraic adjunctions. For example, the complex numbers cannot be because every $f(x)$ is reducible in this field. Such a field is algebraically complete. Steinitz also proved that for every field K there is a unique algebraically complete field K' which is algebraic over K in the sense that all other algebraically complete fields over K (containing K) contain a subfield equivalent to K'.

Steinitz considered also the problem of determining the fields in which the Galois theory of equations holds. To say that Galois theory holds in a field means the following: A Galois field $\overline{K}$ over a given field K is an algebraic field in which every irreducible $f(x)$ in K either remains irreducible or decomposes into a product of linear factors. To every Galois field $\overline{K}$ there exists a set of automorphisms, each of which transforms the elements of $\overline{K}$ into other elements of $\overline{K}$ and such that $\alpha \pm \beta$ and $\alpha\beta$ correspond to $\alpha' \pm \beta'$ and $\alpha'\beta'$ while all elements of K remain invariant (correspond to themselves). The set of automorphisms forms a group G, the Galois group of $\overline{K}$ with respect to K. The main theorem of Galois theory asserts that there is a unique correspondence between the subgroups of G and the subfields of $\overline{K}$ such that to any subgroup G' of G there corresponds a subfield K' consisting of all elements left invariant by G' and conversely. Galois theory is said to hold for those fields in which this theorem holds. Steinitz's result is essentially that Galois theory holds in those fields of finite rank that can be obtained from a given field by a series of adjunctions of roots of irreducible $f(x)$ which have no equal roots. Fields in which all irreducible $f(x)$ have no equal roots are called separable. (Steinitz said *vollkommen* or complete.)

The theory of fields also includes, as Steinitz's classification indicates, finite fields of characteristic p. A simple example of the latter is the set of all residues (remainders) modulo a prime p. The concept of a finite field is due to Galois. In 1830 he published a definitive paper, "Sur la théorie des nombres."[39] Galois wished to solve congruences

$$F(x) \equiv 0 \bmod p,$$

where p is a prime and $F(x)$ is a polynomial of degree n. He took $F(x)$ to be irreducible (modulo p), so that the congruence did not have integral or irrational roots. This obliged him to consider other solutions, which were suggested by the imaginary numbers. Galois denoted one of the roots of $F(x)$ by i (which is not $\sqrt{-1}$). He then considered the expression

$$(4) \qquad a_0 + a_1 i + a_2 i^2 + \cdots + a_{n-1} i^{n-1},$$

where the a_i are whole numbers. When these coefficients assume separately all the least positive residues, modulo p, this expression can take on only p^ν values. Let α be one of these non-zero values, of which there are $p^\nu - 1$. The powers of α also have the form (4). Hence these powers cannot be all different. There must then be at least one power $\alpha^m = 1$, where m is the smallest of such values. Then there will be m different values

$$(5) \qquad 1, \alpha, \alpha^2, \ldots, \alpha^{m-1}.$$

If we multiply these m quantities by an expression β of the same form, we obtain a new group of quantities different from (5) and from each other.

39. *Bulletin des Sciences Mathématiques de Férussac*, 13, 1830, 428–35 = *Œuvres*, 1897, 15–23.

Multiplying the set (5) by γ will produce more such quantities, until we obtain all of the form (4). Hence m must divide $p^\nu - 1$ or $\alpha^{p^\nu - 1} = 1$ and so $\alpha^{p^\nu} = \alpha$. The p^ν values of the form (4) constitute a finite field. Galois had shown in this concrete situation that the number of elements in a Galois field of characteristic p is a power of p.

E. H. Moore[40] showed that any finite abstract field is isomorphic to a Galois field of order p^n, p a prime. There is such a field for every prime p and positive integer n. The characteristic of each field is p. Joseph H. M. Wedderburn (1882–1948), a professor at Princeton University,[41] and Dickson proved simultaneously that any finite field is necessarily commutative (in the multiplication operation). A great deal of work has been done to determine the structure of the additive groups contained in Galois fields and the structure of the fields themselves.

4. Rings

Though the structures rings and ideals were well known and utilized in Dedekind's and Kronecker's work on algebraic numbers, the abstract theory is entirely a product of the twentieth century. The word ideal had already been adopted (Chap. 34, sec. 4). Kronecker used the word "order" for ring; the latter term was introduced by Hilbert.

Before discussing the history, it may be well to be clear about the modern meanings of the concepts. An abstract ring is a collection of elements that form a commutative group with respect to one operation and are subject to a second operation applicable to any two elements. The second operation is closed and associative but may or may not be commutative. There may or may not be a unit element. Moreover the distributive law $a(b + c) = ab + ac$ and $(b + c)a = ba + ca$ holds.

An ideal in a ring R is a sub-ring M such that if a belongs to M and r is any element of R, then ar and ra belong to M. If only ar belongs to M, then M is called a right ideal. If only ra belongs to M, then M is a left ideal. If an ideal is both right and left, it is called a two-sided ideal. The unit ideal is the entire ring. The ideal (a) generated by one element a consists of all elements of the form

$$ra + na,$$

where r belongs to R and n is any whole number. If R has a unit element, then $ra + na = ra + nea = (r + ne)a = r'a$, where r' is now any element of R. The ideal generated by one element is called a principal ideal. Any ideal other than 0 and R is a proper ideal. Similarly if $a_1, a_2, \ldots, a_m$ are m given elements of a ring R with unit element the set of all sums $r_1 a_1 + \cdots + r_m a_m$

40. *N.Y. Math. Soc. Bull.*, 3, 1893, 73–78.
41. *Amer. Math. Soc. Trans.*, 6, 1905, 349–52.

with coefficients r in R is an ideal of R denoted by $(a_1, a_2, \ldots, a_m)$. It is the smallest ideal containing $a_1, a_2, \ldots, a_m$. A commutative ring R is called Noetherian if every ideal has this form.

An ideal M in a ring, since it is a subgroup of the additive group of the ring, divides the ring into residue classes. Two elements of R, a and b, are congruent relative to M if $a - b$ belongs to M or $a \equiv b \pmod{M}$. Under a homomorphism T from a ring R to a ring R', which calls for $T(ab) = (Ta) \cdot (Tb)$, $T(a + b) = Ta + Tb$ and $T1 = 1'$, the elements of R which correspond to the zero element of R' constitute an ideal called the kernel of R, and R' is isomorphic to the ring of residue classes of R modulo the kernel. Conversely, given an ideal L in R one may form a ring R modulo L and a homomorphism of R into R modulo L which has L as its kernel. R modulo L or R/L is called a quotient ring.

The definition of a ring does not call for the existence of an inverse to each element under multiplication. If an inverse (except for 0) and the unit element both exist, the ring is called a division ring (division algebra) and it is in effect a non-commutative (or skew) field. Wedderburn's result already noted (1905) showed that a finite division ring is a commutative field. Up to 1905 the only division algebras known were commutative fields and quaternions. Then Dickson created a number of new ones, commutative and non-commutative. In 1914 he[42] and Wedderburn[43] gave the first examples of non-commutative fields with centers (the set of all elements which commute with each other) of rank n^2.[44]

In the late nineteenth century a great variety of concrete linear associative algebras were created (Chap. 32, sec. 6). These algebras, abstractly considered, are rings, and when the theory of abstract rings was formulated it absorbed and generalized the work on these concrete algebras. This theory of linear associative algebras and the whole subject of abstract algebra received a new impulse when Wedderburn, in his paper "On Hypercomplex Numbers,"[45] took up results of Elie Cartan (1869–1951)[46] and generalized them. The hypercomplex numbers are, we may recall, numbers of the form

$$(6) \qquad x = x_1 e_1 + x_2 e_2 + \cdots + x_n e_n,$$

where the e_i are qualitative units and the x_i are real or complex numbers. Wedderburn replaced the x_i by members of an arbitrary field F. He called

42. *Amer. Math. Soc. Trans.*, 15, 1914, 31–46.

43. *Amer. Math. Soc. Trans.*, 15, 1914, 162–66.

44. In 1958 Michel Kervaire (1927–) in *Proceedings of the National Academy of Sciences*, 44, 1958, 280–83 and John Milnor (1931–) in *Annals of Math.* (2), 68, 1958, 444–49, both using a result of Raoul Bott (1923–), proved that the only possible division algebras with real coefficients, if one does not assume the associative and commutative laws of multiplication, are the real and complex numbers, quaternions, and the Cayley numbers.

45. *Proc. London Math. Soc.*, (2), 6, 1907, 77–118.

46. *Ann. Fac. Sci. de Toulouse*, 12B, 1898, 1–99 = *Œuvres*, Part II, Vol. 1, 7–105.

these generalized linear associative algebras simply algebras. To treat these generalized algebras he had to abandon the methods of his predecessors because an arbitrary field F is not algebraically closed. He also adopted and perfected Benjamin Peirce's technique of idempotents.

In Wedderburn's work, then, an algebra consists of all linear combinations of the form (6) with coefficients now in a field F. The number of e_i, called basal units, is finite and is the order of the algebra. The sum of two such elements is given by

$$\sum x_i e_i + \sum y_i e_i = \sum (x_i + y_i) e_i.$$

The scalar product of an element x of the algebra and an element a of the field F is defined as

$$a \sum x_i e_i = \sum a x_i e_i$$

and the product of two elements of the algebra is defined by

$$\left(\sum x_i e_i \right) \left(\sum y_i e_i \right) = \sum_{i,j} x_i y_i e_i e_j,$$

which is completed by a table expressing all products $e_i e_j$ as some linear combination of the e_i with coefficients in F. The product is required to satisfy the associative law. One can always add a unit element (modulus) 1 such that $x \cdot 1 = 1 \cdot x = x$ for every x and then the elements of the algebra include the field F of the coefficients.

Given an algebra A, a subset B of elements which itself forms an algebra is called a sub-algebra. If x belongs to A and y to B and yx and xy both belong to B, then B is called an invariant sub-algebra. If an algebra A is the sum of two invariant sub-algebras with no common element, then A is called reducible. It is also said to be the direct sum of the sub-algebras.

A simple algebra is one having no invariant sub-algebra. Wedderburn also used and modified Cartan's notion of a semi-simple algebra. To define this notion, Wedderburn made use of the notion of nilpotent elements. An element x is nilpotent if $x^n = 0$ for some integer n. An element x is said to be properly nilpotent if xy and also yx are nilpotent for every y in an algebra A. It can be shown that the set of properly nilpotent elements of an algebra A form an invariant sub-algebra. Then a semi-simple algebra A is one having no nilpotent invariant sub-algebra.

Wedderburn proved that every semi-simple algebra can be expressed as a direct sum of irreducible algebras and each irreducible algebra is equivalent to the direct product of a matrix algebra and a division algebra (primitive algebra in Wedderburn's terminology). This means that each element of the irreducible algebra can be taken to be a matrix whose elements are members of the division algebra. Since semi-simple algebras can be reduced to the direct sum of several simple algebras, this theorem amounts to the determination of all semi-simple algebras. Still another result uses the notion of a total matrix algebra, which is just the algebra of all n by n matrices. An algebra for which the coefficient field F is the complex numbers and which contains no properly nilpotent elements is equivalent to a direct sum of total matrix algebras. This sample of results obtained by Wedderburn may give some indication of the work done on the generalized linear associative algebras.

The theory of rings and ideals was put on a more systematic and axiomatic basis by Emmy Noether, one of the few great women mathematicians, who in 1922 became a lecturer at Göttingen. Many results on rings and ideals were already known when she began her work, but by properly formulating the abstract notions she was able to subsume these results under the abstract theory. Thus she reexpressed Hilbert's basis theorem (Chap. 39, sec. 2) as follows: A ring of polynomials in any number of variables over a ring of coefficients that has an identity element and a finite basis, itself has a finite basis. In this reformulation she made the theory of invariants a part of abstract algebra.

A theory of ideals for polynomial domains had been developed by Emanuel Lasker (1868–1941)[47] in which he sought to give a method of deciding whether a given polynomial belongs to an ideal generated by r polynomials. In 1921[48] Emmy Noether showed that this ideal theory for polynomials can be deduced from Hilbert's basis theorem. Thereby a common foundation was created for the ideal theory of integral algebraic numbers and integral algebraic functions (polynomials). Noether and others penetrated much farther into the abstract theory of rings and ideals and applied it to rings of differential operators and other algebras. However, an account of this work would take us too far into special developments.

5. Non-associative Algebras

Modern ring theory, or, more properly, an extension of ring theory, also includes non-associative algebras. The product operation is non-associative and non-commutative. The other properties of linear associative algebras are applicable. There are today several important non-associative algebras. Historically the most important is the type called a Lie algebra. It is customary in such algebras to denote the product of two elements a and b by $[a, b]$. In place of the associative law the product operation satisfies two conditions,

$$[a, b] = -[b, a] \quad \text{and} \quad [a, [b, c]] + [b, [c, a]] + [c, [a, b]] = 0.$$

47. *Math. Ann.*, 60, 1905, 20–116.
48. *Math. Ann.*, 83, 1921, 24–66.

The second property is called the Jacobi identity. Incidentally, the vector product of two vectors satisfies the two conditions.

An ideal in a Lie algebra L is a sub-algebra L_1 such that the product of any element of L and any element of L_1 is in L_1. A simple Lie algebra is one that has no nontrivial ideals. It is semi-simple if it has no Abelian ideals.

Lie algebras arose out of Lie's efforts to study the structure of his continuous transformation groups. To do this Lie introduced the notion of infinitesimal transformations.[49] Roughly speaking, an infinitesimal transformation is one that moves points an infinitesimal distance. Symbolically it is represented by Lie as

$$(7) \qquad x_i' = x_i + \delta t X_i(x_1, x_2, \cdots, x_n),$$

where δt is an infinitesimally small quantity, or

$$(8) \qquad \delta x_i = \delta t X_i(x_1, x_2, \cdots, x_n).$$

The δt is a consequence of a small change in the parameters of the group. Thus suppose a group of transformations is given by

$$x_1 = \phi(x, y, a) \quad \text{and} \quad y_1 = \psi(x, y, a).$$

Let a_0 be the value of the parameter for which ϕ and ψ are the identity transformation so that

$$x = \phi(x, y, a_0), \qquad y = \psi(x, y, a_0).$$

If a_0 is changed to $a_0 + \delta a$, then by Taylor's theorem

$$x_1 = \phi(x, y, a_0) + \frac{\partial \phi}{\partial a} \delta a + \cdots$$

$$y_1 = \psi(x, y, a_0) + \frac{\partial \psi}{\partial a} \delta a + \cdots$$

so that neglecting higher powers of δa gives

$$\delta x = x_1 - x = \frac{\partial \phi}{\partial a} \delta a, \qquad \delta y = y_1 - y = \frac{\partial \psi}{\partial a} \delta a.$$

For the fixed a_0, $\partial \phi / \partial a$ and $\partial \psi / \partial a$ are functions of x and y so that

$$\frac{\partial \phi}{\partial a} = \xi(x, y), \qquad \frac{\partial \psi}{\partial a} = \eta(x, y)$$

and

$$(9) \qquad \delta x = \xi(x, y) \, \delta a, \qquad \delta y = \eta(x, y) \, \delta a.$$

If δa is δt we get the form (7) or (8). The equations (9) represent an infinitesimal transformation of the group.

49. *Archiv for Mathematik Naturvidenskab*, 1, 1876, 152–93 = *Ges. Abh.*, 5, 42–75.

If $f(x, y)$ is an analytic function of x and y, the effect of an infinitesimal transformation on it is to replace $f(x, y)$ by $f(x + \xi\, \delta a, y + \eta\, \delta a)$, and by applying Taylor's theorem one finds that to first order

$$\delta f = \left(\xi \frac{\partial f}{\partial x} + \eta \frac{\partial f}{\partial y} \right) \delta a.$$

The operator

$$\xi \frac{\partial}{\partial x} + \eta \frac{\partial}{\partial y}$$

is another way of *representing* the infinitesimal transformation (9) because knowledge of one gives the other. Such operators can be added and multiplied in the usual sense of differential operators.

The number of independent infinitesimal transformations or the number of corresponding independent operators is the number of parameters in the original group of transformations. The infinitesimal transformations or the corresponding operators, now denoted by $X_1, X_2, \ldots, X_n$, do determine the Lie group of transformations. But, equally important, they are themselves generators of a group. Though the product $X_i X_j$ is not a linear operator, the expression called the alternant of X_i and X_j,

$$X_i X_j - X_j X_i,$$

is a linear operator and is denoted by $[X_i, X_j]$. With this product operation, the group of operators becomes a Lie algebra.

Lie had begun the work on finding the structure of his simple finite (continuous) groups with r parameters. He found four main classes of algebras. Wilhelm K. J. Killing (1847–1923)[50] found that these classes are correct for all simple algebras but that in addition there are five exceptional cases of 14, 52, 78, 133, and 248 parameters. Killing's work contained gaps and Elie Cartan undertook to fill them.

In his thesis, *Sur la structure des groupes de transformations finis et continus*,[51] Cartan gave the complete classification of all simple Lie algebras over the field of complex values for the variables and parameters. Like Killing, Cartan found that they fall into four general classes and the five exceptional algebras. Cartan constructed the exceptional algebras explicitly. In 1914[52] Cartan determined all the simple algebras with real values for the parameters and variables. These results are still basic.

The use of representations to study Lie algebras has been pursued much as in the case of abstract groups. Cartan, in his thesis and in a paper of 1913,[53]

50. *Math. Ann.*, 31, 1888, 252–90, and in Vols. 33, 34, and 36.
51. 1894; 2nd ed., Vuibert, Paris, 1933 = *Œuvres*, Part I, Vol. 1, 137–286.
52. *Ann. de l'Ecole Norm. Sup.*, 31, 1914, 263–355 = *Œuvres*, Part I, Vol. 1, 399–491.
53. *Bull. Soc. Math. de France*, 41, 1913, 53–96 = *Œuvres*, Part I, Vol. 1, 355–98.

found irreducible representations of the simple Lie algebras. A key result was obtained by Hermann Weyl.[54] Any representation of a semi-simple Lie algebra (over an algebraically closed field of characteristic 0) is completely reducible.

6. The Range of Abstract Algebra

Our few indications of the accomplishments in the field of abstract algebra certainly do not give the full picture of what was created even in the first quarter of this century. It may, however, be helpful to indicate the vast area that was opened up by the conscious turn to abstraction.

Up to about 1900 the various algebraic subjects that had been studied, whether matrices, the algebras of forms in two, three, or n variables, hypernumbers, congruences, or the theory of solution of polynomial equations, had been based on the real and complex number systems. However, the abstract algebraic movement introduced abstract groups, rings, ideals, division algebras, and fields. Beyond investigating the properties of such abstract structures and relationships as isomorphism and homomorphism, mathematicians now found it possible to take almost any algebraic subject and raise questions about it by replacing the real and complex numbers with any one of the abstract structures. Thus in place of matrices with complex elements, one can study matrices with elements that belong to a ring or field. Similarly one can take problems of the theory of numbers and, replacing the positive and negative integers and 0 by a ring, reconsider every question that has previously been investigated for the integers. One can even consider functions and power series with coefficients in an arbitrary field.

Such generalizations have indeed been made. We have noted that Wedderburn in his 1907 work generalized previous work on linear associative algebras (hypernumbers) by replacing the real or complex coefficients by any field. One can replace the field by a ring and investigate the theorems that hold under this change. Even the theory of equations with coefficients in arbitrary or finite fields has been studied.

As another example of the modern tendency to generalize, consider quadratic forms. These were important with integer coefficients in the study of the representation of integers as sums of squares and with real coefficients as the representation of conic and quadric surfaces. In the twentieth century quadratic forms are studied with any and every field as coefficients. As more abstract structures are introduced, these can be used as the base or coefficient field of older algebraic theories, and the process of generalization goes on indefinitely. This use of abstract concepts calls for the use of abstract algebraic techniques; thus many formerly distinct subjects were absorbed in

54. *Mathematische Zeitschrift*, 23, 1925, 271–309, and 24, 1926, 328–95 = *Ges. Abh.*, 2, 543–647.

abstract algebra. This is the case with large parts of number theory, including algebraic numbers.

However, abstract algebra has subverted its own role in mathematics. Its concepts were formulated to unify various seemingly diverse and dissimilar mathematical domains as, for example, group theory did. Having formulated the abstract theories, mathematicians turned away from the original concrete fields and concentrated on the abstract structures. Through the introduction of hundreds of subordinate concepts, the subject has mushroomed into a welter of smaller developments that have little relation to each other or to the original concrete fields. Unification has been succeeded by diversification and specialization. Indeed, most workers in the domain of abstract algebra are no longer aware of the origins of the abstract structures, nor are they concerned with the application of their results to the concrete fields.

Bibliography

Artin, Emil: "The Influence of J. H. M. Wedderburn on the Development of Modern Algebra," *Amer. Math. Soc. Bull.*, 56, 1950, 65–72.

Bell, Eric T.: "Fifty Years of Algebra in America, 1888–1938," *Amer. Math. Soc. Semicentennial Publications*, II, 1938, 1–34.

Bourbaki, N.: *Eléments d'histoire des mathématiques*, Hermann, 1960, pp. 110–28.

Cartan, Elie: "Notice sur les travaux scientifiques," *Œuvres complètes*, Gauthier-Villars, 1952–55, Part I, Vol. 1, pp. 1–98.

Dicke, Auguste: *Emmy Noether, 1882–1935*, Birkhäuser Verlag, 1970.

Dickson, L. E.: "An Elementary Exposition of Frobenius's Theory of Group-Characters and Group-Determinants," *Annals of Math.*, 4, 1902, 25–49.

———: *Linear Algebras*, Cambridge University Press, 1914.

———: *Algebras and Their Arithmetics* (1923), G. E. Stechert (reprint), 1938.

Frobenius, F. G.: *Gesammelte Abhandlungen*, 3 vols., Springer-Verlag, 1968.

Hawkins, Thomas: "The Origins of the Theory of Group Characters," *Archive for History of Exact Sciences*, 7, 1971, 142–70.

MacLane, Saunders: "Some Recent Advances in Algebra," *Amer. Math. Monthly*, 46, 1939, 3–19. Also in Albert, A. A., ed.: *Studies in Modern Algebra*, The Math. Assn. of Amer., 1963, pp. 9–34.

———: "Some Additional Advances in Algebra," in Albert, A. A., ed.: *Studies in Modern Algebra*, The Math. Assn. of Amer., 1963, pp. 35–58.

Ore, Oystein: "Some Recent Developments in Abstract Algebra," *Amer. Math. Soc. Bull.*, 37, 1931, 537–48.

———: "Abstract Ideal Theory," *Amer. Math. Soc. Bull.*, 39, 1933, 728–45.

Steinitz, Ernst: *Algebraische Theorie der Körper*, W. de Gruyter, 1910; 2nd rev. ed., 1930; Chelsea (reprint), 1950. The first edition is the same as the article in *Jour. für Math.*, 137, 1910, 167–309.

Wiman, A.: "Endliche Gruppen linearer Substitutionen," *Encyk. der Math. Wiss.*, B. G. Teubner, 1898–1904, I, Part 1, 522–54.

Wussing, H. L.: *Die Genesis des abstrakten Gruppenbegriffes*, VEB Deutscher Verlag der Wissenschaften, 1969.

50
The Beginnings of Topology

I believe that we lack another analysis properly geometric or linear which expresses location directly as algebra expresses magnitude.   G. W. LEIBNIZ

1. *The Nature of Topology*

A number of developments of the nineteenth century crystallized in a new branch of geometry, now called topology but long known as analysis situs. To put it loosely for the moment, topology is concerned with those properties of geometric figures that remain invariant when the figures are bent, stretched, shrunk, or deformed in any way that does not create new points or fuse existing points. The transformation presupposes, in other words, that there is a one-to-one correspondence between the points of the original figure and the points of the transformed figure, and that the transformation carries nearby points into nearby points. This latter property is called continuity, and the requirement is that the transformation and its inverse both be continuous. Such a transformation is called a homeomorphism. Topology is often loosely described as rubber-sheet geometry, because if the figures were made of rubber, it would be possible to deform many figures into homeomorphic figures. Thus a rubber band can be deformed into and is topologically the same as a circle or a square, but it is not topologically the same as a figure eight, because this would require the fusion of two points of the band.

Figures are often thought of as being in a surrounding space. For the purposes of topology two figures can be homeomorphic even though it is not possible to transform topologically the entire space in which one figure lies into the space containing the second figure. Thus if one takes a long rectangular strip of paper and joins the two short ends, he obtains a cylindrical band. If instead one end is twisted through 360° and then the short ends are joined, the new figure is topologically equivalent to the old one. But it is not possible to transform the three-dimensional space into itself topologically and carry the first figure into the second one.

Topology, as it is understood in this century, breaks down into two somewhat separate divisions: point set topology, which is concerned with geometrical figures regarded as collections of points with the entire collection often regarded as a space; and combinatorial or algebraic topology, which treats geometrical figures as aggregates of smaller building blocks just as a wall is a collection of bricks. Of course notions of point set topology are used in combinatorial topology, especially for very general geometric structures.

Topology has had numerous and varied origins. As with most branches of mathematics, many steps were made which only later were recognized as belonging to or capable of being subsumed under one new subject. In the present case the possibility of a distinct significant study was at least outlined by Klein in his Erlanger Programm (Chap. 38, sec. 5). Klein was generalizing the types of transformation studied in projective and algebraic geometry and he was already aware through Riemann's work of the importance of homeomorphisms.

2. Point Set Topology

The theory of point sets as initiated by Cantor (Chap. 41, sec. 7) and extended by Jordan, Borel, and Lebesgue (Chap. 44, secs. 3 and 4) is not *eo ipso* concerned with transformations and topological properties. On the other hand, a set of points regarded as a space is of interest in topology. What distinguishes a space as opposed to a mere set of points is some concept that binds the points together. Thus in Euclidean space the distance between points tells us how close points are to each other and in particular enables us to define limit points of a set of points.

The origins of point set topology have already been related (Chap. 46, sec. 2). Fréchet in 1906, stimulated by the desire to unify Cantor's theory of point sets and the treatment of functions as points of a space, which had become common in the calculus of variations, launched the study of abstract spaces. The rise of functional analysis with the introduction of Hilbert and Banach spaces gave additional importance to the study of point sets as spaces. The properties that proved to be relevant for functional analysis are topological largely because limits of sequences are important. Further, the operators of functional analysis are transformations that carry one space into another.[1]

As Fréchet pointed out, the binding property need not be the Euclidean distance function. He introduced (Chap. 46, sec. 2) several different concepts that can be used to specify when a point is a limit point of a sequence of points. In particular he generalized the notion of distance by introducing

1. The definitions of the basic properties of point sets, such as compactness and separability, have had different meanings for different authors and are still not standardized. We shall use the present commonly understood meanings.

THE BEGINNINGS OF TOPOLOGY

the class of metric spaces. In a metric space, which can be two-dimensional
Euclidean space, one speaks of the neighborhood of a point and means all
those points whose distance from the point is less than some quantity ε,
say. Such neighborhoods are circular. One could use square neighborhoods
as well. However, it is also possible to suppose that the neighborhoods,
certain subsets of a given set of points, are specified in some way, even
without the introduction of a metric. Such spaces are said to have a neighborhood topology. This notion is a generalization of a metric space. Felix
Hausdorff (1868–1942), in his *Grundzüge der Mengenlehre* (Essentials of Set
Theory, 1914), used the notion of a neighborhood (which Hilbert had
already used in 1902 in a special axiomatic approach to Euclidean plane
geometry) and built up a definitive theory of abstract spaces on this notion.

Hausdorff defines a topological space as a set of elements x together
with a family of subsets U_x associated with each x. These subsets are called
neighborhoods and must satisfy the following conditions:

(a) To each point x there is at least one neighborhood U_x which
contains the point x.

(b) The intersection of two neighborhoods of x contains a neighborhood
of x.

(c) If y is a point in U_x there exists a U_y such that $U_y \subseteq U_x$.

(d) If $x \neq y$, there exist U_x and U_y such that $U_x \cdot U_y = 0$.

Hausdorff also introduced countability axioms:

(a) For each point x, the set of U_x is at most countable.

(b) The set of all distinct neighborhoods is countable.

The groundwork in point set topology consists in defining several basic
notions. Thus a limit point of a set of points in a neighborhood space is
one such that every neighborhood of the point contains other points of the
set. A set is open if every point of the set can be enclosed in a neighborhood
that contains only points of the set. If a set contains all its limit points, then
it is closed. A space or a subset of a space is called compact if every infinite
subset has a limit point. Thus the points on the usual Euclidean line are
not a compact set because the infinite set of points corresponding to the
positive integers has no limit point. A set is connected if, no matter how it
is divided into disjoined sets, at least one of these contains limit points of
the other. The curve of $y = \tan x$ is not connected but the curve of $y = \sin 1/x$
plus the interval $(-1, 1)$ of the Y-axis is connected. Separability, introduced
by Fréchet in his 1906 thesis, is another basic concept. A space is called
separable if it has a denumerable subset whose closure, the set plus its limit
points, is the space itself.

The notions of continuous transformations and homeomorphism can
now also be introduced. A continuous transformation usually presupposes

that to each point of one space there is associated a unique point of the second or image space and that given any neighborhood of an image point there is a neighborhood of the original point (or each original point if there are many) whose points map into the neighborhood in the image space. This concept is no more than a generalization of the $\varepsilon - \delta$ definition of a continuous function, the ε specifying the neighborhood of a point in the image space, and the δ a neighborhood of the original point. A homeomorphism between two spaces S and T is a one-to-one correspondence that is continuous both ways; that is, the transformations from S to T and from T to S are continuous. The basic task of point set topology is to discover properties that are invariant under continuous transformations and homeomorphisms. All of the properties mentioned above are topological invariants.

Hausdorff added many results to the theory of metric spaces. In particular he added to the notion of completeness, which Fréchet had introduced in his 1906 thesis. A space is complete if every sequence $\{a_n\}$ that satisfies the condition that given ε, there exists an N, such that $\left|a_n - a_m\right| < \varepsilon$ for all m and n greater than N, has a limit point. Hausdorff proved that every metric space can be extended to a complete metric space in one and only one way. The introduction of abstract spaces raised several questions that prompted much research. For example, if a space is defined by neighborhoods, is it necessarily metrizable; that is, is it possible to introduce a metric that preserves the structure of the space so that limit points remain limit points? This question was raised by Fréchet. One result, due to Paul S. Urysohn (1898–1924), states that every completely separable normal topological space can be metrized.[2] A normal space is one in which two disjoint closed sets can each be enclosed in an open set and the two open sets are disjoined. A related result of some importance is also due to Urysohn. He proved[3] that every separable metric space, that is, every metric space in which a countable subset is dense in the space, is homeomorphic to a subset of the Hilbert cube; the cube consists of the space of all infinite sequences $\{x_i\}$ such that $0 \leqslant x_i \leqslant 1/i$ and in which distance is defined by $d = \sqrt{\sum_{i=1}^{\infty} (x_i - y_i)^2}$.

The question of dimension, as already noted, was raised by Cantor's demonstration of a one-to-one correspondence of line and plane (Chap. 41, sec. 7) and by Peano's curve, which fills out a square (Chap. 42, sec. 5). Fréchet (already working with abstract spaces) and Poincaré saw the need for a definition of dimension that would apply to abstract spaces and yet grant to line and plane the dimensions usually assumed for them. The definition that had been tacitly accepted was the number of coordinates needed to fix the points of a space. This definition was not applicable to general spaces.

2. *Math. Ann.*, 94, 1925, 262–95.
3. *Math. Ann.*, 94, 1925, 309–15.

In 1912[4] Poincaré gave a recursive definition. A continuum (a closed connected set) has dimension n if it can be separated into two parts whose common boundary consists of continua of dimension $n - 1$. Luitzen E. J. Brouwer (1881–1967) pointed out that the definition does not apply to the cone with two nappes, because the nappes are separated by a point. Poincaré's definition was improved by Brouwer,[5] Urysohn,[6] and Karl Menger (1902–).[7]

The Menger and Urysohn definitions are similar and both are credited with the now generally accepted definition. Their concept assigns a local dimension. Menger's formulation is this: The empty set is defined to be of dimension -1. A set M is said to be n-dimensional at a point P if n is the smallest number for which there are arbitrarily small neighborhoods of P whose boundaries in M have dimension less than n. The set M is called n-dimensional if its dimension in all of its points is less than or equal to n but is n in at least one point.

Another widely accepted definition is due to Lebesgue.[8] A space is n-dimensional if n is the least number for which coverings by closed sets of arbitrarily small diameter contain points common to $n + 1$ of the covering sets. Euclidean spaces have the proper dimension under any of the latter definitions and the dimension of any space is a topological invariant.

A key result in the theory of dimension is a theorem due to Menger (*Dimensionstheorie*, 1928, p. 295) and A. Georg Nöbeling (1907–).[9] It asserts that every n-dimensional compact metric space is homeomorphic to some subset of the $(2n + 1)$-dimensional Euclidean space.

Another problem raised by the work of Jordan and Peano was the very definition of a curve (Chap. 42, sec. 5). The answer was made possible by the work on dimension theory. Menger[10] and Urysohn[11] defined a curve as a one-dimensional continuum, a continuum being a closed connected set of points. (The definition requires that an open curve such as the parabola be closed by a point at infinity.) This definition excludes space-filling curves and renders the property of being a curve invariant under homeomorphisms.

The subject of point set topology has continued to be enormously active. It is relatively easy to introduce variations, specializations, and generalizations of the axiomatic bases for the various types of spaces. Hundreds of concepts have been introduced and theorems established, though the ultimate value of these concepts is dubious in most cases. As in other

4. *Revue de Métaphysique et de Morale*, 20, 1912, 483–504.
5. *Jour. für Math.*, 142, 1913, 146–52.
6. *Fundamenta Mathematicae*, 7, 1925, 30–137 and 8, 1926, 225–359.
7. *Monatshefte für Mathematik und Physik*, 33, 1923, 148–60 and 34, 1926, 137–61.
8. *Fundamenta Mathematicae*, 2, 1921, 256–85.
9. *Math. Ann.*, 104, 1930, 71–80.
10. *Monatshefte für Mathematik und Physik*, 33, 1923, 148–60 and *Math. Ann.*, 95, 1926, 277–306.
11. *Fundamenta Mathematicae*, 7, 1925, 30–137, p. 93 in part.

fields, mathematicians have not hesitated to plunge freely and broadly into point set topology.

3. The Beginnings of Combinatorial Topology

As far back as 1679 Leibniz, in his *Characteristica Geometrica*, tried to formulate basic geometric properties of geometrical figures, to use special symbols to represent them, and to combine these properties under operations so as to produce others. He called this study analysis situs or *geometria situs*. He explained in a letter to Huygens of 1679[12] that he was not satisfied with the coordinate geometry treatment of geometric figures because, beyond the fact that the method was not direct or pretty, it was concerned with magnitude, whereas "I believe we lack another analysis properly geometric or linear which expresses location [*situs*] directly as algebra expresses magnitude." Leibniz's few examples of what he proposed to build still involved metric properties even though he aimed at geometric algorithms that would furnish solutions of purely geometric problems. Perhaps because Leibniz was vague about the kind of geometry he sought, Huygens was not enthusiastic about his idea and his symbolism. To the extent that he was at all clear, Leibniz envisioned what we now call combinatorial topology.

A combinatorial property of geometric figures is associated with Euler, though it was known to Descartes in 1639 and, through the latter's unpublished manuscript, to Leibniz in 1675. If one counts the number of vertices, edges, and faces of any closed convex polyhedron—for example, a cube—then $V - E + F = 2$. This fact was published by Euler in 1750.[13] In 1751 he submitted a proof.[14] Euler's interest in this relation was to use it to classify polyhedra. Though he had discovered a property of all closed convex polyhedra, Euler did not think of invariance under continuous transformation. Nor did he define the class of polyhedra for which the relation held.

In 1811 Cauchy[15] gave another proof. He removed the interior of a face and stretched the remaining figure out on a plane. This gives a polygon for which $V - E + F$ should be one. He established the latter by triangulating the figure and then counting the changes as the triangles are removed one by one. This proof, inadequate because it supposes that any closed convex polyhedron is homeomorphic with a sphere, was accepted by nineteenth-century mathematicians.

Another well-known problem, which was a curiosity at the time but

12. Leibniz, *Math. Schriften*, 1 Abt., Vol. 2, 1850, 19–20 = Gerhardt, *Der Briefwechsel von Leibniz mit Mathematikern*, 1, 1899, 568 = Chr. Huygens, *Œuv. Comp.*, 8, No. 2192.
13. *Novi Comm. Acad. Sci. Petrop.*, 4, 1752–53, 109–40, pub. 1758 = *Opera*, (1), 26, 71–93.
14. *Novi Comm. Acad. Sci. Petrop.*, 4, 1752–53, 140–60, pub. 1758 = *Opera*, (1), 26, 94–108.
15. *Jour. de l'Ecol. Poly.*, 9, 1813, 68–86 and 87–98 = *Œuvres*, (2), 1, 7–38.

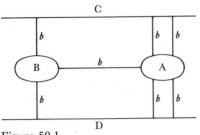

Figure 50.1

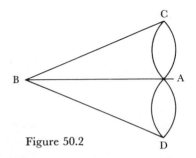

Figure 50.2

whose topological nature was later appreciated, is the Koenigsberg bridge problem. In the Pregel, a river flowing through Koenigsberg, there are two islands and seven bridges (marked b in Fig. 50.1). The villagers amused themselves by trying to cross all seven bridges in one continuous walk without recrossing any one. Euler, then at St. Petersburg, heard of the problem and found the solution in 1735.[16]. He simplified the representation of the problem by replacing the land by points A, B, C, D and the bridges joining them by line segments or arcs and obtained Figure 50.2. The question Euler then framed was whether it was possible to describe this figure in one continuous motion of the pencil without recrossing any arcs. He proved it was not possible in the above case and gave a criterion as to when such paths are or are not possible for given sets of points and arcs.

Gauss gave frequent utterance[17] to the need for the study of basic geometric properties of figures but made no outstanding contribution. In 1848 Johann B. Listing (1806–82), a student of Gauss in 1834 and later professor physics at Göttingen, published *Vorstudien zur Topologie*, in which he discussed what he preferred to call the geometry of position but, since this term was used for projective geometry by von Staudt, he used the term topology. In 1858 he began a new series of topological investigations that were published under the title *Der Census räumlicher Complexe* (Survey of Spatial Complexes).[18] Listing sought qualitative laws for geometrical figures. Thus he attempted to generalize the Euler relation $V - E + F = 2$.

The man who first formulated properly the nature of topological investigations was Möbius, who was an assistant to Gauss in 1813. He had classified the various geometrical properties, projective, affine, similarity, and congruence and then in 1863 in his "Theorie der elementaren Verwandschaft" (Theory of Elementary Relationships)[19] he proposed studying the

16. *Comm. Acad. Sci. Petrop.*, 8, 1736, 128–40, pub. 1741. An English translation of this paper can be found in James R. Newman: *The World of Mathematics*, Simon and Schuster, 1956, Vol. 1, 573–80.
17. *Werke*, 8, 270–86.
18. *Abh. der Ges. der Wiss. zu Gött.*, 10, 1861, 97–180, and as a book in 1862.
19. *Königlich Sächsischen Ges. der Wiss. zu Leipzig*, 15, 1863, 18–57 = *Werke*, 2, 433–71.

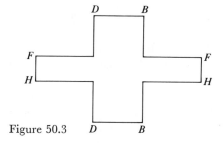

Figure 50.3

relationship between two figures whose points are in one-to-one correspond-ence and such that neighboring points correspond to neighboring points. He began by studying the *geometria situs* of polyhedra. He stressed that a polyhedron can be considered as a collection of two-dimensional polygons, which, since each piece can be triangulated, would make a polyhedron a collection of triangles. This idea proved to be basic. He also showed[20] that some surfaces could be cut up and laid out as polygons with proper identifi-cation of sides. Thus a double ring could be represented as a polygon (Fig. 50.3), provided that the edges that are lettered alike are identified.

In 1858 he and Listing independently discovered one-sided surfaces, of which the Möbius band is best known (Fig. 50.4). This figure is formed by taking a rectangular strip of paper, twisting it at one short edge through 180°, then joining this edge to the opposite edge. Listing published it in *Der Census*; the figure is also described in a publication by Möbius.[21] As far as the band is concerned, its one-sidedness may be characterized by the fact that it can be painted by a continuous sweep of the brush so that the entire surface is covered. If an untwisted band is painted on one side then the brush must be moved over an edge to get onto the other face. One-sidedness may also be defined by means of a perpendicular to the surface. Let it have a definite direction. If it can be moved arbitrarily over the surface and have the same direction when it returns to its original position, the surface is said to be two-sided. If the direction is reversed, the surface is one-sided.

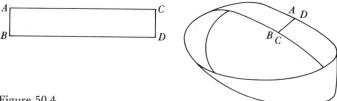

Figure 50.4

20. *Werke*, 2, 518–59.
21. *Königlich Sächsischen Ges. der Wiss. zu Leipzig*, 17, 1865, 31–68 = *Werke*, 2, 473–512; see also p. 519.

On the Möbius band the perpendicular will return to the point on the "opposite side" with reverse direction.

Still another problem that was later seen to be topological in nature is called the map problem. The problem is to show that four colors suffice to color all maps so that countries with at least an arc as a common boundary are differently colored. The conjecture that four colors will always suffice was made in 1852 by Francis Guthrie (d. 1899), a little-known professor of mathematics, at which time his brother Frederick communicated it to De Morgan. The first article devoted to it was Cayley's;[22] in it he says that he could not obtain a proof. The proof was attempted by a number of mathematicians, and though some published proofs were accepted for a time, they have been shown to be fallacious and the problem is still open.

The greatest impetus to topological investigations came from Riemann's work in complex function theory. In his thesis of 1851 on complex functions and in his study of Abelian functions,[23] he stressed that to work with functions some theorems of analysis situs were indispensable. In these investigations he found it necessary to introduce the connectivity of Riemann surfaces. Riemann defined connectivity in the following manner: "If upon the surface F [with boundaries] there can be drawn n closed curves $a_1, a_2, \ldots, a_n$ which neither individually nor in combination completely bound a part of this surface F, but with whose aid every other closed curve forms the complete boundary of a part of F, the surface is said to be $(n + 1)$-fold connected." To reduce the connectivity of a surface (with boundaries) Riemann states,

> By means of a crosscut [*Querschnitt*], that is, a line lying in the interior of the surface and going from a boundary point to a boundary point, an $(n + 1)$-fold connected surface can be changed into an n-fold connected one, F'. The parts of the boundary arising from the cutting play the role of boundary even during the further cutting so that a crosscut can pass through no point more than once but can end in one of its earlier points. ... To apply these considerations to a surface without boundary, a closed surface, we must change it into a bounded one by the specialization of an arbitrary point, so that the first division is made by means of this point and a crosscut beginning and ending in it, hence by a closed curve.

Riemann gives the example of an anchor ring or torus (Fig. 50.5) which is three-fold connected [genus 1 or one-dimensional Betti number 2] and can be changed into a simply connected surface by means of a closed curve abc and a crosscut $ab'c'$.

Riemann had thus classified surfaces according to their connectivity and, as he himself realized, had introduced a topological property. In

22. *Proceedings of the Royal Geographical Society*, 1, 1879, 259–61 = *Coll. Math. Papers*, 11, 7–8.
23. *Jour. für Math.*, 54, 1857, 105–10 = *Werke*, 91–96; see also *Werke*, 479–82.

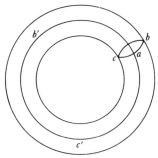

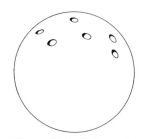

Figure 50.5   Figure 50.6. Sphere with p holes

terms of genus, the term used by the algebraic geometers of the latter part
of the nineteenth century, Riemann had classified closed surfaces by means
of their genus p, $2p$ being the number of closed curves [loop cuts or *Rück-
kehrschnitte*] needed to make the surface simply connected and $2p + 1$ to
cut the surface into two distinct pieces. He regarded as intuitively evident
that if two closed (orientable) Riemann surfaces are topologically equivalent
they have the same genus. He also observed that all closed (algebraic)
surfaces of genus zero, that is, simply connected, are topologically (and conformally and birationally) equivalent. Each can be mapped topologically
on a sphere.

    Since the structure of Riemann surfaces is complicated and a topo-
logically equivalent figure has the same genus, some mathematicians sought
simpler structures. William K. Clifford showed[24] that the Riemann surface
of an n-valued function with w branch-points can be transformed into a
sphere with p holes where $p = (w/2) - n + 1$ (Fig. 50.6). There is the
likelihood that Riemann knew and used this model. Klein suggested another
topological model, a sphere with p handles (Fig. 50.7).[25]

    The study of the topological equivalence of closed surfaces was made
by many men. To state the chief result it is necessary to note the concept of
orientable surfaces. An orientable surface is one that can be triangulated and
each (curvilinear) triangle can be oriented so that any side common to two
triangles has opposite orientations induced on it. Thus the sphere is orientable
but the projective plane (see below) is non-orientable. This fact was discovered
by Klein.[26] The chief result, clarified by Klein in this paper, is that two
orientable closed surfaces are homeomorphic if and only if they have the
same genus. For orientable surfaces with boundaries, as Klein also pointed

24. *Proc. Lon. Math. Soc.*, 8, 1877, 292–304 = *Math. Papers*, 241–54.
25. *Über Riemanns Theorie der algebraischen Funktionen und ihrer Integrale*, B. G. Teubner,
1882; Dover reprint in English, 1963. Also in Klein's *Ges. Math. Abh.*, 3, 499–573.
26. *Math. Ann.*, 7, 1874, 549–57 = *Ges. Math. Abh.*, 2, 63–77.

Figure 50.7. Sphere with p handles

out, the equality of the number of boundary curves must be added to the above condition. The theorem had been proved by Jordan.[27]

The complexity of even two-dimensional closed figures was emphasized by Klein's introduction in 1882 (see sec. 23 of the reference in note 25) of the surface now called the Klein bottle (Fig. 50.8). The neck enters the bottle without intersecting it and ends smoothly joined to the base along C. Along D the surface is uninterrupted and yet the tube enters the surface. It has no edge, no inside, and no outside; it is one-sided and has a one-dimensional connectivity number of 3 or a genus of 1. It cannot be constructed in three dimensions.

The projective plane is another example of a rather complex closed surface. Topologically the plane can be represented by a circle with diametrically opposite points identified (Fig. 50.9). The infinitely distant line is represented by the semicircumference CAD. The surface is closed and its connectivity number is 1 or its genus is 0. It can also be formed by pasting the edge of a circle along the edge of a Möbius band (which has just a single edge), though again the figure cannot be constructed in three dimensions without having points coincide that should be distinct.

Still another impetus to topological research came from algebraic geometry. We have already related (Chap. 39, sec. 8) that the geometers

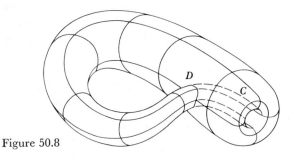

Figure 50.8

27. *Jour. de Math.*, (2), 11, 1866, 105–9 = *Œuvres*, 4, 85–89.

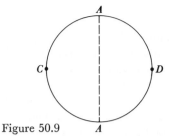

Figure 50.9

had turned to the study of the four-dimensional "surfaces" that represent the domain of algebraic functions of two complex variables and had introduced integrals on these surfaces in the manner analogous to the theory of algebraic functions and integrals on the two-dimensional Riemann surfaces. To study these four-dimensional figures, their connectivity was investigated and the fact learned that such figures cannot be characterized by a single number as the genus characterizes Riemann surfaces. Some investigations by Emile Picard around 1890 revealed that at least a one-dimensional and a two-dimensional connectivity number would be needed to characterize such surfaces.

The need to study the connectivity of higher-dimensional figures was appreciated by Enrico Betti (1823–92), a professor of mathematics at the University of Pisa. He decided that it was just as useful to take the step to n dimensions. He had met Riemann in Italy, where the latter had gone for several winters to improve his health, and from him Betti learned about Riemann's own work and the work of Clebsch. Betti[28] introduced connectivity numbers for each dimension from 1 to $n - 1$. The one-dimensional connectivity number is the number of closed curves that can be drawn in the geometrical structure that do not divide the surface into disjoined regions. (Riemann's connectivity number was 1 greater.) The two-dimensional connectivity number is the number of closed surfaces in the figure that collectively do not *bound* any three-dimensional region of the figure. The higher-dimensional connectivity numbers are similarly defined. The closed curves, surfaces, and higher-dimensional figures involved in these definitions are called cycles. (If a surface has edges, the curves must be crosscuts; that is, a curve that goes from a point on one edge to a point on another edge. Thus the one-dimensional connectivity number of a finite hollow tube (without ends) is 1, because a crosscut from one edge to the other can be drawn without disconnecting the surface.) For the four-dimensional structures used to represent complex algebraic functions $f(x, y, z) = 0$, Betti showed that the one-dimensional connectivity number equals the three-dimensional connectivity number.

28. *Annali di Mat.*, (2), 4, 1870–71, 140–58.

4. The Combinatorial Work of Poincaré

Toward the end of the century, the only domain of combinatorial topology that had been rather fully covered was the theory of closed surfaces. Betti's work was just the beginning of a more general theory. The man who made the first systematic and general attack on the combinatorial theory of geometrical figures, and who is regarded as the founder of combinatorial topology, is Henri Poincaré (1854–1912). Poincaré, professor of mathematics at the University of Paris, is acknowledged to be the leading mathematician of the last quarter of the nineteenth and the first part of this century, and the last man to have had a universal knowledge of mathematics and its applications. He wrote a vast number of research articles, texts, and popular articles, which covered almost all the basic areas of mathematics and major areas of theoretical physics, electromagnetic theory, dynamics, fluid mechanics, and astronomy. His greatest work is *Les Méthodes nouvelles de la mécanique céleste* (3 vols., 1892–99). Scientific problems were the motivation for his mathematical research.

Before he undertook the combinatorial theory we are about to describe, Poincaré contributed to another area of topology, the qualitative theory of differential equations (Chap. 29, sec. 8). That work is basically topological because it is concerned with the form of the integral curves and the nature of the singular points. The contribution to combinatorial topology was stimulated by the problem of determining the structure of the four-dimensional "surfaces" used to represent algebraic functions $f(x, y, z) = 0$ wherein x, y, and z are complex. He decided that a systematic study of the analysis situs of general or n-dimensional figures was necessary. After some notes in the *Comptes Rendus* of 1892 and 1893, he published a basic paper in 1895,[29] followed by five lengthy supplements running until 1904 in various journals. He regarded his work on combinatorial topology as a systematic way of studying n-dimensional geometry, rather than as a study of topological invariants.

In his 1895 paper Poincaré tried to approach the theory of n-dimensional figures by using their analytical representations. He did not make much progress this way, and he turned to a purely geometric theory of manifolds, which are generalizations of Riemann surfaces. A figure is a closed n-dimensional manifold if each point possesses neighborhoods that are homeomorphic to the n-dimensional interior of an $n - 1$ sphere. Thus the circle (and any homeomorphic figure) is a one-dimensional manifold. A spherical surface or a torus is a two-dimensional manifold. In addition to closed manifolds there are manifolds with boundaries. A cube or a solid torus is a three-manifold with boundary. At a boundary point a neighborhood is only part of the interior of a two-sphere.

29. *Jour. de l'Ecole Poly.*, (2), 1, 1895, 1–121 = *Œuvres*, 6, 193–288.

The method Poincaré finally adopted appears in his first supplement.[30] Though he used curved cells or pieces of his figures and treated manifolds, we shall formulate his ideas in terms of complexes and simplexes, which were introduced later by Brouwer. A simplex is merely the n-dimensional triangle. That is, a zero-dimensional simplex is a point; a one-dimensional simplex is a line segment; a two-dimensional simplex is a triangle; a three-dimensional one is a tetrahedron; and the n-dimensional simplex is the generalized tetrahedron with $n + 1$ vertices. The lower-dimensional faces of a simplex are themselves simplexes. A complex is any finite set of simplexes such that any two simplexes meet, if at all, in a common face, and every face of every simplex of the complex is a simplex of the complex. The simplexes are also called cells.

For the purposes of combinatorial topology each simplex or cell of every dimension is given an orientation. Thus a two-simplex E^2 (a triangle) with vertices a_0, a_1, and a_2 can be oriented by choosing an order, say $a_0a_1a_2$. Then any order of the a_i which is obtainable from this by an even number of permutations of the a_i is said to have the same orientation. Thus E^2 is given by $(a_0a_1a_2)$ or $(a_2a_0a_1)$ or $(a_1a_2a_0)$. Any order derivable from the basic one by an odd number of permutations represents the oppositely oriented simplex. Thus $-E^2$ is given by $(a_0a_2a_1)$ or $(a_1a_0a_2)$ or $(a_2a_1a_0)$.

The boundary of a simplex consists of the next lower-dimensional simplexes contained in the given one. Thus the boundary of a two-simplex consists of three one-simplexes. However, the boundary must be taken with the proper orientation. To obtain the oriented boundary one adopts the following rule: The simplex E^k

$$a_0a_1a_2\cdots a_k$$

induces the orientation

(1) $$(-1)^i(a_0a_1\cdots a_{i-1}a_{i+1}\cdots a_k)$$

on each of the $(k-1)$-dimensional simplexes on its boundary. The E_i^{k-1} may have the orientation given by (1), in which case the incidence number, which represents the orientation of E_i^{k-1} relative to E^k, is 1, or it may have the opposite orientation, in which case its incidence number is -1. Whether the incidence number is 1 or -1, the basic fact is that the boundary of the boundary of E^k is 0.

Given any complex one can form a linear combination of its k-dimensional, oriented simplexes. Thus if E_i^k is an oriented k-dimensional simplex,

(2) $$C^k = c_1E_1^k + c_2E_2^k + \cdots + c_lE_l^k,$$

where the c_i are positive or negative integers, is such a combination and is called a chain. The numbers c_i merely tell us how many times a given simplex

30. *Rendiconti del Circolo Matematico di Palermo*, 13, 1899, 285–343 = *Œuvres*, 6, 290–337.

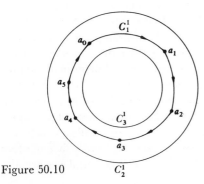

Figure 50.10 C_2^1

is to be counted, a negative number implying also a change in the orientation of the simplex. Thus if our figure is a tetrahedron determined by four points $(a_0a_1a_2a_3)$, we could form the chain $C^3 = 5E_1^3$. The boundary of any chain is the sum of all the next lower-dimensional simplexes on all simplexes of the chain, each taken with the proper incidence number and with the multiplicity appearing in (2). Since the boundary of a chain is the sum of the boundaries of each of the simplexes appearing in the chain, the boundary of the boundary of the chain is zero.

A chain whose boundary is zero is called a cycle. That is, some chains are cycles. Among cycles some are boundaries of other chains. Thus the boundary of the simplex E_1^3 is a cycle that bounds E_1^3. However, if our original figure were not the three-dimensional simplex but merely the surface, we would still have the same cycle but it would not bound in the figure under consideration, namely, the surface. As another example, the chain (Fig. 50.10)

$$C_1^1 = (a_0a_1) + (a_1a_2) + (a_2a_3) + \cdots + (a_5a_0)$$

is a cycle because the boundary of a_1a_2, for example, is $a_2 - a_1$ and the boundary of the entire chain is zero. But C_1^1 does not bound any two-dimensional chain. This is intuitively obvious because the complex in question is the circular ring and the inner hole is not part of the figure.

It is possible for two separate cycles not to bound but for their sum or difference to bound a region. Thus the sum (or difference) of C_2^1 and C_3^1 (Fig. 50.10) bounds the area in the entire ring. Two such cycles are said to be dependent. In general the cycles $C_1^k, \ldots, C_r^k$ are said to be dependent if

$$\sum_{i=1}^{r} c_i C_i^k$$

bounds and not all the c_i are zero.

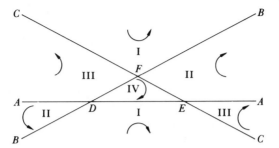

Figure 50.11

Poincaré introduced next the important quantities he called the Betti numbers (in honor of Enrico Betti). For each dimension of possible simplexes in a complex, the number of independent cycles of that dimension is called the Betti number of that dimension (Poincaré actually used a number that is 1 more than Betti's connectivity number.) Thus in the case of the ring, the zero-dimensional Betti number is 1, because any point is a cycle but two points bound the sequence of line segments joining them. The one-dimensional Betti number is 1 because there are one-cycles that do not bound but any two such one-cycles (their sum or difference) do bound. The two-dimensional Betti number of the ring is zero because no chain of two simplexes is a cycle. To appreciate what these numbers mean, one can compare the ring with the circle itself (including its interior). In the latter figure every one-dimensional cycle bounds so that the one-dimensional Betti number is zero.

In this 1899 paper, Poincaré also introduced what he called torsion coefficients. It is possible in more complicated structures—for example, the projective plane—to have a cycle that does not bound but two times this cycle does bound. Thus if the simplexes are oriented as shown in Figure 50.11, the boundary of the four triangles is twice the line BB. (We must remember that AB and BA are the same line segment.) The number 2 is called a torsion coefficient and the corresponding cycle BB a torsion cycle. There may be a finite number of such independent torsion cycles.

It is clear from even these very simple examples that the Betti numbers and torsion coefficients of a geometric figure do somehow distinguish one figure from another, as the circular ring is distinguished from the circle.

In his first supplement (1899) and in the second,[31] Poincaré introduced a method for computing the Betti numbers of a complex. Each simplex E_q of dimension q has simplexes of dimension $q - 1$ on its boundary. These have incidence numbers of $+1$ or -1. A $(q - 1)$-dimensional simplex that does not lie on the boundary of E_q is given the incidence number zero. It

31. *Proc. Lon. Math. Soc.*, 32, 1900, 277–308 = *Œuvres*, 6, 338–70.

is now possible to set up a rectangular matrix that shows the incidence numbers ε_{ij}^q of the jth $(q - 1)$-dimensional simplex with respect to the ith q-dimensional simplex. There is such a matrix T_q for each dimension q other than zero. T_q will have as many rows as there are q-dimensional simplexes in the complex and as many columns as there are $(q - 1)$-dimensional simplexes. Thus T_1 gives the incidence relations of the vertices to the edges. T_2 gives the incidence relations of the one-dimensional simplexes to the two-dimensional ones; and so forth. By elementary operations on the matrices it is possible to make all the elements not on the main diagonal zero and those on this diagonal positive integers or zero. Suppose that γ_q of these diagonal elements are 1. Then Poincaré shows that the q-dimensional Betti number p_q (which is one more than Betti's connectivity number) is

$$p_q = \alpha_q - \gamma_{q+1} - \gamma_q + 1$$

where α_q is the number of q-dimensional simplexes.

Poincaré distinguished complexes with torsion from those without. In the latter case all the numbers in the main diagonal for all q are 0 or 1. Larger values indicate the presence of torsion.

He also introduced the characteristic $N(K^n)$ of an n-dimensional complex K^n. If this has α_k k-dimensional simplexes, then by definition

$$N(K^n) = \sum_{k=0}^{n} (-1)^k \alpha_k.$$

This quantity is a generalization of the Euler number $V - E + F$. Poincaré's result on this characteristic is that if p_k is the kth Betti number of K^n then[32]

$$N(K^n) = \sum_{k=0}^{n} (-1)^k p_k.$$

This result is called the Euler-Poincaré formula.

In his 1895 paper Poincaré introduced a basic theorem, known as the duality theorem. It concerns the Betti numbers of a closed manifold. An n-dimensional closed manifold, as already noted, is a complex for which each point has a neighborhood homeomorphic to a region of n-dimensional Euclidean space. The theorem states that in a closed orientable n-dimensional manifold the Betti number of dimension p equals the Betti number of dimension $n - p$. His proof, however, was not complete.

In his efforts to distinguish complexes, Poincaré introduced (1895) one other concept that now plays a considerable role in topology, the fundamental group of a complex, also known as the Poincaré group or the first

32. These p_k are 1 less than Poincaré's. We use here the statement that is familiar today.

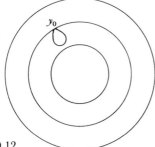

Figure 50.12

homotopy group. The idea arises from considering the distinction between simply and multiply connected plane regions. In the interior of a circle all closed curves can be shrunk to a point. However, in a circular ring some closed curves, those that surround the inner circular boundary, cannot be shrunk to a point, whereas those closed curves that do not surround the inner boundary can be shrunk to a point.

The more precise notion is best approached by considering the closed curves that start and end at a point y_0 of the complex. Then those curves that can be deformed by continuous motion in the space of the complex onto each other are homotopic to each other and are regarded as one class. Thus the closed curves starting and ending at y_0 (Fig. 50.12) in the circular ring and not enclosing the inner boundary are one class. Those which, starting and ending at y_0, do enclose the inner boundary are another class. Those which, starting and ending at y_0 enclose the inner boundary n times, are another class.

It is now possible to define an operation of one class on another, which geometrically amounts to starting at y_0 and traversing any curve of one class, then traversing any curve of the second class. The order in which the curves are chosen and the direction in which a curve is traversed are distinguished. The classes then form a group, called the fundamental group of the complex with respect to the base point y_0. This non-commutative group is now denoted by $\pi_1(K, y_0)$ where K is the complex. In reasonable complexes the group does not actually depend upon the point. That is, the groups at y_0 and y_1, say, are isomorphic. For the circular ring the fundamental group is infinite cyclic. The simply connected domain usually referred to in analysis, e.g. the circle and its interior, has only the identity element for its fundamental group. Just as the circle and the circular ring differ in their homotopy groups, so it is possible to describe higher-dimensional complexes that differ markedly in this respect.

Poincaré bequeathed some major conjectures. In his second supplement he asserted that any two closed manifolds that have the same Betti numbers and torsion coefficients are homeomorphic. But in the fifth supplement[33] he gave an example of a three-dimensional manifold that has the Betti numbers and torsion coefficients of the three-dimensional sphere (the surface of a four-dimensional solid sphere) but is not simply connected. Hence he added simple connectedness as a condition. He then showed that there are three-dimensional manifolds with the same Betti numbers and torsion coefficients but which have different fundamental groups and so are not homeomorphic. However, James W. Alexander (1888–1971), a professor of mathematics at Princeton University and later at the Institute for Advanced Study, showed[34] that two three-dimensional manifolds may have the same Betti numbers, torsion coefficients, and fundamental group and yet not be homeomorphic.

In his fifth supplement (1904), Poincaré made a somewhat more restricted conjecture, namely, that every simply connected, closed, orientable three-dimensional manifold is homeomorphic to the sphere of that dimension. This famous conjecture has been generalized to read: Every simply connected, closed, n-dimensional manifold that has the Betti numbers and torsion coefficients of the n-dimensional sphere is homeomorphic to it. Neither Poincaré's conjecture nor the generalized one has been proven.[35]

Another famous conjecture, called the *Hauptvermutung* (most important conjecture) of Poincaré, asserts that if T_1 and T_2 are simplicial (not necessarily straightedged) subdivisions of the same three-manifold, then T_1 and T_2 have subdivisions that are isomorphic.[36]

5. Combinatorial Invariants

The problem of establishing the invariance of combinatorial properties is to show that any complex homeomorphic to a given complex in the *point set sense* has the same combinatorial properties as the given complex. The proof that the Betti numbers and torsion coefficients are combinatorial invariants was first given by Alexander.[37] His result is that if K and K_1 are any simplicial subdivisions (not necessarily rectilinear) of any two homeomorphic (as point sets) polyhedra P and P_1, then the Betti numbers and

33. *Rendiconti del Circolo Matematico di Palermo*, 18, 1904, 45–110 = *Œuvres*, 6, 435–98.
34. *Amer. Math. Soc. Trans.*, 20, 1919, 339–42.
35. The generalized conjecture has been proved for $n \geq 5$ by Stephen Smale (*Amer. Math. Soc. Bull.*, 66, 1960, 373–75), John R. Stallings (*ibid.*, 485–88) and E. C. Zeeman (*ibid.*, 67, 1961, 270).
36. The *Hauptvermutung* has been shown to be true for finite simplicial complexes (which are more general than manifolds) of dimension less than three and to be false for such complexes of dimension greater than five. It is correct for manifolds of dimension less than or equal to three and an open problem for manifolds of dimension four or greater. See John Milnor, *Annals of Math.*, (2), 74, 1961, 575–90.
37. *Amer. Math. Soc. Trans.*, 16, 1915, 148–54.

torsion coefficients of P and P_1 are equal. The converse is not true, so that the equality of the Betti numbers and torsion coefficients of two complexes does not guarantee the homeomorphism of the two complexes.

Another invariant of importance was contributed by L. E. J. Brouwer. Brouwer became interested in topology through problems in function theory. He sought to prove that there are $3g - 3$ classes of conformally equivalent Riemann surfaces of genus $g > 1$ and was led to take up the related topological problems. He proved[38] the invariance of the dimension of a complex in the following sense: If K is an n-dimensional, simplicial subdivision of a polyhedron P, then every simplicial subdivision of P and every such division of any polyhedron homeomorphic to P is also an n-dimensional complex.

The proofs of this theorem, as well as Alexander's theorem, are made by using a method due to Brouwer,[39] namely, simplicial approximations of continuous transformations. Simplicial transformations (simplexes into simplexes) are themselves no more than the higher-dimensional analogues of continuous transformations, while the simplicial approximation of continuous transformations is analogous to linear approximation applied to continuous functions. If the domain in which the approximation is made is small, then the approximation serves to represent the continuous transformation for the purposes of the invariance proofs.

6. Fixed Point Theorems

Beyond serving to distinguish complexes from one another, combinatorial methods have produced fixed point theorems that are of geometric significance and also have applications to analysis. By introducing notions (which we shall not take up) such as the class of a mapping of one complex into another[40] and the degree of a mapping,[41] Brouwer was able to treat first what are called singular points of vector fields on a manifold. Consider the circle S^1, the spherical surface S^2, and the n-sphere $\sum_{i=1}^{n+1} x_i^2 = 1$ in Euclidean $(n + 1)$-dimensional space. On S^1 it is possible to have a tangent vector at each point and such that the lengths and directions of these vectors vary continuously around the circle while no vector has length zero. There is, one says, a continuous tangent vector field without singularities on S^1. However, no such field can exist on S^2. Brouwer proved[42] that what happens on S^2 must happen on every even-dimensional sphere; that is, no continuous vector field can exist on even-dimensional spheres, so that there must be at least one singular point.

38. *Math. Ann.*, 70, 1910/11, 161–65, and 71, 1911/12, 305–13.
39. *Math. Ann.*, 71, 1911/12, 97–115.
40. *Proceedings Koninklijke Akademie von Wettenschappen te Amsterdam*, 12, 1910, 785–94.
41. *Math. Ann.*, 71, 1911/12, 97–115.
42. *Math. Ann.*, 71, 1911/12, 97–115.

Closely related to the theory of singular points is the theory of continuous transformations of complexes into themselves. Of special interest in such transformations are the fixed points. If we denote by $f(x)$ the transform of a point x under such a transformation, then a fixed point is one for which $f(x) = x$. We may for any point x introduce a vector from x to $f(x)$. In the case of a fixed point the vector is indeterminate and the point is a singular point. The basic theorem on fixed points is due to Brouwer.[43] It applies to the n-dimensional simplex (or a homeomorph of it) and states that every continuous transformation of the n-simplex into itself possesses at least one fixed point. Thus a continuous transformation of a circular disc into itself must possess at least one fixed point. In the same paper Brouwer proved that every one-to-one continuous transformation of a sphere of even dimension into itself that can be deformed into the identity transformation possesses at least one fixed point.

Shortly before his death in 1912, Poincaré showed[44] that periodic orbits would exist in a restricted three-body problem if a certain topological theorem holds. The theorem asserts the existence of at least two fixed points in an annular region between two circles under a topological transformation of the region that carries each circle into itself, moving one in one direction and the other in the opposite direction while preserving the area. This last "theorem" of Poincaré was proved by George D. Birkhoff.[45]

Fixed point theorems were generalized to infinite-dimensional function spaces by Birkhoff and Oliver D. Kellogg in a joint paper[46] and applied to prove the existence of solutions of differential equations by Jules P. Schauder (1899–1940),[47] and by Schauder and Jean Leray (1906–) in a joint paper.[48] A key theorem used for such applications states that if T is a continuous mapping of a closed, convex, compact set of a Banach space into itself, then T has a fixed point.

The use of fixed point theorems to establish the existence of solutions of differential equations is best understood from a rather simple example. Consider the differential equation

$$\frac{dy}{dx} = F(x, y)$$

in the interval $0 \leq x \leq 1$ and the initial condition $y = 0$ at $x = 0$. The solution $\phi(x)$ satisfies the equation

$$\phi(x) = \int_0^x F(x, \phi(x)) \, dx.$$

43. *Math. Ann.*, 71, 1911/12, 97–115.
44. *Rendiconti del Circolo Matematico di Palermo*, 33, 1912, 375–407 = *Œuvres*, 6, 499–538.
45. *Amer. Math. Soc. Trans.*, 14, 1913, 14–22 = *Coll. Math. Papers*, 1, 673–81.
46. *Amer. Math. Soc. Trans.*, 23, 1922, 96–115 = Birkhoff, *Coll. Math. Papers*, 3, 255–74.
47. *Studia Mathematica*, 2, 1930, 170–79.
48. *Ann. de l'Ecole Norm. Sup.*, 51, 1934, 45–78.

We introduce the general transformation

$$g(x) = \int_0^x F(x, f(x)) \, dx$$

where $f(x)$ is an arbitrary function. This transformation associates f with g and can be shown to be continuous on the space of continuous functions $f(x)$ defined on $[0, 1]$. The solution ϕ that we seek is a fixed point of that function space. If one can show that the function space satisfies the conditions that validate a fixed point theorem, then the existence of ϕ is established. The fixed point theorems applicable to function spaces do precisely that. The method illustrated by this simple example enables us to establish the existence of solutions of nonlinear partial differential equations that are common in the calculus of variations and hydrodynamics.

7. Generalizations and Extensions

The ideas of Poincaré and Brouwer were seized upon by a number of men who have extended the scope of topology so much that it is now one of the most active fields of mathematics. Brouwer himself extended the Jordan curve theorem.[49] This theorem (Chap. 42, sec. 5) can be stated as follows: Let S^2 be a two-sphere (surface) and J a closed curve (topologically an S^1) on S^2. Then the zero-dimensional Betti number of $S^2 - J$ is 2. Since this Betti number is the number of components, J separates S^2 into two regions. Brouwer's generalization states that an $(n - 1)$-dimensional manifold separates the n-dimensional Euclidean space R_n into two regions. Alexander[50] generalized Poincaré's duality theorem and indirectly the Jordan curve theorem. Alexander's theorem states that the r-dimensional Betti number of a complex K in the n-dimensional sphere S^n is equal to the $(n - r - 1)$-dimensional Betti number of the complementary domain $S^n - K$, $r \neq 0$ and $r \neq n - 1$. For $r = 0$ the Betti number of K equals 1 plus the $(n - 1)$-dimensional Betti number of $S^n - K$ and for $r = n - 1$ the Betti number of K equals the zero-dimensional Betti number of $S^n - K$ minus 1.[51] This theorem generalizes the Jordan curve theorem, for if we take K to be the $(n - 1)$-dimensional sphere S^{n-1} then the theorem states that the $(n - 1)$-dimensional Betti number of S^{n-1}, which is 1, equals the zero-dimensional Betti number of $S^n - S^{n-1}$ minus 1 so that the zero-dimensional Betti number of $S^n - S^{n-1}$ is 2 and S^{n-1} divides S^n into two regions.

The definitions of the Betti numbers have been modified and generalized

49. *Math. Ann.*, 71, 1911/12, 314–19.
50. *Amer. Math. Soc. Trans.*, 23, 1922, 333–49.
51. Alexander stated his theorem under the condition that the coefficients of his chains were integers modulo 2 (see next paragraph). Our statement is for the usual integral coefficients.

in various ways. Veblen and Alexander[52] introduced chains and cycles
modulo 2; that is, in place of oriented simplexes one uses unoriented ones
but the integral coefficients are taken modulo 2. The boundaries of chains
are also counted in the same way. Alexander then introduced[53] coefficients
modulo m for chains and cycles. Solomon Lefschetz (1884–) suggested
rational numbers as coefficients.[54] A still broader generalization was made
by Lev S. Pontrjagin (1908–1960),[55] in taking the coefficients of chains to be
elements of an Abelian group. This concept embraces the previously mentioned classes of coefficients as well as another class also utilized, namely,
the real numbers modulo 1. All these generalizations, though they do lead
to more general theorems, have not enhanced the power of Betti numbers
and torsion coefficients to distinguish complexes.

   Another change in the formulation of basic combinatorial properties,
made during the years 1925 to 1930 by a number of men and possibly
suggested by Emmy Noether, was to recast the theory of chains, cycles, and
bounding cycles into the language of group theory. Chains of the same
dimension can be added to each other in the obvious way, that is, by adding
coefficients of the same simplex; and since cycles are chains they too can
be added and their sum is a cycle. Thus chains and cycles form groups.
Given a complex K, to each k-dimensional chain there is a $(k-1)$-dimen-
sional boundary chain and the sum of two chains has as its boundary the
sum of the separate boundary chains. Hence the relation of chain to boundary
establishes a homomorphism of the group $C^k(K)$ of k-dimensional chains
into a subgroup $H^{k-1}(K)$ of the group of $(k-1)$-dimensional chains.
The set of all k-cycles $(k > 0)$ is a subgroup $Z^k(K)$ of $C^k(K)$ and goes under
this homomorphism into the identity element or 0 of $C^{k-1}(K)$. Since every
boundary chain is a cycle, $H^{k-1}(K)$ is a subgroup of $Z^{k-1}(K)$.

   With these facts we may make the following definition: For any $k \geq 0$,
the factor group of $Z^k(K)$, that is, the k-dimensional cycles, modulo the
subgroup $H^k(K)$ of the bounding cycles, is called the kth homology group
of K and denoted by $B^k(K)$. The number of linearly independent generators
of this factor group is called the kth Betti number of the complex and denoted
by $p^k(K)$. The kth homology group may also contain finite cyclic groups
and these correspond to the torsion cycles. In fact the orders of these finite
groups are the torsion coefficients. With this group-theoretic formulation
of the homology groups of a complex, many older results can be reformulated
in group language.

   The most significant generalization of the early part of this century
was to introduce homology theory for general spaces, such as compact

52. *Annals of Math.*, (2), 14, 1913, 163–78.
53. *Amer. Math. Soc. Trans.*, 28, 1926, 301–29.
54. *Annals of Math.*, (2), 29, 1928, 232–54.
55. *Annals of Math.*, (2), 35, 1934, 904–14.

metric spaces as opposed to figures that are complexes to start with. The basic schemes were introduced by Paul S. Alexandroff (1896–),[56] Leopold Vietoris (1891–),[57] and Eduard Čech (1893–1960).[58] The details will not be given, since they involve a totally new approach to homology theory. However, we should note that this work marks a step toward the fusion of point set and combinatorial topology.

Bibliography

Bouligand, Georges: *Les Définitions modernes de la dimension*, Hermann, 1935.

Dehn, M., and P. Heegard: "Analysis Situs," *Encyk. der Math. Wiss.*, B. G. Teubner, 1907–10, III AB3, 153–220.

Franklin, Philip: "The Four Color Problem," *Scripta Mathematica*, 6, 1939, 149–56, 197–210.

Hadamard, J.: "L'Œuvre mathématique de Poincaré, *Acta Math.*, 38, 1921, 203–87.

Manheim, J. H.: *The Genesis of Point Set Topology*, Pergamon Press, 1964.

Osgood, William F.: "Topics in the Theory of Functions of Several Complex Variables," *Madison Colloquium*, American Mathematical Society, 1914, pp. 111–230.

Poincaré, Henri: *Œuvres*, Gauthier-Villars, 1916–56, Vol. 6.

Smith, David E.: *A Source Book in Mathematics*, Dover (reprint), 1959, Vol. 2, 404–10.

Tietze, H., and L. Vietoris: "Beziehungen zwischen den verschiedenen Zweigen der Topologie," *Encyk. der Math. Wiss.*, B. G. Teubner, 1914–31, III AB13, 141–237.

Zoretti, L., and A. Rosenthal: "Die Punktmengen," *Encyk. der Math. Wiss.*, B. G. Teubner, 1923–27, II C9A, 855–1030.

56. *Annals of Math.*, (2), 30, 1928/29, 101–87.
57. *Math. Ann.*, 97, 1927, 454–72.
58. *Fundamenta Mathematicae*, 19, 1932, 149–83.

51
The Foundations of Mathematics

> Logic is invincible because in order to combat logic it is
> necessary to use logic.   PIERRE BOUTROUX

> We know that mathematicians care no more for logic than
> logicians for mathematics. The two eyes of exact science are
> mathematics and logic: the mathematical sect puts out the
> logical eye, the logical sect puts out the mathematical eye,
> each believing that it can see better with one eye than with
> two.   AUGUSTUS DE MORGAN

1. *Introduction*

By far the most profound activity of twentieth-century mathematics has been
the research on the foundations. The problems thrust upon the mathematicians, and others that they voluntarily assumed, concern not only the nature
of mathematics but the validity of deductive mathematics.

Several activities converged to bring foundational problems to a head
in the first part of the century. The first was the discovery of contradictions,
euphemistically called paradoxes, notably in set theory. One such contradiction, the Burali-Forti paradox, has already been noted (Chap. 41, sec. 9).
A number of others were discovered in the first few years of this century.
Clearly the discovery of contradictions disturbed the mathematicians
deeply. Another problem that had gradually been recognized and that
emerged into the open early in this century was the consistency of mathematics (Chap. 43, sec. 6). In view of the paradoxes of set theory consistency
had to be established especially in this area.

During the latter part of the nineteenth century a number of men had
begun to reconsider the foundations of mathematics and, in particular, the
relationship of mathematics to logic. Research in this area (about which
we shall say more later) suggested to some mathematicians that mathematics
could be founded on logic. Others questioned the universal application of
logical principles, the meaningfulness of some existence proofs, and even the
reliance upon logical proof as the substantiation of mathematical results.

Controversies that had been smoldering before 1900 broke out into open fire when the paradoxes and the consistency problem added fuel. Thereupon the question of the proper foundation for all mathematics became vital and of widespread concern.

2. The Paradoxes of Set Theory

Following hard upon the discovery by Cantor and Burali-Forti of the paradox involving ordinal numbers came a number of other paradoxes or antinomies. Actually the word paradox is ambiguous, for it can refer to a seeming contradiction. But what the mathematicians actually encountered were unquestionably contradictions. Let us see first what they were.

One paradox was put in popular form by Bertrand Russell (1872–1970) in 1918, as the "barber" paradox. A village barber, boasting that he has no competition, advertises that of course he does not shave those people who shave themselves, but does shave all those who do not shave themselves. One day it occurs to him to ask whether he should shave himself. If he should shave himself, then by the first half of his assertion, he should not shave himself; but if he does not shave himself, then in accordance with his boast, he must shave himself. The barber is in a logical predicament.

Still another paradox was formulated by Jules Richard (1862–1956).[1] A simplified version of it was given by G. G. Berry and Russell and published by the latter.[2] The simplified paradox, also known as Richard's, reads as follows: Every integer can be described in words requiring a certain number of letters. For example, the number 36 can be described as thirty-six or as four times nine. The first description uses nine letters and the second thirteen. There is no one way to describe any given number, but this is not essential. Now let us divide all the positive whole numbers into two groups, the first to include all those that can be described (in at least one way) in 100 letters or fewer and the second group to include all those numbers that require a minimum of 101 letters, no matter how described. Only a finite number of numbers can have a description in 100 or fewer letters, for there are at most 27^{100} expressions with 100 or fewer letters (and some of these are meaningless). There is then a smallest integer in the second group. It can be described by the phrase, "the least integer not describable in one hundred or fewer letters." But this phrase requires fewer than 100 letters. Hence, the least integer not describable in 100 or fewer letters can be described in fewer than 100 letters.

Let us consider another form of this paradox, first stated by Kurt Grelling (1886–1941) and Leonard Nelson (1882–1927) in 1908 and

1. *Revue Générale des Sciences*, 16, 1905, 541.
2. *Proc. Lon. Math. Soc.*, (2), 4, 1906, 29–53.

published in an obscure journal.[3] Some words are descriptive of themselves. For example, the word "polysyllabic" is polysyllabic. On the other hand, the word "monosyllabic" is not monosyllabic. We shall call those words that are not descriptive of themselves heterological. In other words the word X is heterological if X is not itself X. Now let us replace X by the word "heterological." Then the word "heterological" is heterological if heterological is not heterological.

Cantor pointed out, in a letter to Dedekind of 1899, that one could not speak of the set of all sets without ending in a contradiction (Chap. 41, sec. 9). This is essentially what is involved in Russell's paradox (*The Principles of Mathematics*, 1903, p. 101). The class of all men is not a man. But the class of all ideas is an idea; the class of all libraries is a library; and the class of all sets with cardinal number greater than 1 is such a set. Hence, some classes are not members of themselves and some are. This description of classes includes all and the two types are mutually exclusive. Let us now denote by M the class of all classes that are members of themselves and by N the class of all classes that are not members of themselves. Now N is itself a class and we ask whether it belongs to M or to N. If N belongs to N then N is a member of itself and so must belong to M. On the other hand, if N is a member of M, since M and N are mutually exclusive classes, N does not belong to N. Hence N is not a member of itself and should, by virtue of the definition of N, belong to N.

The cause of all these paradoxes, as Russell and Whitehead point out, is that an object is defined in terms of a class of objects that contains the object being defined. Such definitions are also called impredicative and occur particularly in set theory. This type of definition is also used, as Zermelo noted in 1908, to define the lower bound of a set of numbers and to define other concepts of analysis. Hence classical analysis contains paradoxes.

Cantor's proof of the nondenumerability of the set of real numbers (Chap. 41, sec. 7) also uses such an impredicative set. A one-to-one correspondence is assumed to hold between the set of *all* positive integers and the set M of all real numbers. Then to each integer k, there corresponds the set $f(k)$. Now $f(k)$ does or does not contain k. Let N be the set of all k such that k does not belong to $f(k)$. This set N (taken in some order) is a real number. Hence there should be, by the initial one-to-one correspondence, an integer n such that n corresponds to N. Now if n belongs to N, it should not by the definition of N. If n does not belong to N, then by the definition of N it should belong to N. The definition of the set N is impredicative because k belongs to N if and only if there exists a set K in M such that $K = f(k)$ and k does not belong to K. Thus in defining N we make use of the totality M of sets which contains N as a member. That is, to define N, N must already be in the set M.

3. *Abhandlungen der Friesschen Schule*, 2, 1908, 301–24.

It is rather easy to fall unwittingly into the trap of introducing impredicative definitions. Thus if one defines the class of all classes that contain more than five members, he has defined a class that contains itself. Likewise the statement, the set S of all sets definable in twenty-five or fewer words, defines S impredicatively.

These paradoxes jolted the mathematicians, while they compromised not only set theory but large portions of classical analysis. Mathematics as a logical structure was in a sad state, and mathematicians looked back longingly to the happier days before the paradoxes were recognized.

3. The Axiomatization of Set Theory

It is perhaps not surprising that the mathematicians' first recourse was to axiomatize Cantor's rather freely formulated and, as some are wont to say today, naive set theory. The axiomatization of geometry and the number system had resolved logical problems in those areas, and it seemed likely that axiomatization would clarify the difficulties in set theory. The task was first undertaken by the German mathematician Ernst Zermelo who believed that the paradoxes arose because Cantor had not restricted the concept of a set. Cantor in 1895[4] had defined a set as a collection of distinct objects of our intuition or thought. This was rather vague, and Zermelo therefore hoped that clear and explicit axioms would clarify what is meant by a set and what properties sets should have. Cantor himself was not unaware that his concept of a set was troublesome. In a letter to Dedekind of 1899[5] he had distinguished between consistent and inconsistent sets. Zermelo thought he could restrict his sets to Cantor's consistent ones and these would suffice for mathematics. His axiom system[6] contained fundamental concepts and relations that are defined only by the statements in the axioms themselves. Among such concepts was the notion of a set itself and the relation of belonging to a set. No properties of sets were to be used unless granted by the axioms. The existence of an infinite set and such operations as the union of sets and the formation of subsets were also provided for in the axioms. Notably Zermelo included the axiom of choice (Chap. 41, sec. 8).

Zermelo's plan was to admit into set theory only those classes that seemed least likely to generate contradictions. Thus the null class, any finite class, and the class of natural numbers seemed safe. Given a safe class, certain classes formed from it, such as any subclass, the union of safe classes, and the class of all subsets of a safe class, should be safe classes. However, he avoided complementation for, while x might be a safe class,

4. *Math. Ann.*, 46, 1895, 481–512 = *Ges. Abh.*, 282–356.
5. *Ges. Abh.*, 443–48.
6. *Math. Ann.*, 65, 1908, 261–81.

the complement of x, that is, all non-x, in some large universe of objects might not be safe.

Zermelo's development of set theory was improved by Abraham A. Fraenkel (1891–1965).[7] Additional changes were made by von Neumann.[8] The hope of avoiding the paradoxes rests in the case of the Zermelo-Fraenkel system on restricting the types of sets that are admitted while admitting enough to serve the foundations of analysis. Von Neumann's idea was a little more daring. He makes the distinction between classes and sets. Classes are sets so large that they are not contained in other sets or classes, whereas sets are more restricted classes and may be members of a class. Thus sets are the safe classes. As von Neumann pointed out, it was not the admission of the classes that led to contradictions but their being treated as members of other classes.

Zermelo's formal set theory, as modified by Fraenkel, von Neumann, and others, is adequate for developing the set theory required for practically all of classical analysis and avoids the paradoxes to the extent that, as yet, no one has discovered any within the theory. However, the consistency of the axiomatized set theory has not been demonstrated. Apropos of the open question of consistency Poincaré remarked, "We have put a fence around the herd to protect it from the wolves but we do not know whether some wolves were not already within the fence."

Beyond the problem of consistency, the axiomatization of set theory used the axiom of choice, which is needed to establish parts of standard analysis, topology, and abstract algebra. This axiom was considered objectionable by a number of mathematicians, among them Hadamard, Lebesgue. Borel, and Baire, and in 1904, when Zermelo used it to prove the well-ordering theorem (Chap. 41, sec. 8), a host of objections flooded the journals.[9] The questions of whether this axiom was essential and whether it was independent of the others were raised and remained unanswered for some time (see sec. 8).

The axiomatization of set theory, despite the fact that it left open such questions as consistency and the role of the axiom of choice, might have put mathematicians at ease with respect to the paradoxes and have led to a decline in the interest in foundations. But by this time several schools of thought on the foundations of mathematics, no doubt stirred into life by the paradoxes and the problem of consistency, had become active and contentious. To the proponents of these philosophies the axiomatic method as practiced by Zermelo and others was not satisfactory. To some it was objec-

7. *Math. Ann.*, 86, 1921/22, 230–37, and many later papers.
8. *Jour. für Math.*, 154, 1925, 219–40, and later papers.
9. The views of these men are expressed in a famous exchange of letters. See the *Bull. Soc. Math. de France*, 33, 1905, 261–73. Also in E. Borel: *Leçons sur la théorie des fonctions*, Gauthier-Villars, 4th ed., 1950, 150–58.

tionable because it presupposed the logic it used, whereas by this time logic itself and its relation to mathematics was under investigation. Others, more radical, objected to the reliance upon any kind of logic, particularly as it was applied to infinite sets. To understand the arguments that various schools of thought propounded, we must go back in time somewhat.

4. The Rise of Mathematical Logic

One development that caused new controversies as well as dissatisfaction with the axiomatization of set theory concerns the role of logic in mathematics; it arose from the nineteenth-century mathematization of logic. This development has its own history.

The power of algebra to symbolize and even mechanize geometrical arguments had impressed both Descartes and Leibniz, among others (Chap. 13, sec. 8), and both envisioned a broader science than the algebra of numbers. They contemplated a general or abstract science of reasoning that would operate somewhat like ordinary algebra but be applicable to reasoning in all fields. As Leibniz put it in one of his papers, "The universal mathematics is, so to speak, the logic of the imagination," and ought to treat "all that which in the domain of the imagination is susceptible of exact determination." With such a logic one might build any edifice of thought from its simple elements to more and more complicated structures. The universal algebra would be part of logic but an algebraicized logic. Descartes began modestly by attempting to construct an algebra of logic; an incomplete sketch of this work is extant.

In pursuit of the same broad goal as Descartes's, Leibniz launched a more ambitious program. He had paid attention to logic throughout his life and rather early became entranced with the scheme of the scientist and theologian Raymond Lull (1235–1315) whose book *Ars Magna et Ultima* offered a naive mechanical method for producing new ideas by combining existing ones but who did have the concept of a universal science of logic that would be applicable to all reasoning. Leibniz broke from scholastic logic and from Lull but became impressed with the possibility of a broad calculus that would enable man to reason in all fields mechanically and effortlessly. Leibniz says of his plan for a universal symbolic logic that such a science, of which ordinary algebra is but a small part, would be limited only by the necessity of obeying the laws of formal logic. One could name it, he said, "algebraico-logical synthesis."

This general science was to provide, first of all, a rational, universal language that would be adapted to thinking. The concepts, having been resolved into primitive distinct and nonoverlapping ones, could be combined in an almost mechanical way. He also thought that symbolism would be necessary in order to keep the mind from getting lost. Here the influence of algebraic symbolism on his thinking is clear. He wanted a symbolic language capable of expressing human thoughts unambiguously and aiding deduction. This symbolic language was his "universal characteristic."

In 1666 Leibniz wrote his *De Arte Combinatoria*,[10] which contains among other matters his early plans for his universal system of reasoning. He then wrote numerous fragments, which were never published but which are available in the edition of his philosophical writings (see note 10). In his first attempt he associated with each primitive concept a prime number; any concept composed of several primitive ones was represented by the product of the corresponding primes. Thus if 3 represents "man" and 7 "rational," 21 would represent "rational man." He then sought to translate the usual rules of the syllogism into this scheme but was not successful. He also tried, at another time, to use special symbols in place of prime numbers, where again complex ideas would be represented by combinations of symbols. Actually Leibniz thought that the number of primitive ideas would be few, but this proved to be erroneous. Also one basic operation, conjunction, for compounding primitive ideas did not suffice.

He also commenced work on an algebra of logic proper. Directly and indirectly Leibniz had in his algebra concepts we now describe as logical addition, multiplication, identity, negation, and the null class. He also called attention to the desirability of studying abstract relations such as inclusion, one-to-one correspondence, many-to-one correspondences, and equivalence relations. Some of these, he recognized, have the properties of symmetry and transitivity. Leibniz did not complete this work; he did not get beyond the syllogistic rules, which he himself recognized do not encompass all the logic mathematics uses. Leibniz described his ideas to l'Hospital and others, but they paid no attention. His logical works remained unedited until the beginning of the twentieth century and so had little direct influence. During the eighteenth and early nineteenth centuries a number of men sketched attempts similar to Leibniz's but got no further than he had.

A more effective if less ambitious step was taken by Augustus De Morgan. De Morgan published *Formal Logic* (1847) and many papers, some of which appeared in the *Transactions of the Cambridge Philosophical Society*. He sought to correct defects of and improve on Aristotelian logic. In his *Formal Logic* he added a new principle to Aristotelian logic. In the latter the premises "Some M's are A's" and "Some M's are B's" permit no conclusion; and, in fact, this logic says that the middle term M must be used universally, that is, "All M's" must occur. But De Morgan pointed out that from "Most M's are A's" and "Most M's are B's," it follows of necessity that "Some A's are B's." De Morgan put this fact in quantitative form. If there are m

10. Pub. 1690 = G. W. Leibniz: *Die philosophischen Schriften*, ed. by C. I. Gerhardt, 1875–90, Vol. 4, 27–102.

of the M's, and a of the M's are in A and b of the M's are in B, then there are at least $(a + b - m)$ A's that are B's. The point of De Morgan's observation is that the terms may be quantified. He was able as a consequence to introduce many more valid forms of the syllogism. Quantification also eliminated a defect in Aristotelian logic. The conclusion "Some A's are B's" which in Aristotelian logic can be drawn from "All A's are B's" implies the existence of A's but they need not exist.

De Morgan also initiated the study of the logic of relations. Aristotelian logic is devoted primarily to the relationship "to be" and either asserts or denies this relationship. As De Morgan pointed out, this logic could not prove that if a horse is an animal, then a horse's tail is an animal's tail. It certainly could not handle a relation such as x loves y. De Morgan introduced symbolism to handle relations but did not carry this subject very far.

In the area of symbolic logic De Morgan is widely known for what are now called De Morgan's laws. As he stated them,[11] the contrary of an aggregate is the compound of the contraries of the aggregates; the contrary of a compound is the aggregate of the contraries of the components. In logical notation these laws read

$$1 - (x + y) = (1 - x)(1 - y).$$
$$1 - xy = (1 - x) + (1 - y).$$

The contribution of symbolism to an algebra of logic was the major step made by George Boole (1815–1864), who was largely self-taught and became professor of mathematics at Queens College in Cork. Boole was convinced that the symbolization of language would rigorize logic. His *Mathematical Analysis of Logic*, which appeared on the same day as De Morgan's *Formal Logic*, and his *An Investigation of the Laws of Thought* (1854) contain his major ideas.

Boole's approach was to emphasize extensional logic, that is, a logic of classes, wherein sets or classes were denoted by $x, y, z, \ldots$ whereas the symbols $X, Y, Z, \ldots$ represented individual members. The universal class was denoted by 1 and the empty or null class by 0. He used xy to denote the intersection of two sets (he called the operation election), that is, the set of elements common to both x and y, and $x + y$ to denote the set consisting of all the elements in x and in y. (Strictly for Boole addition or union applied only to disjoint sets; W. S. Jevons [1835–82] generalized the concept.) The complement of x is denoted by $1 - x$. More generally $x - y$ is the class of x's that are not y's. The relation of inclusion, that is, x is contained in y, he wrote as $xy = x$. The equal sign denoted the identity of the two classes.

Boole believed that the mind grants to us at once certain elementary processes of reasoning that are the axioms of logic. For example, the law

11. *Trans. Camb. Phil. Soc.*, 10, 1858, 173–230.

of contradiction, that A cannot be both B and not B, is axiomatic. This is expressed by

$$x(1 - x) = 0.$$

It is also obvious to the mind that

$$xy = yx$$

and so this commutative property of intersection is another axiom. Equally obvious is the property

$$xx = x.$$

This axiom is a departure from ordinary algebra. Boole also accepted as axiomatic that

$$x + y = y + x$$

and

$$x(u + v) = xu + xv.$$

With these axioms the law of excluded middle could be stated in the form

$$x + (1 - x) = 1;$$

that is, everything is x or not x. Every X is Y becomes $x(1 - y) = 0$. No X is Y reads as $xy = 0$. Some X are Y is denoted by $xy \neq 0$ and some X are not Y by $x(1 - y) \neq 0$.

From the axioms Boole planned to deduce the laws of reasoning by applying the processes permitted by the axioms. As trivial conclusions he had that $1 \cdot x = x$ and $0 \cdot x = 0$. A slightly more involved argument is illustrated by the following. From

$$x + (1 - x) = 1$$

it follows that

$$z[x + (1 - x)] = z \cdot 1$$

and then that

$$zx + z(1 - x) = z.$$

Thus the class of objects z consists of those that are in x and those that are in $1 - x$.

Boole observed that the calculus of classes could be interpreted as a calculus of propositions. Thus if x and y are propositions instead of classes, then xy is the joint assertion of x and y and $x + y$ is the assertion of x or y or both. The statement $x = 1$ would mean that the proposition x is true and $x = 0$ that x is false. $1 - x$ would mean the denial of x. However Boole did not get very far with his calculus of propositions.

Both De Morgan and Boole can be regarded as the reformers of Aristotelian logic and the initiators of an algebra of logic. The effect of their work was to build a science of logic which thenceforth was detached from philosophy and attached to mathematics.

The calculus of propositions was advanced by Charles S. Peirce. Peirce distinguished between a proposition and a propositional function. A proposition, John is a man, contains only constants. A propositional function, x is a man, contains variables. Whereas a proposition is true or false, a propositional function is true for some values of the variable and false for others. Peirce also introduced propositional functions of two variables, for example, x knows y.

The men who had built symbolic logic thus far were interested in logic and in mathematizing that subject. With the work of Gottlob Frege (1848–1925), professor of mathematics at Jena, mathematical logic takes a new direction, the one that is pertinent to our account of the foundations of mathematics. Frege wrote several major works, *Begriffsschrift* (Calculus of Concepts, 1879), *Die Grundlagen der Arithmetik* (The Foundation of Arithmetic, 1884), and *Grundgesetze der Arithmetik* (The Fundamental Laws of Arithmetic; Vol. 1, 1893; Vol. 2, 1903). His works are characterized by precision and thoroughness of detail.

In the area of logic proper, Frege expanded on the use of variables, quantifiers, and propositional functions; most of this work was done independently of his predecessors, including Peirce. In his *Begriffsschrift* Frege gave an axiomatic foundation to logic. He introduced many distinctions that acquired great importance later, for example the distinction between the statement of a proposition and the assertion that it is true. The assertion is denoted by placing the symbol ⊢ in front of the proposition. He also distinguished between an object x and the set $\{x\}$ containing just x, and between an object belonging to a set and the inclusion of one set in another. Like Peirce, he used variables and propositional functions, and he indicated the quantification of his propositional functions, that is, the domain of the variable or variables for which they are true. He also introduced (1879) the concept of material implication: A implies B means either that A is true and B is true or A is false and B is true or A is false and B is false. This interpretation of implication is more convenient for mathematical logic. The logic of relations was also taken up by Frege; thus the relation of order involved in stating, for example, that a is greater than b was important in his work.

Having built up logic on explicit axioms, he proceeded in his *Grundlagen* to his real goal, to *build mathematics as an extension of logic*. He expressed the concepts of arithmetic in terms of the logical concepts. Thus the definitions and laws of number were derived from logical premises. We shall examine this construction in connection with the work of Russell and Whitehead.

Unfortunately Frege's symbolism was quite complex and strange to mathematicians. His work was in fact not well known until it was discovered by
Russell. Rather ironic too is the fact that, just as the second volume of
Grundgesetze was about to go to press, he received a letter from Russell, who
informed him of the paradoxes of set theory. At the close of Volume 2
(p. 253), Frege remarks, "A scientist can hardly meet with anything more
undesirable than to have the foundation give way just as the work is finished.
I was put in this position by a letter from Mr. Bertrand Russell when the
work was nearly through the press."

5. *The Logistic School*

We left the account of the work on the foundations of mathematics at the
point where the axiomatization of set theory had provided a foundation
that avoided the known paradoxes and yet served as a logical basis for the
existing mathematics. We did point out that this approach was not satisfactory to many mathematicians. That the consistency of the real number
system and the theory of sets remained to be proved was acknowledged
by all; consistency was no longer a minor matter. The use of the axiom of
choice was controversial. But beyond these problems was the overall question
of what the proper foundation for mathematics was. The axiomatic movement of the late nineteenth century and the axiomatization of set theory
had proceeded on the basis that the logic employed by mathematics could
be taken for granted. But by the early nineteen-hundreds there were several
schools of thought that were no longer content with this presupposition.
The school led by Frege sought to rebuild logic and to build mathematics
on logic. This plan, as we have already noted, was set back by the appearance
of the paradoxes, but it was not abandoned. It was in fact independently
conceived and pursued by Bertrand Russell and Alfred North Whitehead.
Hilbert, already impressed by the need to establish consistency, began to
formulate his own systematic foundation for mathematics. Still another
group of mathematicians, known as intuitionists, was dissatisfied with the
concepts and proofs introduced in late nineteenth-century analysis. These
men were adherents of a philosophical position that not only could not be
reconciled with some of the methodology of analysis but also challenged the
role of logic. The development of these several philosophies was the major
undertaking in the foundations of mathematics; its outcome was to open up
the entire question of the nature of mathematics. We shall examine each
of these three major schools of thought.

The first of these is known as the logistic school and its philosophy is
called logicism. The founders are Russell and Whitehead. Independently
of Frege, they had the idea that mathematics is derivable from logic and
therefore is an extension of logic. The basic ideas were sketched by Russell

in his *Principles of Mathematics* (1903); they were developed in the detailed work by Whitehead and Russell, the *Principia Mathematica* (3 vols., 1910–13). Since the *Principia* is the definitive version, we shall base our account on it.

This school starts with the development of logic itself, from which mathematics follows without any axioms of mathematics proper. The development of logic consists in stating some axioms of logic, from which theorems are deduced that may be used in subsequent reasoning. Thus the laws of logic receive a formal derivation from axioms. The *Principia* also has undefined ideas, as any axiomatic theory must have since it is not possible to define all the terms without involving an infinite regress of definitions. Some of these undefined ideas are the notion of an elementary proposition, the notion of a propositional function, the assertion of the truth of an elementary proposition, the negation of a proposition, and the disjunction of two propositions.

Russell and Whitehead explain these notions, though, as they point out, this explanation is not part of the logical development. By a proposition they mean simply any sentence stating a fact or a relationship: for example, John is a man; apples are red; and so forth. A propositional function contains a variable, so that substitution of a value for that variable gives a proposition. Thus "X is an integer" is a propositional function. The negation of a proposition is intended to mean, "It is not true that the proposition holds," so that if p is the proposition that John is a man, the negation of p, denoted by $\sim p$, means, "It is not true that John is a man," or "John is not a man." The disjunction of two propositions p and q, denoted by $p \vee q$, means p or q. The meaning of "or" here is that intended in the sentence, "Men or women may apply." That is, men may apply; women may apply; and both may apply. In the sentence, "That person is a man or a woman," "or" has the more common meaning of either one or the other but not both. Mathematics uses "or" in the first sense, though sometimes the second sense is the only one possible. For example, "the triangle is isosceles or the quadrilateral is a parallelogram" illustrates the first sense. We also say that every number is positive or negative. Here additional facts about positive and negative numbers say that both cannot be true. Thus the assertion $p \vee q$ means p and q, $\sim p$ and q, or p and $\sim q$.

A most important relationship between propositions is implication, that is, the truth of one proposition compelling the truth of another. In the *Principia* implication, $p \supset q$, is defined by $\sim p \vee q$, which in turn means $\sim p$ and q, p and q, or $\sim p$ and $\sim q$. As an illustration consider the implication, If X is a man, then X is mortal. Here the state of affairs could be

X is not a man and X is mortal;
X is a man and X is mortal;
X is not a man and X is not mortal.

Any one of these possibilities is allowable. What the implication forbids is

X is a man and X is not mortal.

Some of the postulates of the *Principia* are:

(a) Anything implied by a true elementary proposition is true.
(b) $(p \lor p) \supset p$.
(c) $q \supset (p \lor q)$.
(d) $(p \lor q) \supset (q \lor p)$.
(e) $[p \lor (q \lor r)] \supset [q \lor (p \lor r)]$.
(f) The assertion of p and the assertion $p \supset q$ permits the assertion of q.

The independence of these postulates and their consistency cannot be proved because the usual methods do not apply. From these postulates the authors proceed to deduce theorems of logic and ultimately arithmetic and analysis. The usual syllogistic rules of Aristotle occur as theorems.

To illustrate how even logic itself has been formalized and made deductive, let us note a few theorems of the early part of *Principia Mathematica*:

2.01. $(p \supset \sim p) \supset \sim p$.

This is the principle of *reductio ad absurdum*. In words, if the assumption of p implies that p is false, then p is false.

2.05. $[q \supset r] \supset [(p \supset q) \supset (p \supset r)]$.

This is one form of the syllogism. In words, if q implies r, then if p implies q, p implies r.

2.11. $p \lor \sim p$.

This is the principle of excluded middle: p is true or p is false.

2.12. $p \supset \sim (\sim p)$.

In words, p implies that not-p is false.

2.16. $(p \supset q) \supset (\sim q \supset \sim p)$.

If p implies q, then not-q implies not-p.

Propositions are a step to propositional functions that treat sets by means of properties rather than by naming the objects in a set. Thus the propositional function "x is red" denotes the set of all red objects.

If the members of a set are individual objects, then the propositional functions applying to such members are said to be of type 0. If the members of a set are themselves propositional functions, any propositional function applying to such members is said to be of type 1. And generally propositional functions whose variables are of types less than and equal to n are of type $n + 1$.

The theory of types seeks to avoid the paradoxes, which arise because a collection of objects contains a member that itself can be defined only in terms of the collection. The resolution by Russell and Whitehead of this difficulty was to require that "whatever involves all members of a collection must not itself be a member of the collection." To carry out this restriction in the *Principia*, they specify that a (logical) function cannot have as one of its arguments anything defined in terms of the function itself. They then discuss the paradoxes and show that the theory of types avoids them.

However, the theory of types leads to classes of statements that must be carefully distinguished by type. If one attempts to build mathematics in accordance with the theory of types, the development becomes exceedingly complex. For example, in the *Principia* two objects a and b are equal if for every property $P(x)$, $P(a)$ and $P(b)$ are equivalent propositions (each implies the other). According to the theory of types, P may be of different types because it may contain variables of various orders as well as the individual objects a or b, and so the definition of equality must apply for all types of P; in other words, there is an infinity of relations of equality, one for each type of property. Likewise, an irrational number defined by the Dedekind cut proves to be of higher type than a rational number, which in turn is of higher type than a natural number, and so the continuum consists of numbers of different types. To escape this complexity, Russell and Whitehead introduced the axiom of reducibility, which affirms the existence, for each propositional function of whatever type, of an equivalent propositional function of type zero.

Having treated propositional functions, the authors take up the theory of classes. A class, loosely stated, is the set of objects satisfying some propositional function. Relations are then expressed as classes of couples satisfying propositional functions of two variables. Thus "x judges y" expresses a relation. On this basis the authors are prepared to introduce the notion of cardinal number.

The definition of a cardinal number is of considerable interest. It depends upon the previously introduced relation of one-to-one correspondence between classes. If two classes are in one-to-one correspondence, they are called similar. The relationship of similarity is proven to be reflexive, symmetric, and transitive. All similar classes possess a common property, and this is their number. However, similar classes may have more than one common property. Russell and Whitehead get around this, as had Frege, by defining the number of a class as the class of all classes that are similar to the given class. Thus the number 3 is the class of all three-membered classes and the denotation of all three-membered classes is $\{x, y, z\}$ with $x \neq y \neq z$. Since the definition of number presupposes the concept of one-to-one correspondence, it would seem as though the definition is circular. The authors point out, however, that a relation is one-to-one if, when x

and x' have the relation to y, then x and x' are identical, and when x has the relation to y and y', then y and y' are identical. Hence the concept of one-to-one correspondence does not involve the number 1.

Given the cardinal or natural numbers, it is possible to build up the real and complex number systems, functions, and in fact all of analysis. Geometry can be introduced through numbers. Though the details in the *Principia* differ somewhat, our own examination of the foundations of the number system and of geometry (Chaps. 41 and 42) shows that such constructions are logically possible without additional axioms.

This, then, is the grand program of the logistic school. What it does with logic itself is quite a story, which we are skimming over here briefly. What it does for mathematics, and this we must emphasize, is to found mathematics on logic. No axioms of mathematics are needed; mathematics becomes no more than a natural extension of the laws and subject matter of logic. But the postulates of logic and all their consequences are arbitrary and, moreover, formal. That is, they have no content; they have merely form. As a consequence, mathematics too has no content, but merely form. The physical meanings we attach to numbers or to geometric concepts are not part of mathematics. It was with this in mind that Russell said that mathematics is the subject in which we never know what we are talking about nor whether what we are saying is true. Actually, when Russell started this program in the early part of the century, he (and Frege) thought the axioms of logic were truths. But he abandoned this view in the 1937 edition of the *Principles of Mathematics*.

The logistic approach has received much criticism. The axiom of reducibility aroused opposition, for it is quite arbitrary. It has been called a happy accident and not a logical necessity; it has been said that the axiom has no place in mathematics, and that what cannot be proved without it cannot be regarded as proved at all. Others called the axiom a sacrifice of the intellect. Moreover, the system of Russell and Whitehead was never completed and is obscure in numerous details. Many efforts were made later to simplify and clarify it.

Another serious philosophical criticism of the entire logistic position is that if the logistic view is correct, then all of mathematics is a purely formal, logico-deductive science whose theorems follow from the laws of thought. Just how such a deductive elaboration of the laws of thought can represent wide varieties of natural phenomena such as acoustics, electromagnetics, and mechanics seems unexplained. Further, in the creation of mathematics perceptual or imaginative intuition must supply new concepts, whether or not derived from experience. Otherwise, how could new knowledge arise? But in the *Principia* all concepts reduce to logical ones.

The formalization of the logistic program apparently does not represent mathematics in any real sense. It presents us with the husk, not the kernel.

Poincaré said, snidely (*Foundations of Science*, p. 483), "The logistic theory is not sterile; it engenders contradictions." This is not true if one accepts the theory of types, but this theory, as noted, is artificial. Weyl also attacked logicism; he said that this complex structure "taxes the strength of our faith hardly less than the doctrines of the early Fathers of the Church or of the Scholastic philosophers of the Middle Ages."

Despite the criticisms, the logistic philosophy is accepted by many mathematicians. The Russell-Whitehead construction also made a contribution in another direction. It carried out a thorough axiomatization of logic in entirely symbolic form and so advanced enormously the subject of mathematical logic.

6. The Intuitionist School

A radically different approach to mathematics has been undertaken by a group of mathematicians called intuitionists. As in the case of logicism, the intuitionist philosophy was inaugurated during the late nineteenth century when the rigorization of the number system and geometry was a major activity. The discovery of the paradoxes stimulated its further development.

The first intuitionist was Kronecker, who expressed his views in the 1870s and 80s. To Kronecker, Weierstrass's rigor involved unacceptable concepts, and Cantor's work on transfinite numbers and set theory was not mathematics but mysticism. Kronecker was willing to accept the whole numbers because these are clear to the intuition. These "were the work of God." All else was the work of man and suspect. In his essay of 1887, "Über den Zahlbegriff" (On the Number Concept),[12] he showed how some types of numbers, fractions for example, could be defined in terms of the whole numbers. Fractional numbers as such were acceptable as a convenience of notation. The theory of irrational numbers and of continuous functions he wished to strip away. His ideal was that every theorem of analysis should be interpretable as giving relations among the integers only.

Another objection Kronecker made to many parts of mathematics was that they did not give constructive methods or criteria for determining in a finite number of steps the objects with which they dealt. Definitions should contain the means of calculating the object defined in a finite number of steps, and existence proofs should permit the calculation to any required degree of accuracy of the quantity whose existence is being established. Algebraists were content to say that a polynomial $f(x)$ may have a rational factor, in which case $f(x)$ is reducible. In the contrary case it is irreducible. In his *Festschrift* "Grundzüge einer arithmetischen Theorie der algebraischen Grössen" (Elements of an Arithmetic Theory of Algebraic Quantities)[13]

---

12. *Jour. für Math.*, 101, 1887, 337–55 = *Werke*, 3, 251–74.
13. *Jour. für Math.*, 92, 1882, 1–122 = *Werke*, 2, 237–387.

Kronecker said, "The definition of reducibility is devoid of a sure foundation until a *method* is given by means of which it can be decided whether a given function is irreducible or not."

Again, though the several theories of the irrational numbers give definitions as to when two real numbers *a* and *b* are equal or when *a* > *b* or *b* > *a*, they do not give criteria to determine which alternative holds in a given case. Hence Kronecker objected to such definitions. They are definitions only in appearance. The entire theory of irrationals was unsatisfactory to him and one day he said to Lindemann, who had proved that π is a transcendental irrational, "Of what use is your beautiful investigation regarding π? Why study such problems, since irrational numbers are non-existent?"

Kronecker himself did little to develop the intuitionist philosophy except to criticize the absence of constructive procedures for determining quantities whose existence was merely established. He tried to rebuild algebra but made no efforts to reconstruct analysis. Kronecker produced fine work in arithmetic and algebra which did not conform to his own requirements because, as Poincaré remarked,[14] he temporarily forgot his own philosophy.

Kronecker had no supporters of his philosophy in his day and for almost twenty-five years no one pursued his ideas. However, after the paradoxes were discovered, intuitionism was revived and became a widespread and serious movement. The next strong advocate was Poincaré. His opposition to set theory because it gave rise to paradoxes has already been noted. Nor would he accept the logistic program for rescuing mathematics. He ridiculed attempts to base mathematics on logic on the ground that mathematics would reduce to an immense tautology. He also mocked the (to him) highly artificial derivation of number. Thus in the *Principia* 1 is defined as $\hat{\alpha}\{\exists x \cdot \alpha = i\,{}^{\prime}x\}$. Poincaré said sarcastically that this was an admirable definition to give to people who never heard of the number 1.

In *Science and Method* (*Foundations of Science*, p. 480) he stated,

> Logistic has to be made over, and one is none too sure of what can be saved. It is unnecessary to add that only Cantorism and Logistic are meant; true mathematics, that which serves some useful purpose, may continue to develop according to its own principles without paying any attention to the tempests raging without, and it will pursue step by step its accustomed conquests which are definitive and which it will never need to abandon.

Poincaré objected to concepts that cannot be defined in a finite number of words. Thus a set chosen in accordance with the axiom of choice is not really defined when a choice has to be made from each of a transfinite num-

14. *Acta Math.*, 22, 1899, 17.

ber of sets. He also contended that arithmetic cannot be justified by an axiomatic foundation. Our intuition precedes such a structure. In particular, mathematical induction is a fundamental intuition and not just an axiom that happens to be useful in some system of axioms. Like Kronecker, he insisted that all definitions and proof should be constructive.

He agreed with Russell that the source of the paradoxes was the definition of collections or sets that included the object defined. Thus the set A of all sets contains A. But A cannot be defined until each member of A is defined, and if A is one member the definition is circular. Another example of an impredicative definition is the definition of the maximum value of a continuous function defined over a closed interval as the largest value that the function takes on in this interval. Such definitions were common in analysis and especially in the theory of sets.

Further criticisms of the current logical state of mathematics were developed and discussed in an exchange of letters among Borel, Baire, Hadamard, and Lebesgue.[15] Borel supported Poincaré's assertion that the integers cannot be founded axiomatically. He too criticized the axiom of choice because it calls for a nondenumerable infinity of choices, which is inconceivable to the intuition. Hadamard and Lebesgue went further and said that even a denumerable infinity of arbitrary successive choices is not more intuitive because it calls for an infinity of operations, which it is impossible to conceive as being effectively realized. For Lebesgue the difficulties all reduced to knowing what one means when one says that a mathematical object exists. In the case of the axiom of choice he argued that if one merely "thinks" of a way of choosing, may one then not change his choices in the course of his reasoning? Even the choice of a single object in one set, Lebesgue maintained, raises the same difficulties. One must know the object "exists" which means that one must name the choice explicitly. Thus Lebesgue rejected Cantor's proof of the existence of transcendental numbers. Hadamard pointed out that Lebesgue's objections led to a denial of the existence of the set of all real numbers, and Borel drew exactly the same conclusion.

All of the above objections by intuitionists were sporadic and fragmented. The systematic founder of modern intuitionism is Brouwer. Like Kronecker, much of his mathematical work, notably in topology, was not in accord with his philosophy, but there is no question as to the seriousness of his position. Commencing with his doctoral dissertation, *On the Foundations of Mathematics* (1907), Brouwer began to build up the intuitionist philosophy. From 1918 on he expanded and expounded his views in papers in various journals, including the *Mathematische Annalen* of 1925 and 1926.

Brouwer's intuitionist position stems from a broader philosophy. The fundamental intuition, according to Brouwer, is the occurrence of perceptions in a time sequence. "Mathematics arises when the subject of twoness,

15. See note 9.

which results from the passage of time, is abstracted from all special occurrences. The remaining empty form [the relation of n to $n + 1$] of the common content of all these twonesses becomes the original intuition of mathematics and repeated unlimitedly creates new mathematical subjects." Thus by unlimited repetition the mind forms the concept of the successive natural numbers. This idea that the whole numbers derive from the intuition of time had been maintained by Kant, William R. Hamilton in his "Algebra as a Science of Time," and the philosopher Arthur Schopenhauer.

Brouwer conceives of mathematical thinking as a process of construction that builds its own universe, independent of the universe of our experience and somewhat as a free design, restricted only in so far as it is based upon the fundamental mathematical intuition. This fundamental intuitive concept must not be thought of as an undefined idea, such as occurs in postulational theories, but rather as something *in terms of which* all undefined ideas that occur in the various mathematical systems are to be intuitively conceived, if they are indeed to serve in mathematical thinking.

Brouwer holds that "in this constructive process, bound by the obligation to notice with care which theses are acceptable to the intuition and which are not, lies the only possible foundation for mathematics." Mathematical ideas are imbedded in the human mind *prior to language, logic, and experience.* The intuition, not experience or logic, determines the soundness and acceptability of ideas. It must of course be remembered that these statements concerning the role of experience are to be taken in the philosophical sense, not the historical sense.

The mathematical objects are for Brouwer acquired by intellectual construction, wherein the basic numbers $1, 2, 3, \ldots$ furnish the prototype of such a construction. The possibility of the unlimited repetition of the empty form, the step from n to $n + 1$, leads to infinite sets. However, Brouwer's infinite is the potential infinity of Aristotle, whereas modern mathematics, as founded for example by Cantor, makes extensive use of actually infinite sets whose elements are all present "at once."

In connection with the intuitionist notion of infinite sets, Weyl, who belonged to the intuitionist school, says that

> . . . the sequence of numbers which grows beyond any stage already reached . . . is a manifold of possibilities opening to infinity; it remains forever in the status of creation, but is not a closed realm of things existing in themselves. That we blindly converted one into the other is the true source of our difficulties, including the antinomies—a source of more fundamental nature than Russell's vicious circle principle indicated. Brouwer opened our eyes and made us see how far classical mathematics, nourished by a belief in the absolute that transcends all human possibilities of realization, goes beyond such statements as can claim real meaning and truth founded on evidence.

The world of mathematical intuition is opposed to the world of causal perceptions. In this causal world, not in mathematics, belongs language, which serves there for the understanding of common dealings. Words or verbal connections are used to communicate truths. Language serves to evoke copies of ideas in men's minds by symbols and sounds. But thoughts can never be completely symbolized. These remarks apply also to mathematical language, including symbolic language. Mathematical ideas are independent of the dress of language and in fact far richer.

Logic belongs to language. It offers a system of rules that permit the deduction of further verbal connections and also are intended to communicate truths. However, these latter truths are not such before they are experienced, nor is it guaranteed that they can be experienced. Logic is not a reliable instrument to uncover truths and can deduce no truths that are not obtainable just as well in some other way. Logical principles are the regularity observed a posteriori in the language. They are a device for manipulating language, or they are the theory of representation of language. The most important advances in mathematics are not obtained by perfecting the logical form but by modifying the basic theory itself. Logic rests on mathematics, not mathematics on logic.

Since Brouwer does not recognize any a priori obligatory logical principles, he does not recognize the mathematical task of deducing conclusions from axioms. Mathematics is not bound to respect the rules of logic, and for this reason the paradoxes are unimportant even if we were to accept the mathematical concepts and constructions the paradoxes involve. Of course, as we shall see, the intuitionists do not accept all these concepts and proofs.

Weyl[16] expands on the role of logic:

> According to his [Brouwer's] view and reading of history, classical logic was abstracted from the mathematics of finite sets and their subsets. . . . Forgetful of this limited origin, one afterwards mistook that logic for something above and prior to all mathematics, and finally applied it, without justification, to the mathematics of infinite sets. This is the Fall and original sin of set theory, for which it is justly punished by the antinomies. It is not that such contradictions showed up that is surprising, but that they showed up at such a late stage of the game.

In the realm of logic there are some clear, *intuitively* acceptable logical principles or procedures that can be used to assert new theorems from old ones. These principles are part of the fundamental mathematical intuition. However, not all logical principles are acceptable to the basic intuition and one must be critical of what has been sanctioned since the days of Aristotle. Because mathematicians have applied freely these Aristotelian laws, they

16. *Amer. Math. Monthly*, 53, 1946, 2–13 = *Ges. Abh.*, 4, 268–79.

have produced antinomies. The intuitionists therefore proceed to analyze which logical principles are allowable in order that the usual logic conform to and properly express the correct intuitions.

As a specific example of a logical principle that is applied too freely, Brouwer cites the law of excluded middle. This principle, which asserts that every meaningful statement is true or false, is basic to the indirect method of proof. It arose historically in the application of reasoning to subsets of finite sets and was abstracted therefrom. It was then accepted as an independent a priori principle and was unjustifiably applied to infinite sets. Whereas for finite sets one can decide whether all elements possess a certain property P by testing each one, this procedure is no longer possible for infinite sets. One may happen to know that an element of the infinite set does not possess the property or it may be that by the very construction of the set we know or can prove that every element has the property. In any case, one cannot use the law of excluded middle to prove the property holds.

Hence if one proves that not all elements of an infinite set possess a property, then the conclusion that there exists at least one element which does not have the property is rejected by Brouwer. Thus from the denial that $a^b = b^a$ holds for all numbers the intuitionists do not conclude that there exists an a and b for which $a^b \neq b^a$. Consequently many existence proofs are not accepted by the intuitionists. The law of excluded middle can be used in cases where the conclusion can be reached in a finite number of steps, for example, to decide the question of whether a book contains misprints. In other cases the intuitionists deny the possibility of a decision.

The denial of the law of excluded middle gives rise to a new possibility, undecidable propositions. The intuitionists maintain, with respect to *infinite* sets, that there is a third state of affairs, namely, there may be propositions which are neither provable nor unprovable. As an example of such a proposition, let us define the kth position in the decimal expansion of π to be the position of the first zero which is followed by the integers $1 \ldots 9$. Aristotelian logic says that k either exists or does not exist and mathematicians following Aristotle may then proceed to argue on the basis of these two possibilities. Brouwer would reject all such arguments, for we do not know whether we shall ever be able to prove that it does or does not exist. Hence all reasoning about the number k is rejected by the intuitionists. Thus there are sensible mathematical questions which may never be settled on the basis of the statements contained in the axioms of mathematics. The questions seem to us to be decidable but actually our basis for expecting that they must be decidable is really nothing more than that they involve mathematical concepts.

With respect to the concepts they will accept as legitimate for mathematical discussion, the intuitionists insist on constructive definitions. For Brouwer, as for all intuitionists, the infinite exists in the sense that one can always find a finite set larger than the given one. To discuss any other type of infinite, the intuitionists demand that one give a method of constructing or defining this infinite in a finite number of steps. Thus Brouwer rejects the aggregates of set theory.

The requirement of constructibility is another ground for excluding any concept whose existence is established by indirect reasoning, that is, by the argument that the nonexistence leads to a contradiction. Aside from the fact that the existence proof may use the objectionable law of excluded middle, to the intuitionists this proof is not satisfactory because they want a constructive definition of the object whose existence is being established: The constructive definition must permit determination to any desired accuracy in a finite number of steps. Euclid's proof of the existence of an infinite number of primes (Chap. 4, sec. 7) is nonconstructive; it does not afford the determination of the nth prime. Hence it is not acceptable. Further, if one proved merely the existence of integers x, y, z, and n satisfying $x^n + y^n = z^n$, the intuitionist would not accept the proof. On the other hand, the definition of a prime number is constructive, for it can be applied to determine in a finite number of steps whether a number is prime. The insistence on a constructive definition applies especially to infinite sets. A set constructed by the axiom of choice applied to infinitely many sets would not be acceptable.

Weyl said of nonconstructive existence proofs (*Philosophy of Mathematics and Natural Science*, p. 51) that they inform the world that a treasure exists without disclosing its location. Proof through postulation cannot replace construction without loss of significance and value. He also pointed out that adherence to the intuitionist philosophy means the abandonment of the existence theorems of classical analysis—for example, the Weierstrass-Bolzano theorem. A bounded monotonic set of real numbers does not necessarily have a limit. For the intuitionists, if a function of a real variable exists in their sense then it is *ipso facto* continuous. Transfinite induction and its applications to analysis and most of the theory of Cantor are condemned outright. Analysis, Weyl says, is built on sand.

Brouwer and his school have not limited themselves to criticism but have sought to build up a new mathematics on the basis of constructions they accept. They have succeeded in saving the calculus with its limit processes, but their construction is very complicated. They also reconstructed elementary portions of algebra and geometry. Unlike Kronecker, Weyl and Brouwer do allow some kinds of irrational numbers. Clearly the mathematics of the intuitionists differs radically from what mathematicians had almost universally accepted before 1900.

7. The Formalist School

The third of the principal philosophies of mathematics is known as the formalist school and its leader was Hilbert. He began work on this philosophy in 1904. His motives at that time were to provide a basis for the number system without using the theory of sets and to establish the consistency of arithmetic. Since his own proof of the consistency of geometry reduced to the consistency of arithmetic, the consistency of the latter was a vital open question. He also sought to combat Kronecker's contention that the irrationals must be thrown out. Hilbert accepted the actual infinite and praised Cantor's work (Chap. 41, sec. 9). He wished to keep the infinite, the pure existence proofs, and concepts such as the least upper bound whose definition appeared to be circular.

Hilbert presented one paper on his views at the International Congress of 1904.[17] He did no more on this subject for fifteen years; then, moved by the desire to answer the intuitionists' criticisms of classical analysis, he took up problems of the foundations and continued to work on them for the rest of his scientific career. He published several key papers during the nineteen-twenties. Gradually a number of men took up his views.

Their mature philosophy contains many doctrines. In keeping with the new trend that any foundation for mathematics must take cognizance of the role of logic, the formalists maintain that logic must be treated simultaneously with mathematics. Mathematics consists of several branches and each branch is to have its own axiomatic foundation. This must consist of logical and mathematical concepts and principles. Logic is a sign language that puts mathematical statements into formulas and expresses reasoning by formal processes. The axioms merely express the rules by which formulas follow from one another. All signs and symbols of operation are freed from their significance with respect to content. Thus all meaning is eliminated from the mathematical symbols. In his 1926 paper[18] Hilbert says the objects of mathematical thought are the symbols themselves. The symbols are the essence; they no longer stand for idealized physical objects. The formulas may imply intuitively meaningful statements, but these implications are not part of mathematics.

Hilbert retained the law of excluded middle because analysis depends upon it. He said,[19] "Forbidding a mathematician to make use of the principle of excluded middle is like forbidding an astronomer his telescope or a boxer the use of his fists." Because mathematics deals only with symbolic expressions, all the rules of Aristotelian logic can be applied to these formal expressions. In this new sense the mathematics of infinite sets is possible. Also, by avoiding the explicit use of the word "all," Hilbert hoped to avoid the paradoxes.

To formulate the logical axioms Hilbert introduced symbolism for

17. *Proc. Third Internat. Congress of Math., Heidelberg*, 1904, 174–85 = *Grundlagen der Geom.*, 7th ed., 247–61; English trans. in *Monist*, 15, 1905, 338–52.
18. *Math. Ann.*, 95, 1926, 161–90 = *Grundlagen der Geometrie*, 7th ed., 262–88. See note 20.
19. Weyl, *Amer. Math. Soc. Bull.*, 50, 1944, 637 = *Ges. Abh.*, 4, 157.

concepts and relations such as "and," "or," "negation," "there exists," and the like. Luckily the logical calculus (symbolic logic) had already been developed (for other purposes) and so, Hilbert says, he has at hand what he needs. All the above symbols are the building blocks for the ideal expressions —the formulas.

To handle the infinite, Hilbert uses, aside from ordinary noncontroversial axioms, the transfinite axiom

$$A(\tau A) \rightarrow A(a).$$

This, he says, means: If a predicate A applies to the fiducial object τA, it applies to all objects a. Thus suppose A stands for being corruptible. If Aristides the Just is fiducial and corruptible, then everybody is corruptible.

Mathematical proof will consist of this process: the assertion of some formula; the assertion that this formula implies another; the assertion of the second formula. A sequence of such steps in which the asserted formulas or the implications are preceding axioms or conclusions will constitute the proof of a theorem. Also, substitution of one symbol for another or a group of symbols is a permissible operation. Thus formulas are derived by applying the rules for manipulating the symbols of previously established formulas.

A proposition is true if and only if it can be obtained as the last of a sequence of propositions such that every proposition of the sequence is either an axiom in the formal system or is itself derived by one of the rules of deduction. Everyone can check as to whether a given proposition has been obtained by a proper sequence of propositions. Thus under the formalist view proof and rigor are well defined and objective.

To the formalist, then, mathematics proper is a collection of formal systems, each building its own logic along with its mathematics, each having its own concepts, its own axioms, its own rules for deducing theorems such as rules about equality or substitution, and its own theorems. The development of each of these deductive systems is the task of mathematics. Mathematics becomes not a subject about something, but a collection of formal systems, in each of which formal expressions are obtained from others by formal transformations. So much for the part of Hilbert's program that deals with mathematics proper.

However, we must now ask whether the deductions are free of contradictions. This cannot necessarily be observed intuitively. But to show noncontradiction, all we need to show is that one can never arrive at the formal statement $1 = 2$. (Since by a theorem of logic any other false proposition implies this proposition, we may confine ourselves to this one.)

Hilbert and his students Wilhelm Ackermann (1896–1962), Paul Bernays (1888–), and von Neumann gradually evolved, during the years 1920 to 1930, what is known as Hilbert's *Beweistheorie* [proof theory] or meta-

1206   THE FOUNDATIONS OF MATHEMATICS

mathematics, a method of establishing the consistency of any formal system. In metamathematics Hilbert proposed to use a special logic that was to be basic and free of all objections. It employs concrete and finite reasoning of a kind universally admitted and very close to the intuitionist principles. Controversial principles such as proof of existence by contradiction, transfinite induction, and the axiom of choice are not used. Existence proofs must be constructive. Since a formal system can be unending, metamathematics must entertain concepts and questions involving at least potentially infinite systems. However, only finitary methods of proof should be used. There should be no reference either to an infinite number of structural properties of formulas or to an infinite number of manipulations of formulas.

Now the consistency of a major part of classical mathematics can be reduced to that of the arithmetic of the natural numbers (number theory) much as this theory is embodied in the Peano axioms, or to a theory of sets sufficiently rich to yield Peano's axioms. Hence the consistency of the arithmetic of the natural numbers became the center of attention.

Hilbert and his school did demonstrate the consistency of simple formal systems and they believed they were about to realize the goal of proving the consistency of arithmetic and of the theory of sets. In his article "Über das Unendliche"[20] he says,

> In geometry and physical theory the proof of consistency is accomplished by reducing it to the consistency of arithmetic. This method obviously fails in the proof for arithmetic itself. Since our proof theory . . . makes this last step possible, it constitutes the necessary keystone in the structure of mathematics. And in particular what we have twice experienced, first in the paradoxes of the calculus and then in the paradoxes of set theory, cannot happen again in the domain of mathematics.

But then Kurt Gödel (1906–78) entered the picture. Gödel's first major paper was "Über formal unentscheidbare Sätze der *Principia Mathematica* und verwandter Systeme I."[21] Here Gödel showed that the consistency of a system embracing the usual logic and number theory cannot be established if one limits himself to such concepts and methods as can formally be represented in the system of number theory. What this means in effect is that the consistency of number theory cannot be established by the narrow logic permissible in metamathematics. Apropos of this result, Weyl said that God exists since mathematics is consistent and the devil exists since we cannot prove the consistency.

The above result of Gödel's is a corollary of his more startling result.

20. *Math. Ann.*, 95, 1926, 161–90 = *Grundlagen der Geometrie*, 7th ed., 262–88. An English translation can be found in Paul Benacerraf and Hilary Putnam: *Philosophy of Mathematics*, 134–181, Prentice-Hall, 1964.
21. *Monatshefte für Mathematik und Physik*, 38, 1931, 173–98; see the bibliography.

The major result (Gödel's incompleteness theorem) states that if any formal theory T adequate to embrace number theory is consistent and if the axioms of the formal system of arithmetic are axioms or theorems of T, then T is incomplete. That is, there is a statement S of number theory such that neither S nor not-S is a theorem of the theory. Now either S or not-S is true; there is, then, a true statement of number theory which is not provable. This result applies to the Russell-Whitehead system, the Zermelo-Fraenkel system, and Hilbert's axiomatization of number theory. It is somewhat ironic that Hilbert in his address at the International Congress in Bologna of 1928 (see note 22) had criticized the older proofs of completeness through categoricalness (Chap. 42, sec. 3) but was very confident that his own system was complete. Actually the older proofs involving systems containing the natural numbers were accepted as valid only because set theory had not been axiomatized but was used on a naive basis.

Incompleteness is a blemish in that the formal system is not adequate to prove all the assertions frameable in the system. To add insult to injury, there are assertions that are undecidable but are intuitively true in the system. Incompleteness cannot be remedied by adjoining S or $\sim S$ as an axiom, for Gödel proved that *any* system embracing number theory must contain an undecidable proposition. Thus while Brouwer made clear that what is intuitively certain falls short of what is mathematically proved, Gödel showed that the intuitively certain goes beyond mathematical proof.

One of the implications of Gödel's theorem is that no system of axioms is adequate to encompass, not only all of mathematics, but even any one significant branch of mathematics, because any such axiom system is incomplete. There exist statements whose concepts belong to the system, which cannot be proved within the system but can nevertheless be shown to be true by nonformal arguments, in fact by the logic of metamathematics. This implication, that there are limitations on what can be achieved by axiomatization, contrasts sharply with the late nineteenth-century view that mathematics is coextensive with the collection of axiomatized branches. Gödel's result dealt a death blow to comprehensive axiomatization. This inadequacy of the axiomatic method is not in itself a contradiction, but it was surprising, because mathematicians had expected that any true statement could certainly be established within the framework of some axiomatic system. Of course the above arguments do not exclude the possibility of new methods of proof that would go beyond what Hilbert's metamathematics permits.

Hilbert was not convinced that these blows destroyed his program. He argued that even though one might have to use concepts outside a formal system, they might still be finite and intuitively concrete and so acceptable. Hilbert was an optimist. He had unbounded confidence in the power of man's reasoning and understanding. At the talk he gave at the 1928

International Congress[22] he had asserted, ". . . to the mathematical understanding there are no bounds . . . in mathematics there is no Ignorabimus
[we shall not know]; rather we can always answer meaningful questions
. . . our reason does not possess any secret art but proceeds by quite definite
and statable rules which are the guarantee of the absolute objectivity of its
judgment." Every mathematician, he said, shares the conviction that each
definite mathematical problem must be capable of being solved. This
optimism gave him courage and strength, but it barred him from understanding that there could be undecidable mathematical problems.

The formalist program, successful or not, was unacceptable to the
intuitionists. In 1925 Brouwer blasted away at the formalists.[23] Of course, he
said, axiomatic, formalistic treatments will avoid contradictions, but nothing
of mathematical value will be obtained in this way. A false theory is none the
less false even if not halted by a contradiction, just as a criminal act is
criminal whether or not forbidden by a court. Sarcastically he also remarked,
"To the question, where shall mathematical rigor be found, the two parties
give different answers. The intuitionist says, in the human intellect; the
formalist says, on paper." Weyl too attacked Hilbert's program. "Hilbert's
mathematics may be a pretty game with formulas, more amusing even than
chess; but what bearing does it have on cognition, since its formulas admittedly have no material meaning by virtue of which they could express
intuitive truths." In defense of the formalist philosophy, one must point out
that it is only for the purposes of proving consistency, completeness, and
other properties that mathematics is reduced to meaningless formulas. As for
mathematics as a whole, even the formalists reject the idea that it is simply a
game; they regard it as an objective science.

Hilbert in turn charged Brouwer and Weyl with trying to throw over-
board everything that did not suit them and dictatorially promulgating an
embargo.[24] He called intuitionism a treason to science. (Yet in his metamathematics he limited himself to intuitively clear logical principles.)

8. *Some Recent Developments*

None of the proposed solutions of the basic problems of the foundations—the
axiomatization of set theory, logicism, intuitionism, or formalism—achieved
the objective of providing a universally acceptable approach to mathematics.
Developments since Gödel's work of 1931 have not essentially altered the
picture. However, a few movements and results are worth noting. A number

22. *Atti Del Congresso Internazionale Dei Matematici*, I, 135–41 = *Grundlagen der Geometrie*, 7th ed., 313–23.
23. *Jour. für Math.*, 154, 1925, 1.
24. *Abh. Math. Seminar der Hamburger Univ.*, 1, 1922, 157–77 = *Ges. Abh.*, 3, 157–77.

of men have erected compromise approaches to mathematics that utilize features of two basic schools. Others, notably Gerhard Gentzen (1909–45), a member of Hilbert's school, have loosened the restrictions on the methods of proof allowed in Hilbert's metamathematics and, for example, by using transfinite induction (induction over the transfinite numbers), have thereby managed to establish the consistency of number theory and restricted portions of analysis.[25]

Among other significant results, two are especially worth noting. In *The Consistency of the Axiom of Choice and of the Generalized Continuum Hypothesis with the Axioms of Set Theory* (1940, rev. ed., 1951), Gödel proved that if the Zermelo-Fraenkel system of axioms without the axiom of choice is consistent, then the system obtained by adjoining this axiom is consistent; that is, the axiom cannot be disproved. Likewise the continuum hypothesis that there is no cardinal number between $\aleph_0$ and $2^{\aleph_0}$, is consistent with the Zermelo-Fraenkel system (without the axiom of choice). In 1963 Paul J. Cohen (1934–), a professor of mathematics at Stanford University, proved[26] that the latter two axioms are independent of the Zermelo-Fraenkel system; that is, they cannot be proved on the basis of that system. Moreover, even if one retained the axiom of choice in the Zermelo-Fraenkel system, the continuum hypothesis could not be proved. These results imply that we are free to construct new systems of mathematics in which either or both of the two controversial axioms are denied.

All of the developments since 1930 leave open two major problems: to prove the consistency of unrestricted classical analysis and set theory, and to build mathematics on a strictly intuitionistic basis or to determine the limits of this approach. The source of the difficulties in both of these problems is infinity as used in infinite sets and infinite processes. This concept, which created problems even for the Greeks in connection with irrational numbers and which they evaded in the method of exhaustion, has been a subject of contention ever since and prompted Weyl to remark that mathematics is the science of infinity.

The question as to the proper logical basis for mathematics and the rise particularly of intuitionism suggest that, in a larger sense, mathematics has come full circle. The subject started on an intuitive and empirical basis. Rigor became a goal with the Greeks, and though more honored in the breach until the nineteenth century, it seemed for a moment to be achieved. But the efforts to pursue rigor to the utmost have led to an impasse in which there is no longer any agreement on what it really means. Mathematics remains alive and vital, but only on a pragmatic basis.

There are some who see hope for resolution of the present impasse. The French group of mathematicians who write under the pseudonym of Nicolas

25. *Math. Ann.*, 112, 1936, 493–565.
26. *Proceedings of the National Academy of Sciences*, 50, 1963, 1143–48; 51, 1964, 105–10.

Bourbaki, offer this encouragement:[27] "There are now twenty-five centuries during which the mathematicians have had the practice of correcting their errors and thereby seeing their science enriched, not impoverished; this gives them the right to view the future with serenity."

Whether or not the optimism is warranted, the present state of mathematics has been aptly described by Weyl:[28] "The question of the ultimate foundations and the ultimate meaning of mathematics remains open; we do not know in what direction it will find its final solution or even whether a final objective answer can be expected at all. 'Mathematizing' may well be a creative activity of man, like language or music, of primary originality, whose historical decisions defy complete objective rationalization."

Bibliography

Becker, Oskar: *Grundlagen der Mathematik in geschichtlicher Entwicklung*, Verlag Karl Alber, 1956, 317–401.

Beth, E. W.: *Mathematical Thought: An Introduction to the Philosophy of Mathematics*, Gordon and Breach, 1965.

Bochenski, I. M.: *A History of Formal Logic*, University of Notre Dame Press, 1962; Chelsea (reprint), 1970.

Boole, George: *An Investigation of the Laws of Thought* (1854), Dover (reprint), 1951.

————: *The Mathematical Analysis of Logic* (1847), Basil Blackwell (reprint), 1948.

————: *Collected Logical Works*, Open Court, 1952.

Bourbaki, N.: *Eléments d'histoire des mathématiques*, 2nd ed., Hermann, 1969, 11–64.

Brouwer, L. E. J.: "Intuitionism and Formalism," *Amer. Math. Soc. Bull.*, 20, 1913/14, 81–96. An English translation of Brouwer's inaugural address as professor of mathematics at Amsterdam.

Church, Alonzo: "The Richard Paradox," *Amer. Math. Monthly*, 41, 1934, 356–61.

Cohen, Paul J., and Reuben Hersh: "Non-Cantorian Set Theory," *Scientific American*, Dec. 1967, 104–16.

Couturat, L.: *La Logique de Leibniz d'après des documents inédits*, Alcan, 1901.

De Morgan, Augustus: *On the Syllogism and Other Logical Writings*, Yale University Press, 1966. A collection of his papers edited by Peter Heath.

Dresden, Arnold: "Brouwer's Contribution to the Foundations of Mathematics," *Amer. Math. Soc. Bull.*, 30, 1924, 31–40.

Enriques, Federigo: *The Historic Development of Logic*, Henry Holt, 1929.

Fraenkel, A. A.: "The Recent Controversies About the Foundations of Mathematics," *Scripta Mathematica*, 13, 1947, 17–36.

Fraenkel, A. A., and Y. Bar-Hillel: *Foundations of Set Theory*, North-Holland, 1958.

Frege, Gottlob: *The Foundations of Arithmetic*, Blackwell, 1953, English and German; also English translation only, Harper and Bros., 1960.

————: *The Basic Laws of Arithmetic*, University of California Press, 1965.

27. *Journal of Symbolic Logic*, 14, 1949, 2–8.
28. *Obituary Notices of Fellows of the Royal Soc.*, 4, 1944, 547–53 = *Ges. Abh.*, 4, 121–29, p. 126 in part.

Gerhardt, C. I., ed.: *Die philosophischen Schriften von G. W. Leibniz*, 1875–80, Vol. 7.

Gödel, Kurt: *On Formally Undecidable Propositions of* Principia Mathematica *and Related Systems*, Basic Books, 1965.

————: "What Is Cantor's Continuum Problem?," *Amer. Math. Monthly*, 54, 1947, 515–25.

————: *The Consistency of the Axiom of Choice and of the Generalized Continuum Hypothesis with the Axioms of Set Theory*, Princeton University Press, 1940; rev. ed., 1951.

Kneale, William and Martha: *The Development of Logic*, Oxford University Press, 1962.

Kneebone, G. T.: *Mathematical Logic and the Foundations of Mathematics*, D. Van Nostrand, 1963. See the Appendix especially on developments since 1939.

Leibniz, G. W.: *Logical Papers*, edited and translated by G. A. R. Parkinson, Oxford University Press, 1966.

Lewis, C. I.: *A Survey of Symbolic Logic*, Dover (reprint), 1960, pp. 1–117.

Meschkowski, Herbert: *Probleme des Unendlichen, Werk und Leben Georg Cantors*, F. Vieweg und Sohn, 1967.

Mostowski, Andrzej: *Thirty Years of Foundational Studies*, Barnes and Noble, 1966.

Nagel, E., and J. R. Newman: *Gödel's Proof*, New York University Press, 1958.

Poincaré, Henri: *The Foundations of Science*, Science Press, 1946, 448–85. This is a reprint in one volume of *Science and Hypothesis*, *The Value of Science*, and *Science and Method*.

Rosser, J. Barkley: "An Informal Exposition of Proofs of Gödel's Theorems and Church's Theorem," *Journal of Symbolic Logic*, 4, 1939, 53–60.

Russell, Bertrand: *The Principles of Mathematics*, George Allen and Unwin, 1903; 2nd ed., 1937.

Scholz, Heinrich: *Concise History of Logic*, Philosophical Library, 1961.

Styazhkin, N. I.: *History of Mathematical Logic from Leibniz to Peano*, Massachusetts Institute of Technology Press, 1969.

Van Heijenoort, Jean: *From Frege to Gödel*, Harvard University Press, 1967. Translations of key papers on logic and the foundations of mathematics.

Weyl, Hermann: "Mathematics and Logic," *Amer. Math. Monthly*, 53, 1946, 2–13 = *Ges. Abh.*, 4, 268–79.

————: *Philosophy of Mathematics and Natural Science*, Princeton University Press, 1949.

Wilder, R. L.: "The Role of the Axiomatic Method," *Amer. Math. Monthly*, 74, 1967, 115–27.

Abbreviations

Journals whose titles have been written out in full in the text are not listed here.

Abh. der Bayer. Akad. der Wiss. Abhandlungen der Königlich Bayerischen Akademie der Wissenschaften (München)

Abh. der Ges. der Wiss. zu Gött. Abhandlungen der Königlichen Gesellschaft der Wissenschaften zu Göttingen

Abh. König. Akad. der Wiss., Berlin Abhandlungen der Königlich Preussischen Akademie der Wissenschaften zu Berlin

Abh. Königlich Böhm. Ges. der Wiss. Abhandlungen der Königlichen Böhmischen Gesellschaft der Wissenschaften

Abh. Math. Seminar der Hamburger Univ. Abhandlungen aus dem Mathematischen Seminar Hamburgischen Universität

Acta Acad. Sci. Petrop. Acta Academiae Scientiarum Petropolitanae

Acta Erud. Acta Eruditorum

Acta Math. Acta Mathematica

Acta Soc. Fennicae Acta Societatis Scientiarum Fennicae

Amer. Jour. of Math. American Journal of Mathematics

Amer. Math. Monthly American Mathematical Monthly

Amer. Math. Soc. Bull. American Mathematical Society, Bulletin

Amer. Math. Soc. Trans. American Mathematical Society, Transactions

Ann. de l'Ecole Norm. Sup. Annales Scientifiques de l'Ecole Normale Supérieure

Ann. de Math. Annales de Mathématiques Pures et Appliquées

Ann. Fac. Sci. de Toulouse Annales de la Faculté des Sciences de Toulouse

Ann. Soc. Sci. Bruxelles Annales de la Société Scientifique de Bruxelles

Annali di Mat. Annali di Matematica Pura ed Applicata

Annals of Math. Annals of Mathematics

Astronom. Nach. Astronomische Nachrichten

Atti Accad. Torino Atti della Reale Accademia delle Scienze di Torino

Atti della Accad. dei Lincei, Rendiconti Atti della Reale Accademia dei Lincei, Rendiconti

Brit. Assn. for Adv. of Sci. British Association for the Advancement of Science

Bull. des Sci. Math. Bulletin des Sciences Mathématiques

Bull. Soc. Math. de France Bulletin de la Société Mathématique de France

Cambridge and Dublin Math. Jour. Cambridge and Dublin Mathematical Journal

Comm. Acad. Sci. Petrop. Commentarii Academiae Scientiarum Petropolitanae

Comm. Soc. Gott. Commentationes Societatis Regiae Scientiarum Gottingensis Recentiores

Comp. Rend. Comptes Rendus

Corresp. sur l'Ecole Poly. Correspondance sur l'Ecole Polytechnique

Encyk. der Math. Wiss. Encyklopädie der Mathematischen Wissenschaften

Gior. di Mat. Giornale di Matematiche

Hist. de l'Acad. de Berlin Histoire de l'Académie Royale des Sciences et des Belles-Lettres de Berlin

Hist. de l'Acad. des Sci., Paris Histoire de l'Académie Royale des Sciences avec les Mémoires de Mathématique et de Physique

Jahres, der Deut. Math.-Verein. Jahresbericht der Deutschen Mathematiker-Vereinigung

Jour. de l'Ecole Poly. Journal de l'Ecole Polytechnique

Jour. de Math. Journal de Mathématiques Pures et Appliquées

Jour. des Sçavans Journal des Sçavans

Jour. für Math. Journal für die Reine und Angewandte Mathematik

Jour. Lon. Math. Soc. Journal of the London Mathematical Society

Königlich Sächsischen Ges. der Wiss. zu Leipzig Berichte über die Verhandlungen der Königlich Sächsischen Gesellschaft der Wissenschaften zu Leipzig

Math. Ann. Mathematische Annalen

Mém. de l'Acad. de Berlin See *Hist. de l'Acad. de Berlin*

Mém. de l'Acad. des Sci., Paris See *Hist. de l'Acad. des Sci., Paris;* after 1795, Mémoires de l'Academie des Sciences de l'Institut de France

Mém. de l'Acad. Sci. de St. Peters. Mémoires de l'Académie Impériale des Sciences de Saint-Petersbourg

Mém. des sav. étrangers Mémoires de Mathématique et de Physique Présentés à l'Académie Royal des Sciences, par Divers Sçavans, et Lus dans ses Assemblées

Mém. divers Savans See *Mém. des sav. étrangers*

Misc. Berolin. Miscellanea Berolinensia; also as *Hist. de l'Acad. de Berlin (q.v.)*

Misc. Taur. Miscellanea Philosophica-Mathematica Societatis Privatae Taurinensis (published by Accademia della Scienze di Torino)

Monatsber. Berliner Akad. Monatsberichte der Königlich Preussischen Akademie der Wissenschaften zu Berlin

N.Y. Math. Soc. Bull. New York Mathematical Society, Bulletin

Nachrichten König. Ges. der Wiss. zu Gött. Nachrichten von der Königlichen Gesellschaft der Wissenschaften zu Göttingen

Nou. Mém. de l'Acad. Roy. des Sci., Bruxelles Nouveaux Mémoires de l'Académie Royale des Sciences, des Lettres, et des Beaux-Arts de Belgique

Nouv. Bull. de la Soc. Philo. Nouveau Bulletin de la Société Philomatique de Paris

Nouv. Mém. de l'Acad. de Berlin Nouveaux Mémoires de l'Académie Royale des Sciences et des Belles-Lettres de Berlin

Nova Acta Acad. Sci. Petrop. Nova Acta Academiae Scientiarum Petropolitanae

Nova Acta Erud. Nova Acta Eruditorum

Novi Comm. Acad. Sci. Petrop. Novi Commentarii Academiae Scientiarum Petropolitanae

Phil. Mag. The Philosophical Magazine

Philo. Trans. Philosophical Transactions of the Royal Society of London

Proc. Camb. Phil. Soc. Cambridge Philosophical Society, Proceedings

Proc. Edinburgh Math. Soc. Edinburgh Mathematical Society, Proceedings

Proc. London Math. Soc. Proceedings of the London Mathematical Society

Proc. Roy. Soc. Proceedings of the Royal Society of London

Proc. Royal Irish Academy Proceedings of the Royal Irish Academy

Quart. Jour. of Math. Quarterly Journal of Mathematics

Scripta Math. Scripta Mathematica

Sitzungsber. Akad. Wiss zu Berlin Sitzungsberichte der Königlich Preussischen Akademie der Wissenschaften zu Berlin

Sitzungsber. der Akad. der Wiss., Wien Sitzungsberichte der Kaiserlichen Akademie der Wissenschaften zu Wien. Mathematisch-Naturwissenschaftlichen Klasse

Trans. Camb. Phil. Soc. Cambridge Philosophical Society, Transactions

Trans. Royal Irish Academy Transactions of the Royal Irish Academy

Zeit. für Math. und Phys. Zeitschrift für Mathematik und Physik

Zeit. für Physik Zeitschrift für Physik

Name Index

Abbati, Pietro, 765
Abel, Niels Henrik, 216, 973-74, 1024, 1053, 1097, 1111, 1137; biography, 644-45; Abelian integrals, 653-55; elliptic functions, 645-50; rigorization of analysis, 947-48, 964-65, 973-74; theory of equations, 754-55
Ackerman, Wilhelm, 1208
Adams, John Couch, 369
Adelard of Bath, 205, 207
Agrippa von Nettesheim, 286
Ahmes, 16, 18, 20
Al-Battânî, 195
Alberti, Leone Battista, 232-33, 285
Al-Bîrûnî, 190, 195, 241
Albuzjani. *See* Abû'l-Wefâ
Alembert, Jean Le Rond d', 623, 867, 1029; algebra, 598; calculus, 425, 465, 954; complex function theory, 626-28; complex numbers, 595; Fourier series, 458-59; mechanics, 616; ordinary differential equations, 477, 484, 494, 500; partial differential equations, 503-5, 510, 512-13, 516-17, 532, 542; rigor, 619
Alexander, James W., 1176-77, 1179-80
Alexander the Great, 4, 101-2
Alexandroff, Paul S., 1181
Alhazen, 193, 196
Al-Karkhî, 192
Al Kashî, 254
Al-Khowârizmî, 192-93
Ampère, André-Marie, 703, 750
Anaxagoras, 38, 147
Anaximander, 27
Anaximenes, 27
Antiphon, 42
Āpastamba, 184
Apollonius, 56, 89-99, 135, 157-58, 168, 300

Archimedes, 37, 105-16, 134, 163-66, 168, 286, 395
Archytas, 27-28, 42, 46, 48-49
Argand, Jean-Robert, 630-31, 776
Aristaeus the Elder, 48
Aristarchus, 156-57
Aristotle, 26-27, 35-37, 51-54, 151-53, 155-56, 162-63, 207, 395, 436, 776, 992, 1028; concept of mathematics, 51, 151-53
Arnauld, Antoine, 252
Aronhold, Siegfried Heinrich, 928
Āryabhata, 184, 187, 189
Arzelà, Cesare, 1046, 1077-78
Ascoli, Giulio, 1077
Augustine, Saint, 204
Autolycus of Pitane, 54

Babbage, Charles, 259, 622
Bäcklund, Albert Victor, 703
Bacon, Francis, 225-26
Bacon, Roger, 145, 207-8, 213, 226
Baire, René, 1079, 1186, 1199
Baldi, Bernadino, 223, 229
Baltzer, Richard, 879
Banach, Stephen, 1088-92
Barrow, Isaac, 252, 281, 339, 346-47, 356-57, 380, 383-84
Bartels, Johann M., 862, 878-79
Bede, Venerable, 202
Beeckman, Isaac, 316, 478
Beltrami, Eugenio, 896, 901-2, 905, 914, 1030, 1122, 1130
Bendixson, Ivar, 737
Benedetti, Giovanni Battista, 223, 229, 234, 478
Berkeley, George, Bishop, 400, 427-28
Bernays, Paul, 1205

Subject Index

xiii